^{17}O NMR Spectroscopy in Organic Chemistry

Editor

David W. Boykin, Ph.D.

Professor and Chair
Department of Chemistry
Georgia State University
Atlanta, Georgia

CRC Press
Taylor & Francis Group
Boca Raton London New York

CRC Press is an imprint of the
Taylor & Francis Group, an **informa** business

CRC Press
Taylor & Francis Group
6000 Broken Sound Parkway NW, Suite 300
Boca Raton, FL 33487-2742

© 1991 by Taylor & Francis Group, LLC
CRC Press is an imprint of Taylor & Francis Group, an Informa business

First issued in paperback 2019

No claim to original U.S. Government works

ISBN 13: 978-0-367-45076-2 (pbk)
ISBN 13: 978-0-8493-4867-9 (hbk)

Visit the Taylor & Francis Web site at
http://www.taylorandfrancis.com

and the CRC Press Web site at
http://www.crcpress.com

Library of Congress Cataloging-in-Publication Data
^{17}O NMR spectroscopy in organic chemistry/editor, David W. Boykin.
   p. cm.
  Includes bibliographical references.
  ISBM 0-8493-4867-6
  1. Nuclear magnetic resonance spectroscopy. 2. Chemistry,
Organic. I. Boykin, David W. (David Withers), 1939-. II. Title: Oxygen-17
NMR spectroscopy in organic chemistry.
QD272.S6A12 1990
547.3'0877—dc20   90-34724
                                                                           CIP

Library of Congress Card Number 90-34724

PREFACE

During the past decade ^{17}O NMR spectroscopy has become an increasingly important method for studying structure, conformation, and electronic distribution in organic molecules. Numerous important functional groups contain oxygen, and the use of ^{17}O NMR spectroscopy as a probe which allows direct observation at a reaction site offers the potential for many new insights. Since oxygen does not occur with the frequency of carbon and hydrogen in organic molecules, ^{17}O NMR spectroscopy will not play the central role in organic structural analysis achieved by its ^1H and ^{13}C counterparts. Nevertheless, the important and often unique role it can play in understanding structure and bonding in oxygen-containing organic molecules is already apparent. Exciting examples of applications of ^{17}O NMR spectroscopy which have appeared include analysis of molecular deformation as a consequence of steric interactions in rigid systems, conformational and stereochemical analysis in flexible ones, dynamic exchange processes, mechanistic studies, and hydrogen bonding investigations.

This monograph has been organized to cover a large range of applications of ^{17}O NMR spectroscopy to organic chemistry. Chapter 1 describes theoretical aspects of chemical shift, quadrupolar and J coupling. Ease of observation of ^{17}O NMR signals is greatly enhanced by enriching a functional group; Chapter 2 describes methods for ^{17}O enrichment. Chapters 3 and 4 examine the effect of steric interactions on ^{17}O chemical shifts of functional groups in flexible and rigid systems. The application of ^{17}O NMR spectroscopy to hydrogen bonding investigations is discussed in Chapter 5. Chapter 6 explores the application of ^{17}O NMR spectroscopy to mechanistic problems in organic and bioorganic chemistry. The substantial field of ^{17}O NMR spectroscopy of oxygen monocoordinated to carbon in alcohols, ethers, and derivatives is reported in Chapter 7. Chapter 8 is a compendium of ^{17}O NMR spectroscopic data on carbonyl-containing functional groups. Chapters 9 and 10 deal with the ^{17}O NMR spectroscopy of oxygen bound to heteroatoms (O, N, P, and S) in organic systems.

In 1983, in the Preface to Volume 2 of his excellent work *NMR of Newly Accessible Nuclei*, Pierre Laszlo accurately noted that the full potential of ^{17}O NMR spectroscopy had yet to be grasped. It is our hope that this volume will provide some glimpse of that potential.

THE EDITOR

David W. Boykin, Ph.D., is Professor of Chemistry and Chair of the Department at Georgia State University, Atlanta, Georgia.

Dr. Boykin received his B.S. degree from the University of Alabama in 1961. He obtained his M.S. and Ph.D. degrees in 1963 and 1965, respectively, from the Department of Chemistry, University of Virginia, Charlottesville. After doing postdoctoral work at the University of Virginia, he was appointed Assistant Professor of Chemistry at Georgia State University. He became Associate Professor of Chemistry in 1968, Professor in 1972, and Chair of the Department in 1974. In 1989, he received the University Alumni Distinguished Award and the CASE Georgia Professor of the Year Award.

His research has been funded by the National Science Foundation, the National Institutes of Health, the Petroleum Research Fund administered by the American Chemical Society, and the U.S. Army Research and Development Command.

Dr. Boykin is the author of more than 100 papers. His current major research interests are applications of ^{17}O NMR spectroscopy to organic chemistry and the design and synthesis of antiviral agents.

CONTRIBUTORS

Alfons L. Baumstark, Ph.D.
Professor
Department of Chemistry
Georgia State University
Atlanta, Georgia

David W. Boykin, Ph.D.
Professor and Chair
Department of Chemistry
Georgia State University
Atlanta, Georgia

Leslie G. Butler, Ph.D.
Associate Professor
Department of Chemistry
Louisiana State University
Baton Rouge, Louisiana

S. Chandrasekaran, Ph.D.
NMR Facilities Manager
Department of Chemistry
Georgia State University
Atlanta, Georgia

Slayton A. Evans, Jr., Ph.D.
Professor
Department of Chemistry
University of North Carolina
Chapel Hill, North Carolina

Naganna M. Goudgaon, Ph.D.
Research Associate
Department of Chemistry
University of Tennessee
Knoxville, Tennessee

George W. Kabalka, Ph.D.
Professor
Department of Chemistry
Director of Basic Research
Biomedical Imaging Center
Department of Radiology
University of Tennessee
Knoxville, Tennessee

Ronald W. Woodard, Ph.D.
Associate Professor
Medicinal Chemistry and Pharmacognosy
College of Pharmacy
University of Michigan
Ann Arbor, Michigan

TABLE OF CONTENTS

Chapter 1

THE NMR PARAMETERS FOR OXYGEN-17

Leslie G. Butler

TABLE OF CONTENTS

I. INTRODUCTION

From an operational point of view, what does a scientist want from an ¹⁷O nuclear magnetic resonance (NMR) spectrum? Electronic and/or geometrical structure about the oxygen atom? Dynamics at the oxygen site? If these are the questions, ¹⁷O NMR spectroscopy, as shown herein with some of the results of a wide variety of exploratory research projects, will offer important insights into chemical problems.

This chapter is directed at the solution-state NMR spectroscopist with $I = {}^{1}/_{2}$ experience who wishes to add ¹⁷O NMR spectroscopy to the repertoire of routinely useful NMR techniques. Two of the common NMR interactions, chemical shielding and J coupling, will be discussed briefly; more weight is given the effect of the quadrupolar interaction upon the solution-state NMR spectrum. Because I am making the assumption that solid-state ¹⁷O NMR spectroscopy will become more common, static and MAS spectral simulations are presented as methods for determining the ¹⁷O quadrupole coupling constant. Finally, an argument will be made for the interpretation of the quadrupole coupling constant so that electronic and geometrical data can be obtained from the ¹⁷O spectrum.

There are three stable oxygen isotopes; because ¹⁶O and ¹⁸O both have $I = 0$, ¹⁷O is the only practical NMR-active nucleus. Table 1 lists some useful data for the ¹⁷O nucleus including a value for the nuclear electric quadrupole moment.[1] At first glance, the very low natural abundance seems to be a major problem. However, the requirement for selective enrichment of many materials can be turned to the advantage of the spectroscopists in the form of spectral simplification. The common chemical shift reference is naturally abundant ¹⁷O in water. Water is not an ideal chemical shift reference because of its relatively large linewidth, but it is acceptable on the basis of the large linewidths for most ¹⁷O resonances (*vide infra*).

II. SOLUTION LINEWIDTHS, CHEMICAL SHIELDING, AND J COUPLING

In fluid solution, the quadrupolar interaction is averaged to zero and the resonant frequency of the ¹⁷O nucleus is determined by the chemical shielding and J coupling interactions. In fluid solution, the $I = {}^{5}/_{2}$ nucleus acts much like an $S = {}^{1}/_{2}$ nucleus: 90° pulses can be defined, the spinlattice relaxation is exponential,[2,3] $T_1 \cong T_2$, and the lineshapes are Lorenztian. There are two features of the $I = {}^{5}/_{2}$ nucleus that should be mentioned. First, recall the multiplicity rule for first-order spectra, $2nI + 1$, where n is the number of spins; a spin J coupled to an ¹⁷O nucleus will be split into a six-line pattern in the limit of high ¹⁷O enrichment. Second, quadrupolar relaxation is almost always the most effective relaxation pathway for the ¹⁷O nucleus; thus, resonances tend to be broad. The quadrupolar relaxation pathway is field independent, and the relaxation rate, $1/T_1$, can be calculated easily.[4,5]

$$\frac{1}{T_1} = \frac{3}{40} \frac{2I + 3}{I^2(2I - 1)} \left(1 + \frac{\eta^2}{3}\right) \left(\frac{e^2qQ}{\hbar}\right)^2 \tau_c \tag{1}$$

For typical values of the quadrupole coupling constant, Equation 1 was used to calculate the expected spin-lattice relaxation rate and the corresponding linewidth; these results are shown in Figure 1. Narrow lines can be expected only for small molecules in non-viscous fluids. Briefly, the rotational correlation time, τ_c, is a linear function of the solvent viscosity.[6] The solvent viscosity is a strong function of temperature; hence, elevated temperatures, and the resulting shorter rotational correlation times, may be employed to reduce the linewidth.

The chemical shift range of ¹⁷O is similar to that of ¹³C. These two nuclei have comparable diamagnetic shielding values, 260.7 and 395.1 ppm for ¹³C and ¹⁷O, respectively.[7] Because

TABLE 1
The Oxygen-17 Nucleus

Natural abundance	0.037%
Nuclear spin	$I = \frac{5}{2}$
Gyromagnetic ratio	$\gamma = -3,626.4$ rad s^{-1} Gauss^{-1}
NMR Larmor frequency	$\nu_L = 13.557$ MHz at 23488 Gauss
Quadrupole moment	$Q = -2.578 \times 10^{-26}$ cm^2

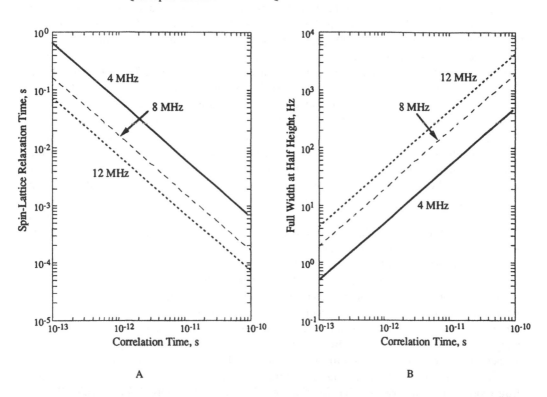

A   B

FIGURE 1. Spin-lattice relaxation times and the corresponding Lorenztian linewidth as a function of the rotational correlation time in fluid solution. The three traces in each figure refer to three different values of the quadrupole coupling constant: $e^2q_{zz}Q/h = 4$ MHz, —; $e^2q_{zz}Q/h = 8$ MHz, ----; $e^2q_{zz}Q/h = 12$ MHz, ... The quadrupolar relaxation rate depends upon the square of the quadrupole coupling constant, thus, the upper trace in Figure 1A corresponds to the smallest value of the quadrupole coupling constant.

of the similarity in observed chemical shifts, the paramagnetic contribution[8] of ^{17}O tracks that of ^{13}C. For example, if we compare ^{13}C and ^{17}O chemical shifts for a formyl group to chemical shifts obtained for reduced organic compounds, namely carbons in alkyls and oxygens in ether sites, we find that the formyl unit is the more deshielded for both ^{13}C and ^{17}O. As in ^{13}C NMR, trends in the chemical shift of ^{17}O nuclei can be used to infer changes in the electronic structure of small molecules. An excellent summary of structural correlations involving the ^{17}O chemical shift has been presented by Kintzinger.[9] Recent work has extended the utility of ^{17}O chemical shifts to such areas as the determination of the intercarbonyl dihedral angle of 1,2-diones.[10] The dione work relies on the variation of σ_p, the paramagnetic contribution to the chemical shielding. Chemical shift calculations using gauge-invariant atomic orbital theory with a minimal basis set do show that, for oxygen in typical organic environments, the paramagnetic contribution to the chemical shielding is much more variable than the diamagnetic contribution.[11] In aromatic systems, correlations have been found between the first ionization potential of the molecule and the ^{17}O chemical shift.[12] In a recent study of aromatic sulfones, the lanthanide shift reagent, $Eu(fod)_3$, was used to remove accidental chemical shift coincidence.[13] In metal carbonyl complexes, ^{17}O chemical shifts have been used to study the binding of the carbonyl ligand to the metal center; the comparison with ^{55}Mn data[14] and with ^{14}N data[15] is interesting. For polyoxoanion metal complexes, there are correlations of oxygen structure, i.e., terminal or bridging, with ^{17}O chemical shifts.[16] Recently, ^{17}O NMR has been used to confirm the structures of a uranyl anion, $(UO_2)_3(CO_3)_6^{6-}$,[17] and oligomeric aquamolybdenum cations;[18] in both works, peak integrations were an important factor in the argument for the proposed structure. In fluid solution where spin-lattice relaxation is rapid and exponential, it should be straightforward to set a delay between pulses long enough to insure that all ^{17}O spins are relaxed, thus permitting a linear relationship between peak area and number of nuclei present in the sample. For the oligomeric aquamolybdenum cations, the rate of oxygen exchange with water was sufficiently slow that a paramagnetic relaxation agent, Mn^{2+}, was used to suppress the bulk water signal. In the case of large peak separations, peak area corrections for the finite width of the rf pulse can be made. For the 1470 ppm shift for neptunium(VII) relative to water, the correction is small (on the order of 2 to 7% for 5 and 10 μs pulses, respectively, at a field of 7 Telsa).[19]

The neptunium(VII) work also illustrates an important application of ^{17}O NMR: the determination of exchange kinetics. For two site exchange processes between solvent water and coordinated oxygen sites, the change in the coordinated oxygen peak linewidth has been used to determine the exchange rate.[20] A caution applies for ^{17}O NMR experiments: the T_2 for $I = {}^5/_2$ nuclei is not a single exponential for solutions with slow rotational correlation times, that is, for cases in which $\omega_o\tau_c \geq 0.1$, where ω_o is the Larmor frequency in radians per second.[3] For very slow exchange processes, simple incorporation of ^{17}O-labeled water into the substrate can be followed by NMR.[21] When one of the sites is paramagnetic, the analysis of Swift and Connick is used.[22] Recent work with paramagnetic metal ions shows an emphasis on activation volumes that are obtained by studying the reaction at various pressures. In this manner, water exchange has been studied for lanthanide(III)[23] and vanadium(IV)[24] complexes. The exchange of acetate ions between solution and manganese(II) complexes has also been studied by variable pressure ^{17}O NMR; a large positive value for the activation volume is ascribed to the bulkiness of the acetic acid molecule.[25]

Observation of J coupling constants continues to be rare. In a recent study of antiarthritic drug oxidation chemistry, a value for $^1J_{PO} = 156$ (5) Hz is reported for $(C_2H_5)_3PO$.[26] A summary of J coupling constants is given by Kintzinger.[9,27]

III. DEFINITION OF THE QUADRUPOLE COUPLING CONSTANT AND ASYMMETRY PARAMETER

All nuclei with spin greater than $^1/_2$ have a nuclear electric quadrupole moment. The nuclear electric quadrupole moment can be viewed as one element of an expansion for describing the nuclear charge distribution. First, a nucleus has an electric charge; the electric charge times the electric potential established by the electrons about the nucleus gives the Coulomb energy. Second, the dot product of an electric dipole moment with an electric field also yields an energy term. However, the requirement for time reversal symmetry of the nuclear wavefunction means that nuclei cannot have electric dipole moments.[28] Also, the electric field at a nuclear site is usually zero, since any electric field would exert a force (nuclear charge times electric field) to accelerate the nucleus to a region where the electric field is zero. Third, the nuclear electric quadrupole moment times the electric field gradient also yields an energy term. Figure 2 illustrates an electric quadrupole moment interacting with an electric field gradient. Notice that the object will have a preferred orientation even though it does not have an electric dipole moment.

Because the nuclear spin angular momentum is quantized, the orientation of the nuclear

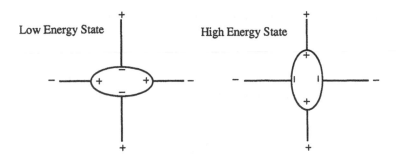

FIGURE 2. Simple example of the alignment of an electric quadrupole with an electric field gradient.

electric quadrupole moment with respect to the electric field gradient is quantized. Classically, the energy associated with this electrostatic interaction is

$$H_Q = \frac{1}{6} \sum_{k,j} V_{kj} Q_{kj} \tag{2}$$

where the indices k,j are over the x,y,z axes, V is the electric field gradient tensor, and Q is the electric quadrupole moment. Introduction of the nuclear spin angular momentum operators yields

$$H_Q = \frac{eQ}{6I(2I - 1)} \sum_{k,j} V_{kj} \left[\frac{3}{2} (I_k I_j + I_j I_k) - \delta_{kj} I^2 \right] \tag{3}$$

where e is the electrostatic charge in esu, Q is the nuclear electric quadrupole moment in cm^2, δ_{kj} is the Kronecker delta function, and I_x, I_y, I_z, and I^2 are the spin angular momentum operators. Equation 3 yields H_Q in ergs; most of the equations for the quadrupolar interaction in the textbooks are in cgs units or, occasionally, atomic units. In Equation 3, the electric field gradient, V, is in the laboratory axis system. It is often useful to transform the electric field gradient into another axis system, the principal axis system, in which the tensor that describes the electric field gradient is diagonal. In the principal axis system, Equation 3 transforms to

$$H_Q = \frac{e^2 q_{zz} Q/\hbar}{4I(2I - 1)} \left[3 I_z^2 - I^2 + \frac{\eta}{2} (I_+^2 + I_-^2) \right] \tag{4}$$

where the two new parameters, eq_{zz} and η, describe the electric field gradient tensor. The incorporation of Planck's constant, $\hbar$, converts H_Q to units of radians per second.

The electric field gradient tensor — used herein as a connection between NMR spectroscopy and molecular orbital theory — is exceedingly valuable to us and somewhat intricate. Therefore, we discuss its properties and interpretation at length. As mentioned above, the electric field gradient tensor can be described either in the laboratory axis system or in the principal axis system; the two are related by two unitary transformations. Here, the unitary transformations are two successive rotations about two orthogonal axes.[29] In the principal axis system, the electric field gradient tensor is diagonal; the convention is to assign labels to the three diagonal elements such that the following relationship applies:

$$|eq_{zz}| \geq |eq_{yy}| \geq |eq_{xx}| \tag{5}$$

Since the electric field gradient tensor is always traceless, that is

$$eq_{zz} + eq_{yy} + eq_{xx} = 0 \qquad (6)$$

there are only two independent parameters in the tensor. The convention is to describe the size of the electric field gradient tensor with eq_{zz} and the shape with the asymmetry parameter, η:

$$\eta = \frac{eq_{xx} - eq_{yy}}{eq_{zz}} \qquad (7)$$

The size of the electric field gradient tensor is referred to as the quadrupole coupling constant and is usually given in frequency units, $e^2q_{zz}Q/h$. The asymmetry parameter is dimensionless and ranges in value from 0 to 1. A value of zero usually indicates local symmetry about the quadrupolar nucleus of C_3 or higher; the asymmetry parameter for ^{17}O in $(CH_3CH_2)_3PO$ should be zero because of the threefold axis of symmetry about the P-O bond. Symmetry conditions also apply to the size of the electric field gradient tensor. Because the tensor is traceless (Equation 6), sites with at least tetrahedral or octahedral symmetry will have a quadrupole coupling constant of zero; examples for ^{35}Cl ($I = {}^3/_2$) having a zero value for the quadrupole coupling constant are $Cl^-(g)$, $NaCl(s)$, and $ClO_4{}^-(aq)$.

Equation 8 shows the relationship between the principal and laboratory electric field gradient tensors in terms of the electric field gradient elements:

$$\begin{bmatrix} -eq_{zz}/2\,(1-\eta) & & \\ & -eq_{zz}/2\,(1+\eta) & \\ & & eq_{zz} \end{bmatrix}^{PA}$$

$$= \begin{bmatrix} eq_{xx} & & \\ & eq_{yy} & \\ & & eq_{zz} \end{bmatrix}^{PA} = U^{adj} \begin{bmatrix} eq_{xx} & eq_{xy} & eq_{xz} \\ eq_{yx} & eq_{yy} & eq_{yz} \\ eq_{zx} & eq_{zy} & eq_{zz} \end{bmatrix}^{Lab} U \qquad (8)$$

The left side of Equation 8 is electric field gradient data as obtained from the NMR experiment; the right side of Equation 8 is the electric field gradient data as would be obtained from molecular orbital calculations. A note about notation: the symbols "PA" and "Lab" will be used to denote principal and laboratory axis systems, respectively; x,y,z will be the axes in each of the two systems. The other two conventions used in the literature, x,y,z for the principal axis system and x',y',z' for the laboratory axis system, and the converse, are difficult for us to follow. Note: Equation 8 will be used later on in a slightly modified form; multiplying through by eQ/h converts the elements to frequency units, which are more familiar to the spectroscopist.

In a way, Equation 8 summarizes the objective of interpreting quadrupole coupling constants: the comparison of experimental data to the results of bonding models. At this juncture, we have the option of pursuing either the acquisition of the experimental data or the interpretation of the results. Let us do the latter by analyzing the results of a molecular orbital calculation for the water molecule.

It is obvious that the electric field gradient tensor should depend upon the position and charge of nuclei and electrons about the quadrupolar nucleus. If we recall the progression of electrostatic interactions discussed above — the electric potential, $1/r$, and the electric field, $1/r^2$ — we would expect to find that the electric field gradient is a $1/r^3$ operator, as is shown here for the eq_{zz} element in the laboratory axis system

$$eq_{zz}^{Lab} = +\sum_{n} Z_n \frac{3 z_n^2 - r_n^2}{r_n^5} - e\left\langle \Psi^* \left| \sum_{i} \frac{3 z_i^2 - r_i^2}{r_i^5} \right| \Psi \right\rangle \qquad (9)$$

where the index n is over the nuclei with charge Z_n in the molecule at distance r_n from the quadrupolar nucleus. The index i is over the electrons in the molecule. The other elements of the electric field gradient tensor are calculated in a similar manner:

$$eq_{xy}^{Lab} = +\sum_{n} Z_n \frac{3 x_n y_n}{r_n^5} - e\left\langle \Psi^* \left| \sum_{i} \frac{3 x_i y_i}{r_i^5} \right| \Psi \right\rangle \qquad (10)$$

Excited states are not involved in the expectation value. Thus electric field gradients are considerably easier to calculate than are chemical shielding or J coupling constants.

Figure 3 shows the orientation of a water molecule in the laboratory axis system; the orientation is arbitrary, though it is a typical example of the orientation one might choose for a molecular orbital calculation. The data listed in Table 2 were taken from an excellent summary of molecular orbital calculation results for small molecules compiled by Snyder and Basch.[30] The molecular orbital calculation was done with a double-zeta basis set and the Hartree-Fock self-consistent field method in a computer program called POLYATOM. This program is similar to the currently popular Gaussian-80 to -88 programs.[31]

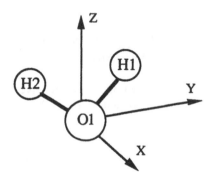

FIGURE 3.   Orientation of water molecule in laboratory axis system.

TABLE 2
Calculated Electric Field Gradient Elements for Water in the Laboratory Frame[a]

Site	$(3x^2 - r^2)/r^5$	$(3y^2 - r^2)/r^5$	$(3z^2 - r^2)/r^5$	$3xy/r^5$	$3xy/r^5$	$3yz/r^5$
H1	−0.3132	0.2950	0.0182	0.0	0.0	0.3806
O1	−2.0487	1.8086	0.2400	0.0	0.0	0.0

[a]   The units are ea_0^{-3}, where $e = 1$ and a_0 is the Bohr radius of 5.2917715×10^{-9} cm.

The conversion from atomic units to cgs units is straightforward. Listed below is the conversion to an electric field gradient in cgs units, then the conversion to frequency units for the oxygen nucleus for convenience in display:

$$eq_{zz}\left[\frac{esu}{cm^3}\right] = eq_{zz}\left[\frac{e}{a_0^3}\right] \times \left(\frac{1\ a_0}{5.2917715 \times 10^{-9}\ cm}\right)^3 \times \left(\frac{4.80325 \times 10^{-10}\ esu}{1\ e}\right)$$

$$(11)$$

and

$$e^2q_{zz}Q/h[Hz] = eq_{zz}\left[\frac{esu}{cm^3}\right] \times (4.80325 \times 10^{-10} \text{ esu}) \times \left(\frac{-2.578 \times 10^{-26} \text{ cm}^2}{6.62619 \times 10^{-27} \text{ erg s}}\right) \tag{12}$$

One cgs conversion factor used in Equation 12 is 1 esu^2 = 1 erg cm. Thus, we have the quadrupolar interaction in the laboratory axis system and in the principal axis system in frequency units as

$$\begin{bmatrix} -1.454 \text{ MHz} & & \\ & -10.955 \text{ MHz} & \\ & & 12.410 \text{ MHz} \end{bmatrix}^{PA}$$

$$= U^{adj} \begin{bmatrix} 12.410 \text{ MHz} & & \\ & -10.955 \text{ MHz} & \\ & & -1.454 \text{ MHz} \end{bmatrix}^{Lab} U \tag{13}$$

The asymmetry parameter is simply

$$\eta = \frac{-1.454 \text{ MHz} - -10.955 \text{ MHz}}{12.410 \text{ MHz}} = 0.766 \tag{14}$$

For completeness, the results for the hydrogen site are given below (Equation 15). The deuterium nuclear electric quadrupole moment is $+2.860 \times 10^{-27}$ cm^2.[32]

$$\begin{bmatrix} -167 \text{ kHz} & & \\ & -210 \text{ kHz} & \\ & & 377 \text{ kHz} \end{bmatrix}^{PA} = U^{adj} \begin{bmatrix} -210 \text{ kHz} & & \\ & 198 \text{ kHz} & 256 \text{ kHz} \\ & 256 \text{ kHz} & 12 \text{ kHz} \end{bmatrix}^{Lab} U \tag{15}$$

Table 3 lists the experimental[33,34] and calculated quadrupole coupling constants and asymmetry parameters for the gas phase water molecule.

TABLE 3
Comparison of Calculated Field Gradient
Parameters to Experimental Results for Water

Site	Calculated	Experimental	Ref.
Oxygen			
$e^2q_{zz}Q/h$	12.410 MHz	10.175 (67) MHz	33
η	0.766	0.75 (1)	
Deuterium			
$e^2q_{zz}Q/h$	377 kHz	318.6 (24) kHz	34
η	0.114	0.06 (.16)	

It is sometimes important to note the orientation of the largest element of the electric field gradient tensor, as expressed in the principal axis system, with respect to the molecular axis system. For example, eq_{zz}^{PA} at the oxygen site is normal to the plane of the water molecule; at the deuterium site, eq_{zz}^{PA} is aligned with the O-H bond vector. As a check of the molecular orbital results, one can check the calculated electric field gradient at the deuterium site. One should expect to find a positive value for $e^2q_{zz}Q/h$ with eq_{zz}^{PA} aligned along the X-D bond. Among the few known exceptions to this rule are μ-hydride in diborane[35] and short, symmetric hydrogen bonds.[36,37] One caution about molecular orbital programs and atomic units: programs can calculate either "eq" ($e = \pm 1$) or "q", hence, differences by a factor of -1 in atomic units can result. Therefore, it is very important to check the value of a deuterium quadrupole coupling constant to confirm the sign of the calculated ^{17}O quadrupole coupling constant.

IV. ACQUISITION OF THE QUADRUPOLE COUPLING CONSTANT AND ASYMMETRY PARAMETER

There are three experimental techniques that have been used on a more or less routine basis to measure ^{17}O quadrupole coupling constants and asymmetry parameters: microwave spectroscopy,[38] adiabatic demagnetization in the laboratory frame (ADLF) spectroscopy,[36,39] and high-field solid-state NMR spectroscopy (both static and magic angle spinning).[40] Each has advantages and significant disadvantages.

With microwave spectroscopy, the orientation and sign of the quadrupole coupling constant can be obtained; this possibility makes microwave spectroscopy a unique technique for ^{17}O NMR spectroscopy.[41] The development of pulsed-beam, Fourier transform microwave spectrometers has yielded a great increase in sensitivity.[42] Two leading references to this field are a review[43] and a determination of the ^{14}N quadrupole coupling constant in cyclo-pentadienylnickel nitrosyl.[44]

ADLF spectroscopy is a field cycling NMR technique; the sample is shuttled between regions of zero and high magnetic field[45] or, alternately, the magnetic field is cycled on and off.[46] By this method, the ^{17}O spectrum at zero magnetic field can be obtained. Also, if the ^1H spin-lattice relaxation time in zero magnetic field is long (>3 s), then phase-alternation enables detection of ^{17}O present in natural abundance.[47] The advantage of acquiring ^{17}O spectra in zero magnetic field is that the transition frequencies depend upon the magnitude of the quadrupole coupling constant and the asymmetry parameter. However, the sign of the quadrupole coupling constant is not usually obtained in a zero field experiment.

The $I = \frac{5}{2}$ spin gives rise to three zero field transitions, which are shown in Figure 4. The spin state energies have, until recently, been obtained by solving Equation 4 as a function of the asymmetry parameter. The analytical solution given in Equations 16 to 18 was used to prepare Figure 4.[48] First, we define the frequency, ν_Q, in hertz as

$$\nu_Q = \frac{3\ e^2q_{zz}Q}{2I\ (2I\ -\ 1)h} \tag{16}$$

then, the spin state energies, in hertz, are given by

$$E_i = -\frac{2}{3}\nu_Q\left[7\left(1 + \frac{\eta^2}{3}\right)\right]^{1/2}\cos\left(\frac{\Theta}{3} + \alpha_i\right) \tag{17}$$

where the parameter Θ is obtained from

$$\cos\Theta = \frac{-10(1 - \eta^2)}{\left[7\left(1 + \frac{\eta^2}{3}\right)\right]^{3/2}} \tag{18}$$

and the variables $\alpha_1 = 0$, $\alpha_2 = 240°$, and $\alpha_3 = 120°$ are unique for each spin state level. The transition $|\pm\frac{1}{2}\rangle \rightarrow |\pm\frac{5}{2}\rangle$ is a $\Delta m = 2$ transition and only gains significant allowedness

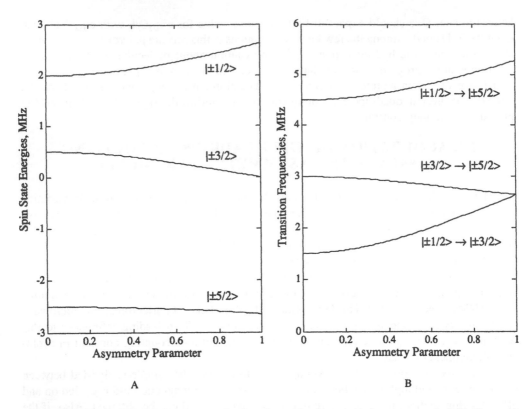

FIGURE 4.   The S = $^5/_2$ spin state energy levels and transition frequencies in zero magnetic field. Figure 4A shows the spin state energy levels for a quadrupole coupling constant, $e^2q_{zz}Q/h = -10$ MHz. Because the nuclear electric quadrupole moment of [17]O is negative, this corresponds to a positive value for the electric field gradient parameter, eq_{zz}. Figure 4B shows the corresponding zero-field transition frequencies. The simple zero-field spectrum does not change with the sign of the quadrupole coupling constant.

at $\eta = 0.4$ and greater. One of the most interesting of the [17]O ADLF spectra is that of $KHCO_3$ which shows pairs of transitions for all three oxygen sites.[49]

High-field solid-state ^{17}O NMR spectroscopy is a relatively new field.[40] The advantage of solid-state NMR spectroscopy for a quadrupolar spin is that information about the quadrupole coupling constant and asymmetry parameter is retained.[29,50] An important contribution of the recent work by Oldfield and co-workers is the demonstration that [17]O spins can be polarized by the ^{1}H spin system just as is done for ^{13}C in the CP/MAS experiment.[40]

In the high-field NMR experiment (100 MHz [^{1}H] and above), crystalline or polymeric materials yield "powder patterns". There are two common types of high-field solid-state NMR experiments: static and magic angle spinning (MAS). The powder patterns for static and MAS spectra have different shapes, so it is useful to do both experiments to check for consistent values of quadrupole coupling constant and asymmetry parameter. Unfortunately, obtaining quadrupole coupling constants and asymmetry parameters from a powder pattern requires computer simulation of a powder pattern and comparison to the experimental spectrum. For both experiments, analytical solutions for the calculation of the powder pattern have been published: static, Equation 10,[51] and MAS, Equation 5.[52] Because the expectation is that applications of high-field solid-state ^{17}O NMR spectroscopy will grow rapidly, Figures 5 to 7 show spectral simulations for a range of experiments, quadrupole coupling constants, and asymmetry parameters. These figures should allow one to select an appropriate instrument for an experiment and to estimate the value of the quadrupole coupling constant and asymmetry parameter from the experimental spectrum.

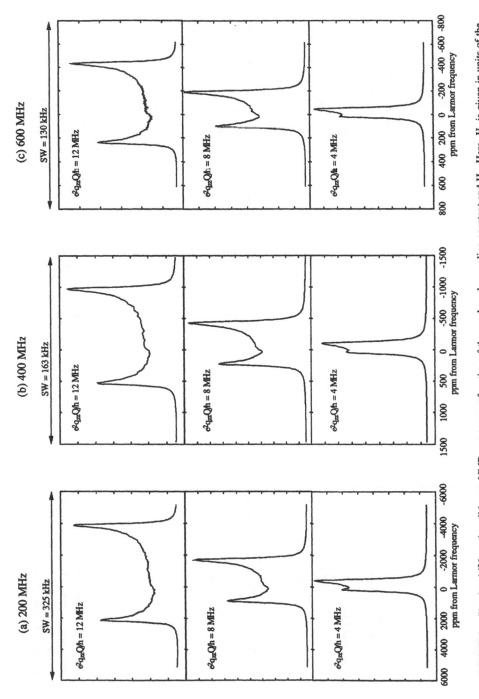

FIGURE 5. Simulated ^{17}O static solid-state NMR spectra as a function of the quadrupole coupling constant and H_o. Here, H_o is given in units of the ^1H Larmor frequency: 200, 400, and 600 MHz corresponding to 4.70, 9.40, and 14.1 Telsa, respectively. Simulations shown for 512 complex points. The number of orientations and the Lorenztian linebroadening was adjusted at each field strength to yield a smooth trace. The orientation step size and linebroadening factors are: (a) 200 MHz, inc = 0.6°, LB = 5000 Hz; (b) 400 MHz, inc = 0.6°, LB = 3000 Hz; and (c) 600 MHz, inc = 0.9°, LB = 2000 Hz.

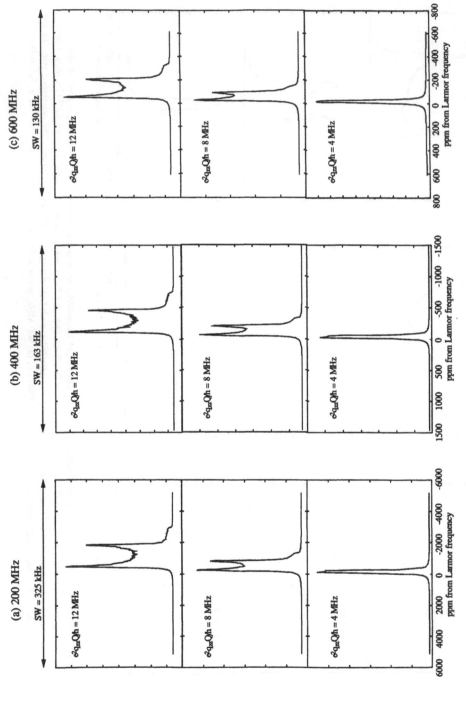

FIGURE 6. Simulated ¹⁷O MAS solid-state NMR spectra as a function of the quadrupole coupling constant and H_o. Here, H_o is given in units of the ¹H Larmor frequency; 200, 400, and 600 MHz corresponding to 4.70, 9.40, and 14.1 Telsa, respectively. Simulations shown for 512 complex points. The number of orientations and the Lorenztian linebroadening was adjusted at each field to yield a smooth trace. The orientation step size and linebroadening factors are: (a) 200 MHz, inc = 0.6°, LB = 2000 Hz; (b) 400 MHz, inc = 0.6°, LB = 1000 Hz; and (c) 600 MHz, inc = 0.9°, LB = 1000 Hz.

(a) Static

(b) MAS

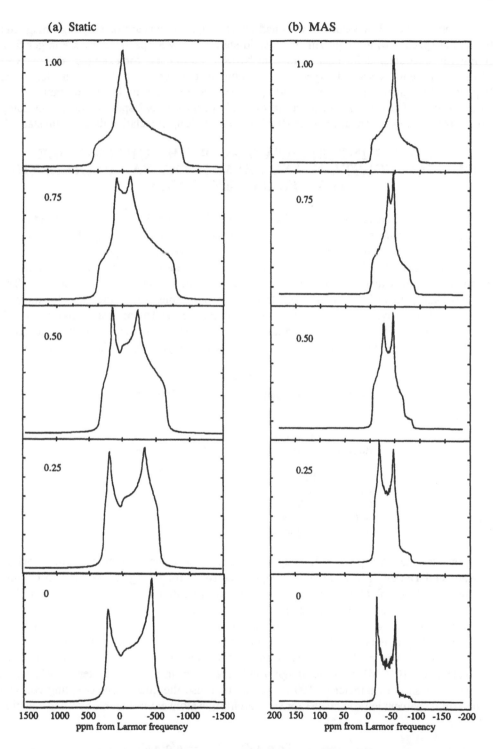

1500 1000 500 0 -500 -1000 -1500
ppm from Larmor frequency

200 150 100 50 0 -50 -100 -150 -200
ppm from Larmor frequency

FIGURE 7. The effect of the asymmetry parameter upon the appearance of the ^{17}O static and MAS solid-state NMR spectrum. These simulations are done for a quadrupole coupling constant of 8 MHz and a magnetic field of 9.40 Tesla, corresponding to a ^{1}H Larmor frequency of 400 MHz. Figure 7A shows the ^{17}O static solid-state NMR spectra as a function of the asymmetry parameter from 0 to 1. The orientation step size and linebroadening factors are 0.9° and 2000 Hz, respectively. Figure 7B shows the ^{17}O MAS solid-state NMR spectra. The orientation step size and linebroadening factors are 0.9° and 1000 Hz, respectively.

Except for very low values of the quadrupole coupling constant or very high magnetic fields, the spectral width requirements in solid-state ^{17}O NMR spectroscopy are larger than commonly encountered in other solid-state NMR experiments, for example, the ^{13}C CP/ MAS experiment. A second experimental feature that differs from $I = \frac{1}{2}$ nuclei is the "apparent" magnetic moment of a quadrupolar spin in a static sample. At constant H_1, the duration of an rf pulse required for a 90° tip angle of an $I = 5/2$ spin is reduced on going from short rotational correlation times (fluid solution) to long correlation times (solid state).[53]

V. DIRECT INTERPRETATION OF THE QUADRUPOLE COUPLING CONSTANT AND ASYMMETRY PARAMETER: THE TOWNES-DAILEY MODEL

In the two previous sections, we have discussed first the definition of the quadrupole coupling constant and the asymmetry parameter as an electrostatic property of a nuclear site, and second, the acquisition of these parameters from the experimental spectrum. The question for this section is this: is it always necessary to perform molecular orbital calculations for comparison to the experimental results, or does a more facile interpretive scheme exist? The Townes-Dailey analysis of halogen quadrupole coupling constants is the basis for most of the current interpretive schemes.[54] Herein we will introduce the Townes-Dailey model through a consideration of ^{35}Cl quadrupole coupling constants. Then, an application of ^{17}O quadrupole coupling constants to the study of bonding in organic carbonyl compounds will be discussed.

TABLE 4
Chlorine Quadrupole Coupling Constants

Molecule	Most important structure	$e^2q_{zz}Q/h$, MHz	Percent ionic character
$Cl^-(g)$	$[Ne]3s^23p^6$	0	100
$NaCl_{(s)}$	Na^+Cl^-	<1	100
$ICl_{(g)}$	I–Cl	−82.5	25
$ClCN_{(g)}$	Cl–C≡N	−83.2	25
$Cl_{(g)}$	$[Ne]3s^23p^5$	−110.4	0

Consider the ^{35}Cl quadrupole coupling constants in Table 4 taken from the work of Townes and Dailey. The quadrupole coupling constant for $Cl^-(g)$ is required to be zero because the electric field gradient has a trace of zero (Equation 6). The quadrupole coupling constant for the neutral chlorine atom is not zero. Why is this so? Because the electron configuration of Cl(g) is $[Ne]3s^23p^5$, there is a lack of one p electron necessary to achieve the spherical symmetry which then leads to a value of zero for the quadrupole coupling constant. Thus, one p electron creates an electric field gradient at the chlorine nucleus equivalent to 110.4 MHz. If we assign a chlorine atom an ionic character of 0%, and a chloride ion an ion character of 100%, we can then use the quadrupole coupling constant to determine percent ionic character for various chlorine atom environments by using equation 19:[55]

$$\text{Percent ionic character} = \left[\frac{e^2q_{zz}Q/h_{\text{free atom}} - e^2q_{zz}Q/h_{\text{molecule}}}{e^2q_{zz}Q/h_{\text{free atom}}} \right] \times 100 \qquad (19)$$

With Equation 19, the percent ionic character listed in Table 4 for the bonds to chlorine have been assigned. As the ionic character increases and the electron configuration about the chlorine nucleus approaches $[Ne]3s^23p^6$, Equation 19 indicates that the quadrupole coupling constant should respond in a linear fashion. The results are consistent with our expectations: for the ICl and ClCN molecules, we would expect the percent ionic character to be small, but not zero.

More elaborate Townes-Dailey models incorporate hybridization of atomic orbitals into valence bonding and lone pair orbitals. Another example of the Townes-Dailey model is shown in the development by Cheng and Brown of a model for determining σ and π bond orbital populations for X-O bonds (X = C,N,P,S) from the ^{17}O quadrupole coupling constant and asymmetry parameter.[49,56,57]

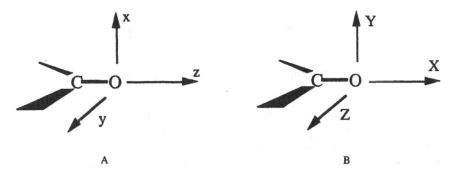

FIGURE 8.   Orientations of the laboratory and principal axes systems for a representative carbonyl. A. Laboratory axis system for labeling atomic orbitals. B. Principal axis system for electric field gradient tensor.

The ADLF experiment for 4-chlorobenzaldehyde shows two transitions: $|\pm^1/_2>\rightarrow|\pm^3/_2>$ at 1900 (1) kHz and $|\pm^3/_2>\rightarrow|\pm^5/_2>$ at 3086 (2) kHz. With the analytical solutions given by Equations 16 to 18, one can determine the parameters: $e^2q_{zz}Q/h$ = 10.648 (2) MHz, and η = 0.437 (2). To determine the σ and π bond orbital populations, the participation of the oxygen valence level atomic orbitals, v_{2s}, v_{2x}, v_{2y}, v_{2z}, in the valence bonding is defined as

Orbital	Orbital population	Name	
$\phi_1 = v_{2x}$	p_π	π bond	
$\phi_2 = v_{2y}$	2	lone pair	(20)
$\phi_3 = \alpha v_{2s} - \sqrt{1 - \alpha^2}\, v_{2z}$	p_σ	σ bond	
$\phi_4 = \sqrt{1 - \alpha^2}\, v_{2s} + \alpha\, v_{2z}$	2	lone pair	

where the quantity α represents the fraction of oxygen s orbital character in the sp-hybridized σ orbital. The laboratory axis system in which the orientation of the oxygen atomic 2p orbitals is defined is shown in Figure 8a. The value of α^2 that best fits the ^{17}O data is 0.25.[49] From Equation 20, we can determine the populations of the oxygen valence level atomic p orbitals. We define Pop(x) as the population of the v_{2x} orbital, and so on. Thus, we have

$$Pop(x) = p_\pi$$
$$Pop(y) = 2 \quad\quad (21)$$
$$Pop(z) = p_\sigma (1 - \alpha^2) + 2 \alpha^2$$

Like the chlorine example discussed earlier, we need to know the effect of a single 2p electron upon the value of the [17]O quadrupole coupling constant; $e^2q_{210} \, Q/h = +20.9$ MHz.[49] The next step is to relate the electric field gradients along the laboratory x,y,z axes to the populations of the atomic p orbitals. The following equations must sum to a value of zero since the electric field gradient tensor is traceless (Equation 6):

$$q_{zz}^{Lab} = \left\{ Pop(z) - \frac{1}{2} \left[Pop(x) + Pop(y) \right] \right\} q_{210}$$

$$q_{yy}^{Lab} = \left\{ Pop(y) - \frac{1}{2} \left[Pop(x) + Pop(z) \right] \right\} q_{210}$$

$$q_{xx}^{Lab} = \left\{ Pop(x) - \frac{1}{2} \left[Pop(y) + Pop(z) \right] \right\} q_{210} \qquad (22)$$

Here, q_{zz}^{Lab} is used to explicitly state that, as yet, the orientation of the principal axis system with respect to the laboratory axis system is unknown. For 4-chlorobenzaldehyde, Cheng and Brown determined that the principal axis system is oriented as shown in Figure 8B.[49] With this information, together with the atomic p orbital populations of Equation 21, Equation 22 can be rewritten as

$$q_{xx}^{PA} = q_{zz}^{Lab} = \left\{ -1 - \frac{p_\pi}{2} + p_\sigma \left(1 - \alpha^2 \right) + 2\,\alpha^2 \right\} q_{210}$$

$$q_{zz}^{PA} = q_{yy}^{Lab} = \left\{ 2 - \frac{p_\pi}{2} - \frac{p_\sigma}{2} \left(1 - \alpha^2 \right) - \alpha^2 \right\} q_{210}$$

$$q_{yy}^{PA} = q_{xx}^{Lab} = \left\{ -1 + p_\pi - \frac{p_\sigma}{2} \left(1 - \alpha^2 \right) - \alpha^2 \right\} q_{210} \qquad (23)$$

Incorporation of the [17]O data, the effect of a single electron in a 2p orbital, $e^2q_{210} \, Q/h = +20.9$ MHz, and the relationship among the principal electric field gradient tensor elements (Equation 8) gives

$$\frac{e^2q_{zz}Q/h}{e^2q_{210}Q/h} = 2 - \frac{p_\pi}{2} - \frac{p_\sigma}{2} (1 - \alpha^2) - \alpha^2$$

$$\frac{\frac{-1}{2} (1 + \eta) \, e^2q_{zz}Q/h}{e^2q_{210}Q/h} = -1 + p_\pi - \frac{p_\sigma}{2} (1 - \alpha^2) - \alpha^2$$

$$\frac{\frac{-1}{2} (1 - \eta) \, e^2q_{zz}Q/h}{e^2q_{210}Q/h} = -1 - \frac{p_\pi}{2} + p_\sigma (1 - \alpha^2) + 2\,\alpha^2 \qquad (24)$$

Solving for p_π and p_σ yields: $p_\pi = 1.416$ and $p_\sigma = 1.420$. In the 28 carbonyl sites studied by Cheng and Brown, p_π ranged from 1.366 for p-benzoquinone to 1.742 for sodium bicarbonate. The total range for p_σ was less, as would be expected for the less polarizable σ bond. There are factors to be noted: first, the orientation of the principal axis system with respect to the laboratory axis system can change.[49,58] Second, the formal charge on oxygen ($= 6 - 4 - p_\pi - p_\sigma$) is excessively large, on the order of $-1 \, e^-$. Nevertheless, the results of the Townes-Dailey model can provide quick analysis of orbital populations without resorting to elaborate molecular orbital calculations and comparative methods.

There are two other applications of the Townes-Dailey model that may be of interest in analysis of [17]O data. Edmonds and co-workers used a Townes-Dailey model to interpret ^{14}N data for a series of tetrahedral nitrogen sites in dimethylammonium chloride, diethylammonium chloride, methylammonium chloride, ethylammonium chloride, glycine, L-proline, L-serine, and acetamide.[59] The ^{14}N data for an extensive series of pyridine complexes with Lewis acids were analyzed with a model that yields the occupancy of the nitrogen donor orbital directed toward the Lewis acid.[60,61] From comparison of donor orbital occupation, it was possible to order the acidities of the Lewis acids.

VI. CONCLUSION

Solution-state and solid-state ^{17}O NMR spectroscopy has a bright future. One anticipates that the best work will come about from analyzing all of the [17]O data: chemical shift, J coupling, and quadrupole coupling constants. Based on the early success with CP/MAS of enriched [17]O samples, this technique is expected to grow rapidly in the range and number of applications.

REFERENCES

1. **Schaefer, H. F., III, Klemm, R. A., and Harris, F. E.,** Atomic hyperfine structure. II. First-order wavefunctions for the ground states of B, C, N, O, and F, *Phys. Rev.,* 181, 137, 1969.
2. **Hubbard, P. S.,** Nonexponential nuclear magnetic relaxation by quadrupole interactions, *J. Chem. Phys.,* 53, 985, 1970.
3. **Bull, T. E., Forsén, S., and Turner, D. L.,** Nuclear magnetic relaxation of spin 5/2 and spin 7/2 nuclei including the effects of chemical exchange, *J. Chem. Phys.,* 70, 3106, 1979.
4. **Abragam, A.,** *The Principles of Nuclear Magnetism,* Oxford, London, 1978, 314.
5. **Fukushima, E. and Roeder, S. B. W.,** *Experimental Pulse NMR,* Addison-Wesley, Reading, MA, 1981, 158.
6. **Berne, B. J. and Pecora R.,** *Dynamic Light Scattering,* John Wiley & Sons, New York, 1976, 149.
7. **Malli, G. and Froese, C.,** Nuclear magnetic shielding constants calculated from numerical Hartree-Fock wavefunctions, *Int. J. Quantum Chem.,* 1S, 95, 1967.
8. **Ando, L. and Webb, G. A.,** *Theory of NMR Parameters,* Academic Press, New York, 1983, chap. 3.
9. **Kintzinger, J.-P.,** Oxygen NMR: characteristic parameters and applications, in *NMR-17. Oxygen-17 and Silicon-29,* Diehl, P., Fluck, E., and Kosfeld, R., Eds., Springer-Verlag, New York, 1981, 1.
10. **Cerfontain, H., Kruk, C., Rexwinkel, R., and Stunnenberg, F.,** Determination of the intercarbonyl dihedral angle of 1,2-diketones by ^{17}O NMR, *Can. J. Chem.,* 65, 2234, 1987.
11. **You, Xiaozeng, and Weixiong, Wu.,** ^{15}N and ^{17}O NMR chemical shift calculations using the MNDO/ GIAO method, *Magn. Reson. Chem.,* 25, 860, 1987.
12. **Jørgensen, K. A.,** A ^{17}O NMR study on the correlation of ionization potentials and [17]O chemical shifts in 4-substituted pyridine-N-oxides and 4-substituted N-(benzylidene)phenylamine-N-oxides, *Chem. Phys.,* 114, 443, 1987.
13. **Kelly, J. W. and Evans, S. A., Jr.,** Oxygen-17 NMR spectral studies of selected aromatic sulfones, *Magn. Reson. Chem.,* 25, 305, 1987.
14. **Onaka, S., Sugawara, T., Kawada, Y., Yokoyama, Y., and Iwamura, H.,** ^{17}O NMR studies on a series of manganese carbonyl derivatives with Sn-Mn bond(s), *Bull. Chem. Soc. Jpn.,* 59, 3079, 1986.
15. **Guy, M. P., Coffer, J. L., Rommel, J. S., and Bennett, D. W.,** ^{13}C, ^{17}O, and ^{14}N NMR spectroscopic studies of a series of mixed isocyanide/carbonyl complexes of tungsten: $W(CO)_{6-n}(CNR)_n$($R = tert$-butyl,p-tolyl;$n = 1$-3), *Inorg. Chem.,* 27, 2942, 1988.
16. **Klemperer, W. G.,** ^{17}O-NMR spectroscopy as a structural probe, *Angew. Chem. Int. Ed. Engl.,* 17, 246, 1978.
17. **Ferri, D., Glaser, J., and Grenthe, I.,** Confirmation of the structure of $(UO_2)_3(CO_3)_6^{6-}$ by ^{17}O NMR, *Inorg. Chim. Acta,* 148, 133, 1988.
18. **Richens, D. T., Helm, L., Pittet, P.-A., and Merbach, A. E.,** Structural elucidation of oligomeric aquamolybdenum cations in solution by ^{17}O NMR, *Inorg. Chim. Acta,* 132, 85, 1987.

19. **Appelman, E. H., Kostka, A. G., and Sullivan, J. C.,** Oxygen-17 NMR of seven-valent neptunium in aqueous solution, *Inorg. Chem.,* 27, 2002, 1988.
20. **Johnson, C. S., Jr.,** Chemical rate processes and magnetic resonance, in *Advances in Magnetic Resonance,* Vol. 1, Waugh, J. S., Ed., Academic Press, New York, 1965, 33.
21. **Dahn, H. and Ung-Truong, M.-N.,** ^{17}O-NMR spectra of cyclopropenones and tropone. Oxygen exchange with water, *Helv. Chim. Acta,* 70, 2130, 1987.
22. **Swift, T. J. and Connick, R. E.,** NMR-relaxation mechanisms of ^{17}O in aqueous solutions of paramagnetic cations and the lifetime of water molecules in the first coordination sphere, *J. Chem. Phys.,* 37, 307, 1962.
23. **Cossy, C., Helm, L., and Merbach, A. E.,** Water exchange kinetics on lanthanide(III) ions: a variable temperature and pressure ^{17}O NMR study, *Inorg. Chim. Acta,* 139, 147, 1987.
24. **Kuroiwa, Y., Harada, M., and Tomiyasu, H.,** High pressure oxygen-17 NMR study on the kinetics of water exchange reaction in pentaquaoxovanadium(IV), *Inorg. Chim. Acta,* 146, 7, 1988.
25. **Ishii, M., Funahashi, S., and Tanaka, M.,** Variable-pressure oxygen-17 NMR studies on acetic acid exchange of managanese(II) perchlorate and manganese(II) acetate, *Inorg. Chem.,* 27, 3192, 1988.
26. **Isab, A. A., Shaw, C. F., III, and Locke, J.,** GC-MS and ^{17}O NMR tracer studies of Et_3PO formation from auranofin and $H_2{}^{17}O$ in the presence of bovine serum albumin: an *in vitro* model for auranofin metabolism, *Inorg. Chem.,* 27, 3406, 1988.
27. **Kintzinger, J.-P.,** Oxygen-17 NMR, in *NMR of Newly Accessible Nuclei,* Vol. 2, Laszlo, P., Ed., Academic Press, New York, 1983, chap. 4.
28. **Preston, M. A.,** *Physics of the Nucleus,* Addison-Wesley, New York, 1962, 69.
29. **Gerstein, B. C. and Dybowski, C. R.,** *Transient Techniques in NMR of Solids,* Academic Press, New York, 1985, 110.
30. **Snyder, L. C. and Basch, H.,** *Molecular Wave Functions and Properties,* John S. Wiley & Sons, New York, 1972, T-18.
31. **Binkley, J. S., Frisch, M. J., DeFrees, D. J., Raghavachari, K., Whitesides, R. A., Schelgel, H. B., Fluder, E. M., and Pople, J. A.,** *GAUSSIAN-82,* Carnegie-Mellon University, Pittsburgh, 1984.
32. **Reid, R. V., Jr. and Vaida, M. L.,** Quadrupole moment of the deuteron, *Phys. Rev. Lett.,* 34, 1064, 1975.
33. **Verhoeven, J., Dymanus, A., and Bluyssen, H.,** Hyperfine structure of $HD^{17}O$ by beam-maser spectroscopy, *J. Chem. Phys.,* 50, 3330, 1969.
34. **Thaddeus, P., Krisher, L. C., and Loubser, J. H. N.,** Hyperfine structure in the microwave spectrum of HDO, HDS, CH_2O, and CHDO: beam-maser spectroscopy on asymmetric-top molecules, *J. Chem. Phys.,* 40, 257, 1964.
35. **Barfield, M., Gottlieb, H. P. W., and Doddrell, D. M.,** Calculations of deuterium quadrupole coupling constants employing semiempirical molecular orbital theory, *J. Chem. Phys.,* 69, 4504, 1978.
36. **Butler, L. G. and Brown, T. L.,** Nuclear quadrupole coupling constants and hydrogen bonding. A molecular orbital study of oxygen-17 and deuterium field gradients in formaldehyde-water hydrogen bonding, *J. Am. Chem. Soc.,* 103, 6541, 1981.
37. **Gready, J. E., Bacskay, G. B., and Hush, N. S.,** Comparison of the effects of symmetric versus asymmetric H bonding on the ^2H and ^{17}O nuclear quadrupole coupling constants: application to formic acid and the hydrogen diformate ion, *Chem. Phys.,* 64, 1, 1982.
38. **Gordy, W. and Cook, R. L.,** *Microwave Molecular Spectra,* John Wiley & Sons, New York, 1984.
39. **Edmonds, D. T.,** Nuclear quadrupole double resonance. *Phys. Rep. C,* 29, 233, 1977.
40. **Walter, T. H., Turner, G. L., and Oldfield, E.,** Oxygen-17 cross-polarization NMR spectroscopy of inorganic solids, *J. Magn. Reson.,* 76, 106, 1988.
41. **Flygare, W. H. and Lowe, J. T.,** Experimental study of the nuclear-quadrupole and spin-rotation interaction of ^{17}O in formaldehyde, *J. Chem. Phys.,* 43, 3645, 1965.
42. **Balle, T. J. and Flygare, W. H.,** Fabry-Perot cavity pulsed Fourier transform microwave spectrometer with a pulsed nozzle particle source, *Rev. Sci. Instrum.,* 52, 33, 1981.
43. **Sheridan, J.,** Recent studies of nuclear quadrupole effects in microwave spectroscopy, *Adv. Nucl. Quadrupole Resonance,* 5, 125, 1983.
44. **Kukolich, S. G., Rund, J. V., Pauley, D. J., and Bumgarner, R. E.,** Nitrogen quadrupole coupling in cyclopentadienylnickel nitrosyl, *J. Am. Chem. Soc.,* 110, 7356, 1988.
45. **Slusher, R. E. and Hahn, E. L.,** Sensitive detection of nuclear quadrupole interactions in solids, *Phys. Rev.,* 166, 332, 1968.
46. **Ader, R. and Shporer, M.,** A double-resonance spectrometer for pure NQR detection, *J. Magn. Reson.,* 47, 483, 1982.
47. **Hsieh, Y., Koo, J. C., and Hahn, E. L.,** Pure nuclear quadrupole resonance of naturally abundant ^{17}O in organic solids, *Chem. Phys. Lett.,* 13, 563, 1972.
48. **Creel, R. B., Brooker, H. R., and Barnes, R. G.,** Exact analytic expressions for NQR parameters in terms of the transition frequencies, *J. Magn. Reson.,* 41, 146, 1980.
49. **Cheng, C. P. and Brown, T. L.,** Oxygen-17 nuclear quadrupole double resonance spectroscopy. I. Introduction. Results for organic carbonyl compounds, *J. Am. Chem. Soc.,* 101, 2327, 1979.

50. **Poole, Jr., C. P. and Farach, H. A.,** *Theory of Magnetic Resonance,* 2nd ed., Wiley-Interscience, New York, 1987.
51. **Baugher, J. F., Taylor, P. C., Oja, T., and Bray, P. J.,** Nuclear magnetic resonance powder patterns in the presence of completely asymmetric quadrupole and chemical shift effects: application to metavanadates, *J. Chem. Phys.,* 50, 4914, 1969.
52. **Kundla, E., Samoson, A., and Lippmaa, E.,** High-resolution NMR of quadrupolar nuclei in rotating solids, *Chem. Phys. Lett.,* 83, 229, 1981.
53. **Fukushima, E. and Roeder, S. B. W.,** *The Experimental Pulse NMR,* Addison-Wesley, Reading, MA, 1981, 110.
54. **Townes, C. H. and Dailey, B. P.,** Determination of electron structure of molecules from nuclear quadrupole effects, *J. Chem. Phys.,* 17, 782, 1949.
55. **Flygare, W. H.,** *Molecular Structure and Dynamics,* Prentice-Hall, Englewood Cliffs, NJ, 1978, 315.
56. **Cheng, C. P. and Brown, T. L.,** Oxygen-17 nuclear quadrupole double resonance spectroscopy, *Symp. Faraday Soc.,* No. 13, 75, 1979.
57. **Cheng, C. P. and Brown, T. L.,** Oxygen-17 nuclear quadrupole double-resonance spectroscopy. III. Results for N-O, P-O, and S-O bonds, *J. Am. Chem. Soc.,* 102, 6418, 1980.
58. **Gready, J. E.,** The relationship between nuclear quadrupole coupling constants and the asymmetry parameter. The interplay of theory and experiment, *J. Am. Chem. Soc.,* 103, 3682, 1981.
59. **Edmonds, D. T., Hunt, M. J., and Mackay, A. L.,** Pure quadrupole resonance of ^{14}N in a tetrahedral environment, *J. Magn. Reson.,* 9, 66, 1973.
60. **Rubenacker, G. V. and Brown, T. L.,** Nitrogen-14 nuclear quadrupole resonance spectra of coordinated pyridine. An extended evaluation of the coordinated nitrogen model, *Inorg. Chem.,* 19, 392, 1980.
61. **Rubenacker, G. V. and Brown, T. L.,** Nitrogen-14 nuclear quadrupole resonance spectra of pyridine-halogen complexes, *Inorg. Chem.,* 19, 398, 1980.

Chapter 2

17O-ENRICHMENT METHODS

George W. Kabalka* and Naganna M. Goudgaon

TABLE OF CONTENTS

---

*   Asterisks following authors' names denote individuals to whom correspondence may be addressed.

I. INTRODUCTION

Isotopically labeled compounds have been utilized in research since Hevesy first proposed the concept of tracers in 1913.[1] Although radioisotopes generally come to mind in discussions focused on tracer techniques, stable isotopes actually received more attention in scientific research in the early part of the twentieth century because of the development of mass spectrometers which had sufficient resolution to identify the isotopic distribution of naturally occurring elements.[2]

Scientific interest in tracer techniques rapidly shifted to radioisotope technologies in the 1940s with the establishment of the Atomic Energy Commission. The subsequent availability of carbon-14, coupled with the development of simple radiation detection equipment, insured that radiotracer techniques would overshadow the use of stable isotopes as tracers for nearly twenty years.[3,4] It is quite possible that the use of radioisotopes in the clinical arena will experience a dramatic expansion during the next decade due to the recent development of relatively inexpensive medical cyclotrons which are used to produce short-lived positron-emitting nuclides such as carbon-11, nitrogen-13, and oxygen-15.[5,6]

Developments in the stable isotopes area have been no less dramatic. The U.S. Atomic Energy Commission initiated a program in 1969 to make stable isotopes readily available at reasonable prices. The availability of the less abundant, stable isotopes of carbon, nitrogen, and oxygen, coupled with recent developments in gas chromatography-mass spectrometry and multinuclear magnetic resonance spectroscopy, has once again pushed stable isotopes to the forefront of tracer research. These advances, along with the absence of sufficiently long-lived radioisotopes of oxygen, ensure that oxygen-17 will play an important role in research in the foreseeable future. Currently, oxygen-17 and oxygen-18 are utilized extensively in tracer studies involving nuclear magnetic resonance[7,8] (NMR) and mass spectrometry.[9,10]

Syntheses involving oxygen isotopes tend to involve rather straight-forward organic reactions. In addition, since isotopes of a given element are chemically equivalent, any reaction involving oxygen-16 can be utilized to incorporate oxygen-17, oxygen-18 and, theoretically, oxygen-15. As a consequence, we have included syntheses of oxygen-18 labeled compounds in this chapter in instances which help to elaborate known routes to oxygen-17 labeled materials or which predicate new routes to oxygen-17 labeled materials. There is one caveat that should be kept in mind when contemplating the synthesis of an oxygen-17 labeled compound and that is the fact that the starting reagents tend to be quite expensive, on the order of $20,000 to $100,000 per mole of the oxygen-17 enriched reagent. A consequence of the expense is the realization that the limiting reagent must be the oxygen-17 enriched material (normally water, carbon dioxide, or molecular oxygen). Ironically, it is these reagents which are traditionally used in excess by the synthetic chemist. Fortunately, it is not difficult to use these molecules as limiting reagents. Simple modifications of existing reactions are required; oxygen gas and carbon dioxide, for example, are added to evacuated systems as opposed to the more traditional procedures which involve bubbling them through the reaction mixture.

II. SYNTHESIS OF 17O-LABELED COMPOUNDS

A. HYDROLYSIS REACTIONS
1. Substitution Reactions
One of the most straightforward routes to simple oxygen-17 labeled materials involves substitution of a labeled hydroxide for a labile leaving group such as a halide (Equation 1). Not surprisingly, a variety of

$$R\text{-}X \xrightarrow{\overset{17}{O}H^{-}} \overset{17}{R\text{-}O}H \tag{1}$$

procedures have been developed. In this section, the classic nucleophilic substitution reactions are summarized, whereas addition-elimination procedures will be highlighted in Section II.A.2.

Chang and le Noble prepared [^{17}O]-*exo*-2-norbornyl brosylate and its ^{17}O-sulfonyl analog to study ion-pair return by means of ^{17}O NMR spectroscopy.[11] [^{17}O]-*Exo*-2-norbornyl brosylate was prepared by heating 2-norbornyl bromide with one equivalent each of 40% [^{17}O]-water, mercuric bromide and 2,6-di-*tert*-butylpyridine in glyme at 75°C in a sealed tube for 48 h. Conversion of the 2-norbornanol to the desired product was accomplished using Winstein's procedure[12] (Equation 2). The sulfonyl-^{17}O analog was also prepared by reacting ^{17}O-enriched *p*-bromobenzenesulfonyl chloride with unlabeled 2-norbornanol (Equation 3). Labeled *p*-bromobenzenesulfonyl chloride was prepared by heating *p*-bromobenzenesulfonyl chloride with one equivalent of [^{17}O]-water, and then reconverting the sulfonic acid so obtained with thionyl chloride (30 min of reflux with a trace of dimethylformamide) (Equation 4).

$$\tag{2}$$

$$\tag{3}$$

$$\tag{4}$$

Gragerov and co-workers[13] synthesized [^{18}O]-ethanol by heating ethyl iodide with Ag$_2$O in ^{18}O-labeled water at 100°C in a sealed tube for 10 h. Similarly, [^{18}O]-methanol was prepared by heating methyl iodide with silver oxide in labeled [^{18}O]-water at 100°C in a sealed tube for 8 h (Equation 5). The process could certainly be utilized to prepare many simple, oxygen-17 labeled alcohols.

$$CH_3CH_2\text{-}I \xrightarrow[\overset{18}{H_2O}]{Ag_2O} CH_3CH_2\overset{18}{O}H \tag{5}$$

In another sealed tube experiment involving oxygen-18 labeled water, a number of ^{18}O-labeled phenolic compounds[14] were prepared. The procedure involves heating a mixture of phenol and ^{18}O-enriched water in a sealed tube at 180°C (Equation 6).

$$\tag{6}$$

Other studies involved oxygen-18 substitution for halogens and alkoxide groups to yield carbonyl derivatives. Koenig prepared *N*-benzoyl-[^{18}O]-*tert*-butylhydroxylamine to investigate the reactivity of radicals formed during thermal decomposition of *N*-benzoyl-*N*-nitroso-

O-tert-butylhydroxylamine.[15] The synthesis of the ¹⁸O-labeled compound started from ben-zotrichloride and labeled water (Equations 7 and 8).

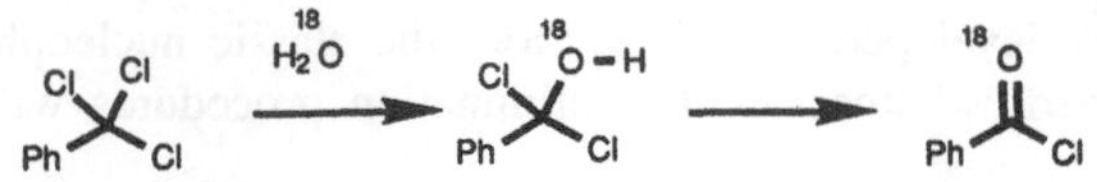

$$(7)$$

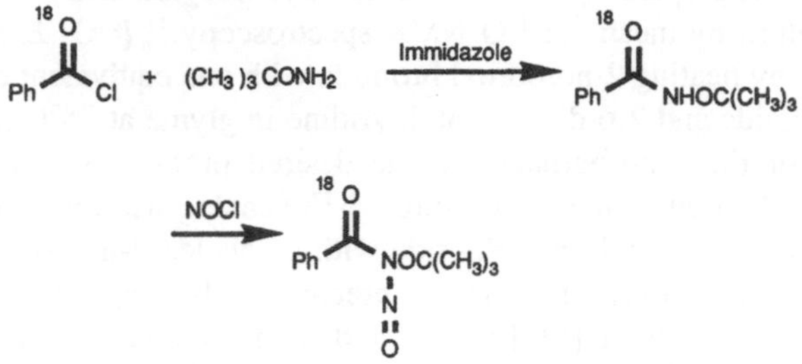

$$(8)$$

In a mechanistically related synthesis, Sawyer developed a simple high yield method[16] for the synthesis of [¹⁸O]-methanol and [¹⁸O]-ethanol. The method involves the reaction of tri-*n*-butyl orthoformate with H₂¹⁸O in the presence of HCl, followed by lithium aluminium hydride reduction of the resulting ¹⁸O-labeled butyl formate generated [¹⁸O]-methanol (Equation 9).

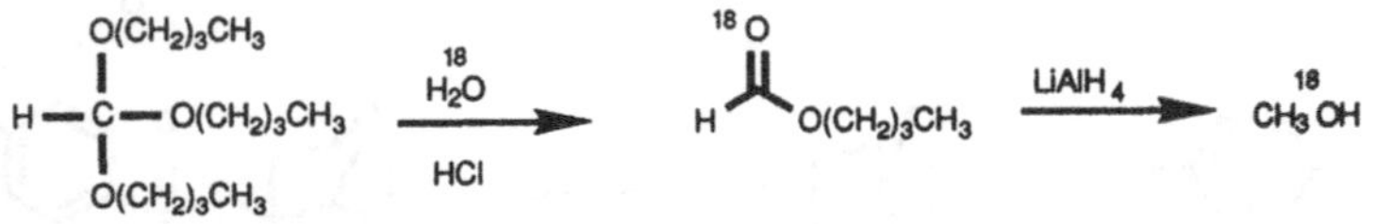

$$(9)$$

Similarly, [¹⁸O]-ethanol was prepared by the acid-catalyzed hydrolysis of 1,1-dipro-poxyethane followed by reduction of the intermediate aldehyde to alcohol (Equation 10).

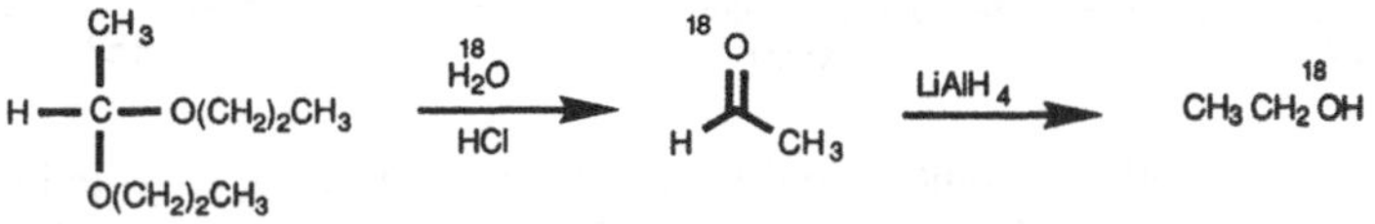

$$(10)$$

The Sawyer and Koenig methods both involve an initial substitution of a labeled hy-droxide moiety for the leaving group, followed by subsequent formation of a carbonyl derivative (Equation 11). Although the mechanisms

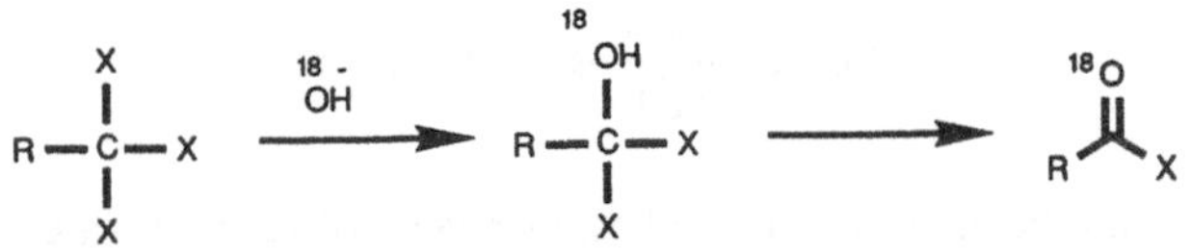

$$(11)$$

vary slightly, the methods are clearly suitable for the formation of a variety of oxygen-17 labeled agents. McClelland utilized an ortho ester to prepare oxygen-18 both a labeled ester and amide.[17] ¹⁸O-Labeled methyl benzoate was prepared by the reaction of trimethyl or-thobenzoate with 90% [¹⁸O]-water in the presence of a small amount of HCl. Treatment of the methyl benzoate obtained with NaNH₂ in liquid NH₃ resulted in the formation of the labeled benzamide (Equation 12).

$$\text{(12)}$$

Gold et al. investigated the reaction of propane-1-diazotic acid and optically pure (*S*)-1-phenyl-ethanediazotic acid in $H_2{}^{18}O$ at neutral (pH 7.0 to 8.5) and basic (pH >14) conditions.[18] They found that the extent of conversion is not dependent on structure, but rather on the degree of proton-transfer-mediated equilibration of the original ${}^{16}OH^-$ counterion with $[{}^{18}O]$-H_2O, prior to the C-N bond cleavage (Equation 13).

$$\text{Ph} - \overset{|}{\underset{CH_3}{CH}} - N \equiv N - X \quad \xrightarrow{\overset{18}{H_2O}} \quad \text{Ph} - \overset{|}{\underset{CH_3}{CH}}\,\overset{18}{OH} \qquad \text{(13)}$$

More recently,[19] Turro, Paczkowski, and Wan prepared oxygen-17 labeled ketones via hydrolysis of the corresponding ethylene glycol acetal with the oxygen-labeled water (Equation 14).

$$\text{(14)}$$

The procedure is closely related to the Sawyer synthesis of labeled alcohols via the intermediacy of a labeled ester.

The substitution reactions are not limited to carbon centers. Gerlt et al. reported the preparation of ${}^{17}O$ labeled mono-, di-, and trimethyl phosphate, AMP, ADP, and ATP. Trimethyl $[{}^{17}O]$-phosphate was synthesized by reacting enriched $P{}^{17}OCl_3$ with excess methanol[20] (Equation 15). Evaporation of the solvent and HCl furnished the labeled phosphate having a 49% ${}^{17}O$ isotopic composition.

$$PCl_5 \quad \xrightarrow{\overset{17}{H_2O}} \quad \overset{17}{O}PCl_3 \quad \xrightarrow{CH_3OH} \quad \overset{17}{O}P(OCH_3)_3 \qquad \text{(15)}$$

Dimethyl $[{}^{17}O]$-phosphate (sodium salt) and monomethyl $[{}^{17}O]$-phosphate (disodium salt) were prepared by the reaction of trimethyl $[{}^{17}O]$-phosphate with excess NaI in refluxing acetone (Equation 16). $[{}^{17}O]$-AMP was prepared

$$\overset{17}{O} \equiv P(OCH_3)_3 \quad \xrightarrow[Me_2CO]{NaI} \quad CH_3O - \overset{\overset{17}{O}}{\underset{\underset{O}{|}}{\overset{||}{P}}} - OCH_3 \quad \xrightarrow[Me_2CO]{Excess\ NaI} \quad {}^-O - \overset{\overset{17}{O}}{\underset{\underset{O}{|}}{\overset{||}{P}}} - OCH_3 \qquad \text{(16)}$$

by reaction of $P{}^{17}OCl_3$ with adenosine dissolved in triethyl phosphate.[21] $[\alpha\text{-}{}^{17}O]$-ADP and $[\beta\text{-}{}^{17}O]$-ADP were prepared according to the procedure of Hoard and Ott[22] from $[{}^{17}O]$-AMP and appropriate equivalents of unlabeled inorganic phosphate. $[\alpha\text{-}{}^{17}O]$-ATP was prepared from $[{}^{17}O]$-AMP and five equivalents of unlabeled pyrophosphate. $[\beta\text{-}{}^{17}O]$-ATP and $[\gamma\text{-}{}^{17}O]$-ATP were synthesized using Ott's procedure[22] by the reaction of labeled ADP with inorganic phosphate.

Gerlt and co-workers also prepared chiral $[{}^{17}O, {}^{18}O]$phosphodiesters to determine the stereochemical course of both enzymatic and nonenzymatic hydrolyses by means of ${}^{17}O$

NMR spectroscopy.[23] The diastereomeric ^{17}O-enriched *p*-anilidates of cyclic dAMP were synthesized using enriched $P^{17}OCl_3$ and were reacted separately with a tenfold excess of 90% enriched $C^{18}O_2$.

2. Addition-Elimination Reactions

The reversible addition of water to the electron deficient carbon of a carbonyl group is a reaction familiar to synthetic chemsits (Equation 17). The reaction can be carried out under a variety of acid and base

$$\underset{}{\overset{}{>}}C=O \; \underset{}{\overset{H_2O}{\rightleftharpoons}} \; HO-\underset{|}{\overset{|}{C}}-OH \tag{17}$$

catalyzed conditions on carbonyl groups, which can be present in the molecule in the acid, ester, acid chloride, and anhydride form. Not surprisingly, because of the reversible nature of the reaction, it has often been utilized to incorporate oxygen isotopes into organic molecules.

a. Hydrate Formation

The formation of hydrates is not generally synthetically useful because most hydrates are unstable unless the carbonyl group is sterically unhindered or particularly electron deficient. Formalin and chloral hydrate (''Mickey Finn'') are two familiar examples of the few stable hydrates available. It is the inherent instability of the hydrates which serve to make them useful intermediates in the syntheses of labeled molecules using oxygen-labeled water (Equation 18).

$$H_2^{17}O \; + \; \underset{R}{\overset{O}{\underset{}{R}}}\!\! \rightleftharpoons \; R-\underset{OH}{\overset{^{17}OH}{C}}-R \; \rightleftharpoons \; \underset{R}{\overset{^{17}O}{\underset{}{R}}} \tag{18}$$

Boykin and co-workers prepared oxygen-17 enriched benzyl alcohols and benzyl acetates for a NMR study focused on substituent effects.[24] Labeled benzyl alcohols were prepared by acid catalyzed exchange of the corresponding benzaldehydes with enriched water, followed by a reduction using sodium borohydride (Equation 19). Enriched benzyl acetates were synthesized by reacting labeled benzyl alcohols with acetic anhydride in pyridine (Equation 20).

$$X-\!\!\!\left\langle \bigcirc \right\rangle\!\!\!-CHO + H_2^{17}O \longrightarrow X-\!\!\!\left\langle \bigcirc \right\rangle\!\!\!-\overset{^{17}O}{CH} \overset{NaBH_4}{\longrightarrow} X-\!\!\!\left\langle \bigcirc \right\rangle\!\!\!-^{17}CH_2OH \tag{19}$$

$$X-\!\!\!\left\langle \bigcirc \right\rangle\!\!\!-^{17}CH_2OH \; + \; (CH_3CO)_2O \overset{Pyridine}{\longrightarrow} X-\!\!\!\left\langle \bigcirc \right\rangle\!\!\!-^{17}CH_2OCOCH_3 \tag{20}$$

Crandall, Centeno, and Borreson utilized this simple exchange reaction to prepare a series of oxygen-17 labeled cyclohexanones[25] (Equation 21).

$$\tag{21}$$

Gorodetsky et al. synthesized ^{17}O-labeled β-diketones and utilized them to study the relative concentrations of the two enol tautomers by means of ^{17}O NMR spectroscopy.[26]

Totally labeled diketo compounds (both carbonyl groups) were prepared by exchange with ^{17}O-enriched water. The compounds were mixed with an excess of [^{17}O]-water (molar ratio 1:4) and diluted with dioxane. The reaction solution was refluxed overnight under nitrogen atmosphere and the labeled compound recovered by distilling off the solvent. Under these conditions the oxygen atoms of acetylacetone, benzoylacetone, acetylcyclohexanone, acetylcyclopentanone, and 1,2-cyclohexadione exchange both oxygen atoms with H$_2^{17}$O (Equation 22).

$$R_1 \overset{O}{\diagdown}\overset{O}{\diagup} R_2 \quad \xrightarrow[\text{Dioxane}]{H_2^{17}O} \quad R_1 \overset{^{17}O}{\diagdown}\overset{^{17}O}{\diagup} R_2 \qquad (22)$$

On the other hand, tropolone did not exchange sufficiently to show a detectable ^{17}O resonance, while formyl and acetyl camphor and benzocyclohexanone incorporated ^{17}O into only one of their two carbonyl groups. For these compounds, hydrochloric acid was added to the reaction mixture to make the solution sufficiently acidic to achieve the exchange.

Dahn et al. utilized a similar reaction to prepare a series of cyclic ketones some of which contained hetero atoms[27] (Equation 23). They investigated the rate of the exchange reaction and found that the oxygen and sulfur containing heterocycles exchanged at a slower rate

$$(CH_2)_n \overset{O}{\underset{X}{\diagdown C \diagup}} (CH_2)_n \quad \underset{\longleftarrow}{\overset{H_2^{17}O}{\longrightarrow}} \quad (CH_2)_n \overset{^{17}O}{\underset{X}{\diagdown C \diagup}} (CH_2)_n \qquad (23)$$

n = 2,3,4

X = O, S, CH$_2$, NR

than their carbocyclic counterparts. Interestingly, the transannular reaction between the nitrogen and the carbonyl group in N-ethyl-azacylodecan-6-one completely inhibits the exchange reaction.

Byrn and Calvin also investigated the exchange reaction between oxygen-labeled water and aldehydes and ketones.[28] They utilized oxygen-18 enriched water in their studies and measured the exchange rates using infrared spectroscopy. They evaluated the exchange rates in terms of steric and electronic considerations and attempted to correlate their results obtained using simple molecules to results obtained in biological molecules such as chlorophyl.

Follmann and Hogenkamp utilized the hydrate sequence[29] in an elegant synthesis of D-ribose-2-^{18}O and D-ribose-3-^{18}O by the use of 1,2:5,6-di-O-isopropylidene-α-D-allofuranose, 2, as a common precursor. Keto sugar, 2, obtained by oxidation of diisopropylidene-D-glucose, 1, is especially attractive for the incorporation of oxygen isotopes as it forms a stable hydrate, 3. Treatment of 2 with ^{18}O-enriched water (96.5%) furnishes a hydrate carrying labeled oxygen in the gem-diol group. Its reduction with sodium borohydride in tetrahydrofuran generates 1,2:5,6-di-O-iso-propylidene-α-D-allofuranose, 4, containing ^{18}O in the 3-hydroxyl group (Equation 24). The authors also prepared ^{18}O-labeled nucleosides and nucleotides by making use of labeled ribose derivatives.

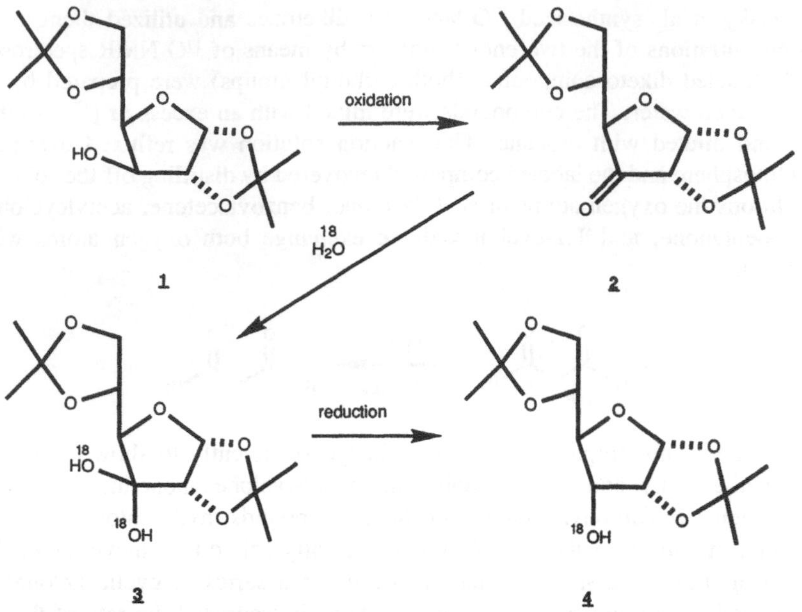

(24)

b. Carboxylic Acid Exchange

Carboxylic acids readily undergo an oxygen-exchange reaction analogous to the hydration reaction of aldehydes and ketones (Equation 25). The label is, of course, equally distributed between the hydroxyl and carbonyl groups. The reaction has been extremely popular for preparing oxygen-17 labeled amino acids.

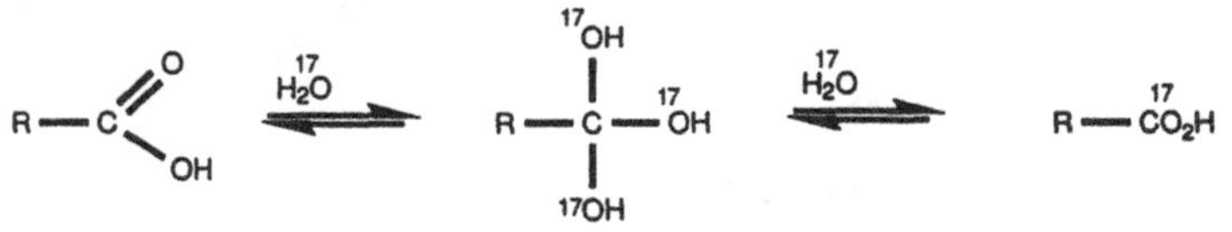

(25)

Gerothanassis prepared labeled amino acids[30,31] to study ^{17}O chemical shifts by NMR spectroscopy. The enrichment method involves isotopic exchange between the carbonyl function of an amino acid and [^{17}O]-H_2O in the presence of a strong acid at elevated temperatures (Equation 26). In a typical experiment, 5 M HCl was added to a mixture of glycine and ^{17}O-enriched H_2O. Glycine, alanine, cystine, glutamic, and aspartic acid were prepared utilizing this straightforward exchange reaction.

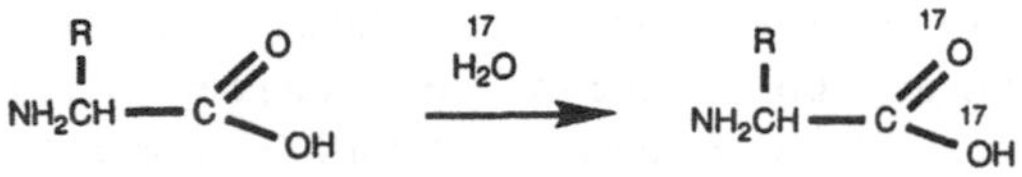

(26)

Baldwin and co-workers prepared both oxygen-17 and oxygen-18 labeled L-α-aminoadipic acid using an exchange reaction[32] (Equation 27). They were used as a precursor in the synthesis of enriched tripeptide, δ-(L-α-aminoadipyl)-L-cysteinyl-D-valine[LLD-ACV*].

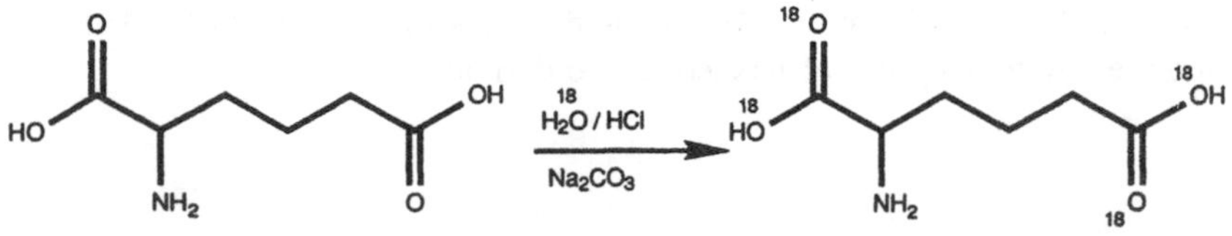

(27)

Jolivet and co-workers[33] prepared oxygen-18 labeled glycolate via exchange with oxygen-18 labeled water. The product was purified by ion-exchange chromatography.

Ponnusamy and Fiat synthesized[34] an oxygen-17 labeled protected L-proline and L-pyroglutamic acid at the carboxyl group by acid catalyzed exchange of oxygen-17 labeled water. The α-amino group of the amino acids were protected by *tert*-butyloxycarbonylation.

Recently, Eckert and Fiat described the synthesis of tyrosine labeled at both the phenolic site and the carboxylic acid site.[35] [17]O-Enriched tyrosine is prepared by selective diazotization of the *p*-amino group of *p*-aminophenylalanine followed by hydrolysis in enriched water (Equation 28).

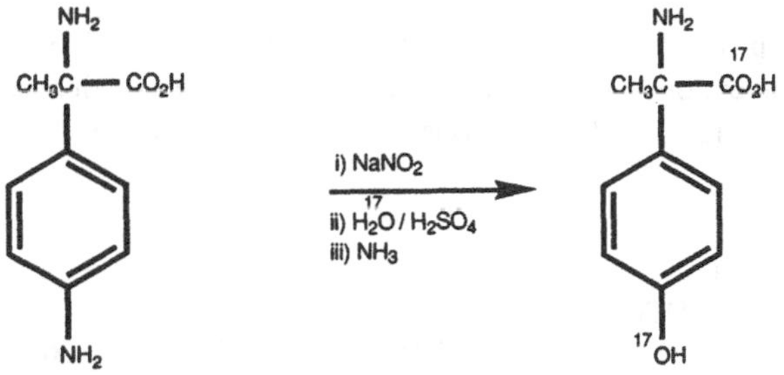

(28)

In a typical experiment *p*-aminophenylalanine was dissolved in enriched water and the solution acidified with H_2SO_4. A solution of $NaNO_2$ in water was added with stirring and the resulting solution heated until the N_2 evolution ceased. On neutralization with NH_3 gas, the tyrosine precipitated.

c. *Ester Hydrolysis*

The hydrolysis of esters under either acid or base (saponification) conditions offers a straightforward route to oxygen-labeled acids (Equation 29). Indeed, ester hydrolysis was utilized as an alternate route to the labeled carboxylic acids discussed in Section II.A.2.b.

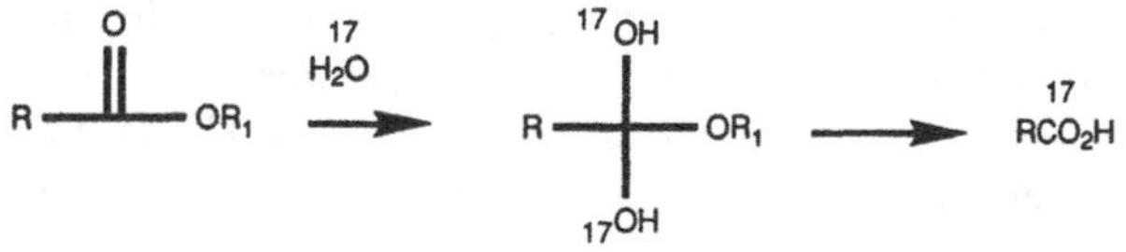

(29)

For example, Lauterwein et al. prepared several [17]O-enriched alkanoic amino acids[36] by hydrolyzing the amino acid methyl ester in basic [[17]O]-H_2O (Equation 30). In the case of asparagine, glutamine, and cystine, the enrichment was performed using an acid-catalyzed exchange. They utilized the labeled acids in an oxygen-17 nuclear magnetic resonance chemical shift investigation.

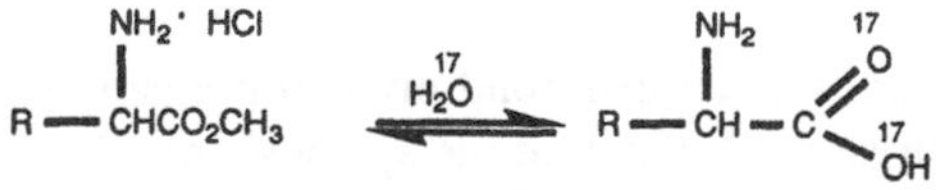

(30)

Steinschneider et al. also prepared enriched amino acids[37] via saponification of the corresponding methyl esters in highly enriched [[17]O]-H_2O (Equation 31).

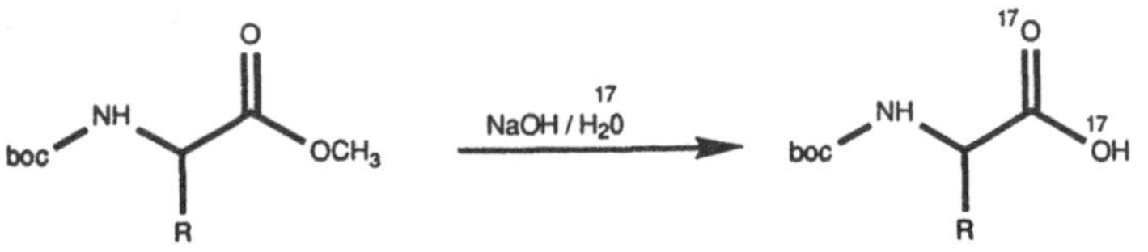

(31)

This approach has been extended to selectively label the γ-COOH group of glutamic acid. However, excessive amounts of labeled water would be required if these methods were to be applied on a preparative scale of the less soluble amino acids. They also explored the acid-catalyzed exchange of oxygen-17 into the α-COOH group of more water soluble unprotected amino acids, subsequently introducing the N-α-boc protecting group in a separate step (Equation 32).

$$
\underset{H_2NCHCO_2H}{\overset{R}{|}} \quad \xrightarrow{\text{6 N HCl / } H_2\overset{17}{O}} \quad \underset{H_2NCHCO_2H}{\overset{R}{|}}{\overset{17}{}} \tag{32}
$$

The same group also reported[38] enrichment with oxygen-17 of several amino acids starting with the conversion to the *O*-methyl ester of their *N*-t-boc derivatives (Equation 33). Oxygen-17 was then introduced in aqueous solution or in dioxane.

$$
\underset{RCHCO_2H}{\overset{\overset{\displaystyle O}{\|}}{N\overset{\displaystyle}{H}COC_4H_9\text{-}t}} \xrightarrow{\text{DCC, } CH_3OH} \underset{RCHCO_2CH_3}{\overset{\overset{\displaystyle O}{\|}}{N\overset{\displaystyle}{H}COC_4H_9\text{-}t}} \xrightarrow[\text{HCl}]{NaOH/H_2\overset{17}{O}} \underset{RCHCO_2H}{\overset{\overset{\displaystyle O}{\|}}{N\overset{\displaystyle}{H}\overset{17}{C}OC_4H_9\text{-}t}}{\overset{17}{}} \tag{33}
$$

$$
\xrightarrow{\text{dry HCl}} \underset{RCHCO_2H}{\overset{NH_2}{|}}{\overset{17}{}}
$$

Baltzer and Becker reported the synthesis of [^{17}O]-benzoic acid by alkaline hydrolysis of methyl benzoate in aqueous acetone containing [^{17}O]-H_2O.[39]

d. Hydrolysis of Carbonyl Analogs

Boykin et al. reported the preparation of oxygen-17 enriched cinnamic acid, methyl cinnamate, *p*-substituted benzoic acids, and methyl benzoates.[40] They investigated the effect of substituents on chemical shifts using ^{17}O NMR spectroscopy. The enrichment method involves the hydrolysis of the appropriate acid chloride with $H_2{}^{17}O$ (Equation 34). Labeled methyl esters were prepared by reaction of the enriched acids with diazomethane (Equation 35).

$$
X\!\!-\!\!\langle\!\!\bigcirc\!\!\rangle\!\!-\!\!COCl \quad + \quad H_2\overset{17}{O} \quad \longrightarrow \quad X\!\!-\!\!\langle\!\!\bigcirc\!\!\rangle\!\!-\!\!\overset{17}{C}O_2H \tag{34}
$$

$$
X\!\!-\!\!\langle\!\!\bigcirc\!\!\rangle\!\!-\!\!\overset{17}{C}O_2H \quad + \quad CH_2N_2 \quad \longrightarrow \quad X\!\!-\!\!\langle\!\!\bigcirc\!\!\rangle\!\!-\!\!\overset{17}{C}O_2CH_3 \tag{35}
$$

Virtually any polarized, unsaturated bond involving a carbon atom can be involved in a hydrolysis reaction. Burgar et al. reported the synthesis of [^{17}O]-cytosine and investigated its use in the study of hydrogen bonding between nucleic acid bases by ^{17}O NMR spectroscopy.[41] They have synthesized ^{17}O-labeled cytosine by direct hydrolysis of 2-chloro-6-aminopyrimidine with ^{17}O-labeled water at 140°C for 2 h (Equation 36). The reaction involves the nucleophilic addition of labeled water to a chloro substituted carbon-nitrogen double bond followed by expulsion of chloride and tautomerization of the resulting enol.

$$(36)$$

Attempts to label cytosine by the direct exchange method with ^{17}O-enriched water were not successful.

2-Chloro-6-aminopyrimidine was treated with ^{17}O-enriched water to furnish a mixture of cytosine and uracil. ^{17}O-Labeled cytosine was separated by the procedure described by Hilbert and Johnson.[42] The enrichment was determined in aqueous solution and was found to be 40% of the maximum possible value.

A more traditional addition-elimination reaction was utilized by Wang et al. to prepare ^{18}O-enriched nucleic acid bases involving a nucleophilic addition of labeled water to the carbon-nitrogen[43] (Equation 37). The incorporation occurs with all compounds studied at C-4 to the extent of 80% and is found to decrease drastically with substitution at C-5 position. The differences in the rates of incorporation were rationalized in terms of steric, electronic, and tautomeric effects.

$$(37)$$

In a typical experiment, a few milligrams of uracil was mixed with a drop of ^{18}O-enriched H_2O and a catalytic amount of $SOCl_2$ in a sealed pyrex glass tubing and heated for 15 h at 100°C. The enriched compound was isolated using common purification methods. Similarly, ^{17}O-enriched bases or nucleic acids can be prepared.

Aoyama et al. investigated the acid-catalyzed oxygen exchange of tertiary nitroso compounds with ^{18}O-enriched water[44] (Equation 38).

$$(38)$$

e. Addition to Heteroatom System

The addition-elimination reactions of labeled water are not, of course, limited to unsaturated systems containing carbon atoms. A variety of nitrogen, phosphorous, and sulfur compounds are available which readily undergo nucleophilic attack by water.

Goldberg and Walseth prepared oxygen-labeled nucleotide phosphonyls by a phosphodiesterase-promoted hydrolysis in the presence of ^{18}O-enriched water[45] (Equation 39).

$$(39)$$

Kobayashi et al. prepared oxygen-18 enriched p-bromobenzenesulfonic and p-toluensulfinic acid to study the rates of oxygen exchange reactions[46] (Equation 40). The enrichment was achieved by dissolving the acid in $H_2^{18}O$ and dioxane, and heating the resulting solution to 90°C for the required time. The product was purified either by distillation under reduced pressure or crystallization.

$$R-\overset{\overset{O}{\|}}{S}-OH \quad \xrightarrow[\Delta]{\overset{18}{H_2O}} \quad R-\overset{\overset{18}{\overset{O}{\|}}}{S}-\overset{18}{OH} \tag{40}$$

King et al. synthesized a series of oxygen-18 labeled sultones[47] (Equation 41). One of the pathways involved an intramolecular ring closure of an oxygen-labeled γ-hydroxy group on a sulfonyl chloride.

$$\overset{18}{HO}CH_2CH_2CH_2SO_2Cl \quad \longrightarrow \tag{41}$$

Dahn and Ung-Truong prepared ^{17}O-labeled mesoionic compounds to study oxygen-17 chemical shifts by NMR spectroscopy. The enrichment was achieved utilizing nitrosation conditions in which the oxygen exchange of a nitrosating agent with water is more rapid than its attack on an amine (Equations 42 and 43).[48]

$$\overset{+}{H_2NO_2} + \overset{17}{H_2O} \quad \rightleftharpoons \quad H_2\overset{17}{N}\overset{+}{O_2} + H_2O \tag{42}$$

$$H_2\overset{17}{N}\overset{+}{O_2} + Ph\text{-}NH\text{-}CH_2COOH \longrightarrow Ph\text{-}\underset{\underset{17\,O}{\overset{\|}{N}}}{N}\text{-}CH_2COOH \longrightarrow Ph-N \tag{43}$$

A typical procedure involves the nitrosation in aqueous HCl solution containing 1.38% of ^{17}O, 10% of ^{18}O and N-phenylglycine, and cyclizing the isolated labeled N-nitroso compound. The tracer incorporation was determined by mass spectra and was found to be ≥99% based on M$^+$ peak.

B. SOLVOLYSIS REACTIONS
1. Substitution Reactions

In theory, solvolysis reactions should provide a convenient route to oxygen-labeled ethers and esters. In reality, the reactions are not widely used presumably due to the difficulty, and/or expense, associated with obtaining the necessary oxygen-labeled precursors. As an example of the potential of these reactions, Dessings et al. synthesized oxygen-17 labeled, 1,4-anhydro-6-azido-2,3-di-O-benzoyl-6-deoxy-β-D-galactopyranose, **6**, from 1-O-acetyl-2,3-di-O-benzoyl-4,6-bis-O-(methylslfonyl)-α-D-glucopyranose, **5**[49] (Equation 44).

5   **6**

2. Addition-Elimination Reactions

As noted in Section II.A.2., the hydrolysis of carboxylic acid derivatives is a convenient route to a variety of labeled carboxylic acids. Parallel syntheses of ester derivatives are possible but few in number.

Kursanov and Kudryavtsev prepared oxygen-18 labeled ethyl propionate to study the mechanism during hydrolysis in basic medium. Labeled ethanol reacted with propionyl chloride to generate the desired EtCO[18]OEt[50] (Equation 45).

$$CH_3CH_2COCl \ + \ \overset{18}{CH_3CH_2OH} \ \longrightarrow \ CH_3CH_2\overset{\overset{O}{\|}}{C}\text{-}O^{18}\text{-}CH_2CH_3 \tag{45}$$

More recently, Boykin and co-workers[51] synthesized oxygen-17 enriched 2,2-dimethylsuccinic anhydride utilizing Bruice's procedure[52] (Equation 45).

$$\tag{46}$$

C. OXIDATION REACTIONS
1. Oxidations

A straightforward approach to oxygen-17 labeled compounds involves simple oxidation reactions between [17O]-O$_2$ and reactive substrates. Suprisingly, this approach has not been extensively investigated. Kabalka et al. reported that [17O]-O$_2$ reacts rapidly with organoboranes to produce the corresponding oxygen-17 labeled alcohols (Equation 47).[53]

$$\tag{47}$$

The reaction proceeds under surprisingly mild conditions and should be applicable to the synthesis of a variety of oxygen-17 labeled alcohols.

2. Oxidative Additions

Providing that appropriately labeled oxidants are available, oxidative additions to unsaturated carbon systems would provide a simple route to oxygen-labeled materials. Thus, one could envision the formation of labeled diols, epoxides, and carbonyl reagents via the reactions of alkenes with oxygen-labeled oxidants, such as osmium tetroxide, permanganate, chromate, etc. Few examples of such reactions are currently available.

Turro et al. did prepare [17]O- and ^{18}O-labeled 9,10-diphenylanthracene endoperoxide, 1,4-dimethylnaphthalene endoperoxide, and 1,4,8-trimethylnaphthalene endoperoxide by irradiation of the appropriate aromatic hydrocarbon and methylene blue under continuous enriched oxygen purging (Equation 48).[54]

$$\tag{48}$$

Baumstark and co-workers reported the preparation of [17]O-enriched α-azohydroperoxides in high yields by oxidation of the corresponding phenylhydrazones with one equivalent of 20 atom % [17]O-enriched molecular oxygen (Equation 49).[55] Labeld α-azohydroperoxides are efficient [17]O-labeling reagents.

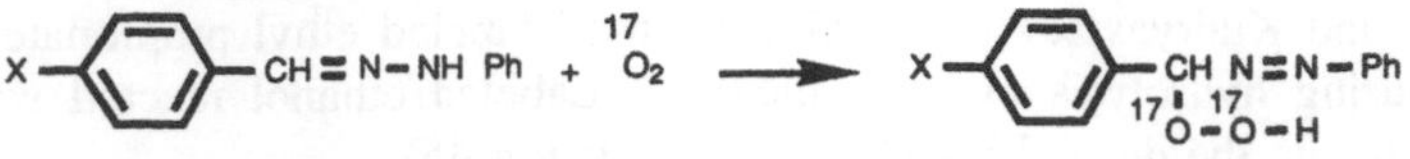

$$X-\text{C}_6\text{H}_4-CH\equiv N-NH\ Ph\ +\ \overset{17}{O_2}\ \longrightarrow\ X-\text{C}_6\text{H}_4-\underset{\overset{17}{O}-\overset{17}{O}-H}{CH\ N\equiv N-Ph} \tag{49}$$

Kobayashi et al. reported the reaction of an oxygen-18 labeled benzoyl peroxide derivative with *p,p'*-dimethoxy-*trans*-stilbene (Equation 50).[56] They found that the label was scrambled, which indicates

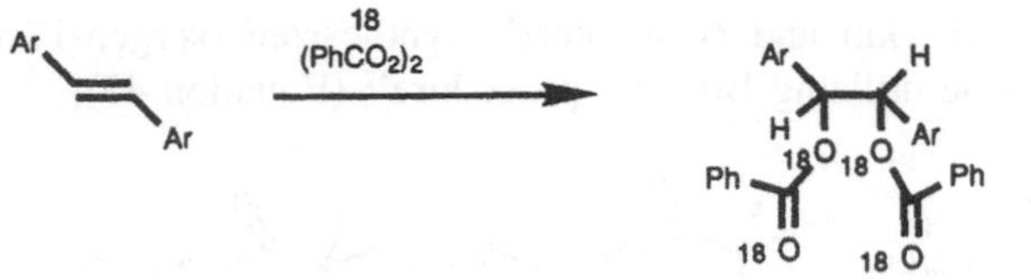

$$\text{(50)}$$

that a four-centered transition state is unlikely during the oxidative addition.

D. MISCELLANEOUS REACTIONS

Essentially any reaction involving the addition of an oxygen-containing fragment to a reactive molecule can be utilized to prepare complex oxygen-labeled agents. In practice, the utility of a specific reaction depends primarily on the availability of the precursor oxygen-labeled fragment. Not surprisingly, syntheses involving isotopically labeled water are abundant, but examples in which other oxygen-labeled species are used are few in number.

1. Addition Reactions

One of the simplest addition reactions of an agent other than water, alcohol, or molecular oxygen would involve the carbonation of a Grignard reagent. Christ et al. utilized the reaction of $[^{17}O]\text{-}CO_2$ to prepare labeled carboxylic acids (Equation 51).[57]

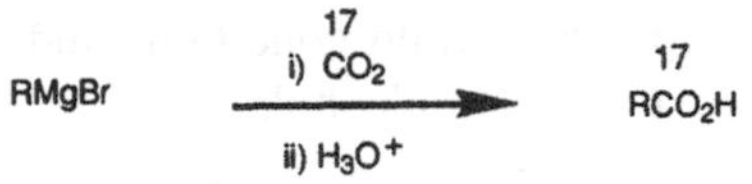

$$RMgBr\ \xrightarrow[\text{ii) } H_3O^+]{\text{i) } \overset{17}{CO_2}}\ R\overset{17}{CO_2}H \tag{51}$$

Another straightforward approach would involve the carbonylation of organoboranes with oxygen-17 labeled carbon monoxide (Equation 52). The

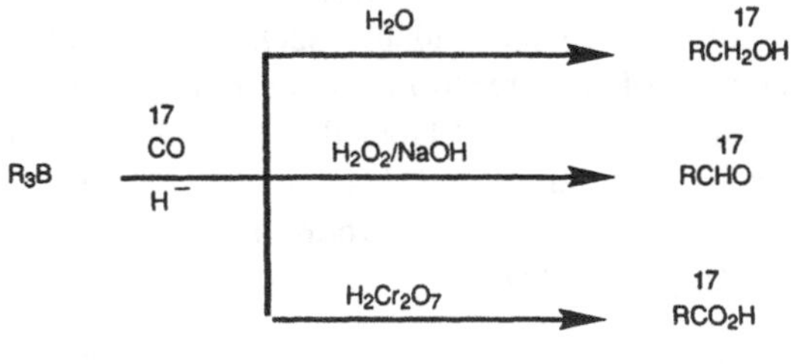

$$\text{(52)}$$

carbonylation reaction has been utilized successfully with carbon-13 labeled carbon monoxide but has not been used to synthesize oxygen-17 labeled agents.[58]

2. Isotopic Enrichment

Kinetic isotope effects have been utilized extensively in organic chemistry as probes of reaction mechanisms. In certain instances the reaction rate differences can be used to isotopically enrich either the products or the starting materials in a given reaction.

Turro et al. achieved a 10 to 20% enrichment of oxygen-17 labeled dibenzylketone by photolyzing dibenzylketone on porous silica gel (Equation 53).[59] The enrichment is a result of the magnetic isotope

$$\text{(53)}$$

effect of oxygen-17. Upon re-encounter (after excitation), radical pairs possessing magnetic nuclei have a higher probability of undergoing combination reactions.

Buchachenko and co-workers also report the use of the magnetic isotope effect to achieve oxygen-17 enrichment via the reaction of hydrocarbons with peroxides[60,61] and molecular oxygen.[62]

Selective reactivity of oxygen-17 in a photodissociation reaction was used by Marling to synthesize oxygen-17 enriched carbon monoxide (Equation 54).[63] Marling reports a 27-fold enrichment of oxygen-17 labeled carbon

$$\text{(54)}$$

monoxide using a neon-22 in laser.

The further development of isotopic enrichment such as these could lower the cost of oxygen-17 labeled precursors and help stimulate interest in oxygen-17 synthetic methodology.

REFERENCES

1. **Hevesy, G. V. and Paneth, F.**, Die Löslichkeit des Bleisulfids und Bleichromates, *Anorg. Chem.*, 82, 323, 1913.
2. **Ott, D. G.**, *Synthesis with Stable Isotopes of Carbon, Nitrogen, and Oxygen*, John Wiley & Sons, New York, 1981.
3. **Raaen, V. F.**, *Carbon-14*, McGraw-Hill, New York, 1968.
4. **Murray, A. and William, D. L.**, *Organic Synthesis with Isotopes*, Interscience, New York, 1958.
5. **Phelps, M. E., Mazziotta, J. C., Schelbert, H. R., Eds.**, *Positron Emission Tomography and Autoradiography: Principles and Applications For the Brain and Heart*, Raven Press, New York, 1986.
6. **English, R. J. and Brown, S. E.**, *Single Photon Emission Tomography*, Society of Nuclear Medicine, New York, 1986.
7. **Boykin, D. W., Balakrishnan, P., and Baumstark, A. L.**, Natural abundance ^{17}O-NMR spectroscopy of heterocyclic N-oxides and di N-oxides. Structural effects, *J. Heterocyclic. Chem.*, 22, 981, 1985.
8. **Rodger, C., Sheppard, N., McFarlane, C., and McFarlane, W.**, *Group VI-Oxygen, Sulphur, Selenium, and Tellurium, in NMR and the Periodic Table*, Harris, R. K. and Mann, B. E., Eds., Academic Press, New York, 1978.
9. **Baillee, T. A. and Prickett, K. S.**, *Applications of Oxygen-18 in Mechanistic Studies of Drug Metabolism, in Synthesis and Applications of Isotopically Labelled Compounds*, Muccino, R. R., Ed., Elsevier, New York, 1986.
10. **Wong, W. W. and Klein, P. D.**, A review of techniques for the preparation of biological samples for mass spectrometric measurements of hydrogen-2/hydrogen-1 and oxygen-18/oxygen-16 isotope ratios, *Mass Spectrom. Rev.*, 5, 313, 1986.
11. **Chang, S. and le Noble, W. J.**, Study of ion-pair return in 2-norbornyl brosylate by means of ^{17}O NMR, *J. Am. Chem. Soc.*, 105, 3708, 1983.
12. **Winstein, S. and Trifan, D.**, Neighboring carbon and hydrogen. X. Solvolysis of *endo*-norbornyl arylsulfonates, *J. Am. Chem. Soc.*, 74, 1147, 1952.
13. **Gragerov, I. P., Rekasheva, A. F., Tarasenko, A. M., Levit, A. F., and Samchenko, I. P.**, Synthesis of some organic compounds labeled with deuterium and oxygen-18, *Zh. Obshch. Khim.*, 31, 1113, 1961.
14. **Oae, S., Kiritani, R., and Tagaki, W.**, The oxygen exchange reaction of phenol in acidic media, *Bull. Chem. Soc. Jpn.*, 39, 1961, 1966.

15. **Koenig, T., Deinzer, M., and Hoobler, J. A.,** Thermal decomposition of *N*-nitrosohydroxylamines. III. *N*-Benzoyl-*N*-nitroso-*O*-*tert*-butyl-hydroxylamine, *J. Am. Chem. Soc.,* 93, 938, 1971.

16. **Sawyer, C. B.,** A simple high yield synthesis of methanol-^{18}O and ethanol-^{18}O, *J. Org. Chem.,* 37, 4225, 1972.

17. **McClelland, R. A.,** Benzamide oxygen exchange concurrent with acid hydrolysis, *J. Am. Chem. Soc.,* 97, 5281, 1975.

18. **Gold, B., Deshpande, A., Linder, W., and Hines, L.,** Reactions of alkanediazotic acids at near neutral and basic pH in [^{18}O]H$_2$O, *J. Am. Chem. Soc.,* 106, 2072, 1984.

19. **Turro, N. J., Paczkowski, M. A., and Wan, P.,** Magnetic isotope effects in photochemical reactions. Observation of carbonyl ^{17}O hyperfine coupling to phenylacetyl and benzoyl radicals. ^{17}O Enrichment studies, *J. Org. Chem.,* 50, 1399, 1985.

20. **Gerlt, J. A., Demou, P. C., and Mehdi, S.,** ^{17}O-NMR Spectral properties of simple phosphate esters and adenine nucleotides, *J. Am. Chem. Soc.,* 104, 2848, 1982.

21. **Townsend, L. B. and Tipson, R. S., Eds.,** *Nucleic Acid Chemistry, Part II,* John S. Wiley & Sons, New York, 1978, 861.

22. **Hoard, D. E. and Ott, D. G.,** Conversion of mono- and oligodeoxyribonucleotides to 5'-triphosphates, *J. Am. Chem. Soc.,* 87, 1785, 1965.

23. **Coderre, J. A., Mehdi, S., Demou, P. C., Weker, R., Traficante, D. D., and Gerlt, J. A.,** Oxygen chiral phosphodiesters. III. Use of ^{17}O NMR spectroscopy to demonstrate configurational differences in the diastereomers of cyclic 2'-deoxyadenosine 3',5'-[^{17}O,^{18}O]-monophosphate, *J. Am. Chem. Soc.,* 103, 1870, 1981.

24. **Balakrishnan, P., Baumstark, A. L., and Boykin, D. W.,** ^{17}O NMR spectroscopy: unusual substituent effects in *p*-substituted benzyl alcohols and acetates, *Tetrahedron Lett.,* 25, 169, 1984.

25. **Crandall, J. K., Centeno, M. A., and Borreson, S.,** Oxygen-17 nuclear magnetic resonance cyclohexanones, *J. Org. Chem.,* 44, 1184, 1979.

26. **Gorodetsky, M., Luz, Z., and Mazur, Y.,** Oxygen-17 NMR studies of the equilibria between the enol forms of β-diketones, *J. Am. Chem. Soc.,* 89, 1183, 1967.

27. **Dahn, H., Schlunke, H. P., and Temler, J.,** Chemische Verschiebungen bei ^{17}O-NMR und Hydrationsgeschwindigkeiten von Cyclischen Ketonen mit Transannularer Wechselwirkung, *Helv. Chim. Acta,* 55, 907, 1972.

28. **Byrn, M. and Calvin, M.,** Oxygen-18 exchange reactions of aldehydes and ketones, *J. Am. Chem. Soc.,* 88, 1916, 1966.

29. **Follmann, H. and Hogenkamp, H. P. C.,** The synthesis of ribose and of adenine nucleotides containing oxygen-18, *J. Am. Chem. Soc.,* 92, 671, 1970.

30. **Gerothanassis, I. P., Hunston, R., and Lauterwein, J.,** ^{17}O NMR of enriched acetic acid, glycine, glutamic acid and aspartic acid in aqueous solution. I. Chemical shift studies, *Helv. Chim. Acta,* 65, 1764, 1982.

31. **Gerothanassis, I. P., Hunston, R., and Lauterwein, J.,** ^{17}O NMR of enriched acetic acid, glycine, glutamic acid, and aspartic acid in aqueous solutions. II. Relaxation studies, *Helv. Chim. Acta,* 65, 1774, 1982.

32. **Adlington, R. M., Aplin, R. T., Baldwin, J. E., Field, L. D., John, M. M., Abraham, E. P., and White, R. L.,** Conversion of ^{17}O/^{18}O-labeled δ-(L-α-aminoadipyl)-L-cysteinyl-D-valine into ^{17}O/^{18}O-labeled isopenicillin N in a cell-free extract of C-acremonium, *J. Chem. Soc., Chem. Commun.,* p. 137, 1982.

33. **Jolivet, P., Gans, P., and Triantaphylides, C.,** Determination of glycolic acid level in higher plants during photorespiration by stable isotope dilution mass spectrometry with double-labeling experiments, *Analytical Biochemistry,* 147, 86, 1985.

34. **Ponnusamy, E. and Fiat, D.,** Synthesis of oxygen-17 labeled thyrotropin releasing hormone, *J. Labelled Compounds Radiopharmaceuticals,* 12, 1135, 1985.

35. **Eckert, H. and Fiat, D.,** Isotopic labeling of tyrosine, followed by ^{17}O NMR, *Int. J. Peptide Protein Res.,* 27, 613, 1986.

36. **Gerothanassis, I. P., Hunston, R. N., and Lauterwein, J.,** ^{17}O NMR chemical shifts of the twenty protein amino acids in aqueous solution, *Magn. Reson. Chem.,* 23, 659, 1985.

37. **Steinschneider, A., Burgar, M. I., Buku, A., and Fiat, D.,** Labeling of amino acids and peptides with isotopic oxygen as followed by ^{17}O NMR, *Int. J. Peptide Protein Res.,* 18, 324, 1981.

38. **Steinschneider, A., Valentine, T. A. M., Burgar, M. I., and Fiat, D.,** A route for oxygen isotope enrichment of α-COOH groups in amino acids, *Int. J. Appl. Radiat. Isotopes,* 32, 120, 1981.

39. **Baltzer, L. and Becker, E. D.,** Solvation of organic oxyanions studied by ^{17}O NMR, *J. Am. Chem. Soc.,* 105, 5730, 1983.

40. **Balakrishnan, P., Baumstark, A. L., and Boykin, D. W.,** ^{17}O NMR spectroscopy: effect of substituents on chemical shifts for *p*-substituted benzoic acids, methyl benzoates, cinnamic acids and methyl cinnamates, *Org. Magn. Reson.,* 22, 753, 1984.

41. **Burgar, M. I., Dhawan, D., and Fiat, D.**, ^{17}O and ^{14}N Spectroscopy of ^{17}O-labeled nucleic acid bases, *Org. Magn. Reson.*, 20, 184, 1982.

42. **Hilbert, G. H. and Johnson, T. B.**, Researches on pyrimidines. CXII. An improved method for the synthesis of cytosine, *J. Am. Chem. Soc.*, 42, 1152, 1930.

43. **Wang, S. Y., Hahn, B. S., Fenslau, C., and Zafiriou, O. C.**, Enrichment of ^{18}O in the nucleic acid bases, *Biochem. Biophys. Res. Commun.*, 48, 1630, 1972.

44. **Aoyama, M., Takahashi, T., Minato, H., and Kobayashi, M. A.**, Comparative study on the oxygen exchange of C-, N-, and O-nitroso compounds with $H_2^{18}O$, *Chem. Lett.*, 245, 1976.

45. **Goldberg, N. D. and Walseth, T. F.**, A second role for second messengers: uncovering the utility of cyclic nucleotide hydrolysis, *Biotechnology*, 3, 235, 1985.

46. **Kobayashi, M., Minato, H., and Ogi, Y.**, Oxygen exchange of benzenesulfinic acids in water, *Bull. Chem. Soc. Jpn.*, 45, 1224, 1972.

47. **King, J. F., Skonieczny, S., Khemani, K. C., and Stothers, J. B.**, Combined ^{17}O-NMR spectra and ^{18}O isotope effects in ^{13}C-NMR spectra for oxygen labeling studies. Carbon-sulfur oxygen migration in the aqueous chlorination of mercapto alcohols, *J. Am. Chem. Soc.*, 105, 6514, 1983.

48. **Dahn, H. and Ung-Truong, M.-N.**, ^{17}O-NMR spectra of mesoionic compounds: the polarized carbonyl group, *Helv. Chim. Acta*, 71, 241, 1988.

49. **Dessinges, A., Castillon, S., Olesker, A., Thang, T. T., and Lukacs, G.**, Oxygen-17 NMR and oxygen-18-induced isotopic shifts in carbon-13 NMR for the elucidation of a controversial reaction mechanism in carbohydrate chemistry, *J. Am. Chem. Soc.*, 106, 450, 1984.

50. **Kursanov, D. N. and Kudryavtsev, R. V.**, A study of the mechanism of hydrolysis by means of heavy oxygen isotope. I. Hydrolysis of ethyl propionate in basic medium, *Zh. Obshch. Khim.*, 26, 1040, 1956.

51. **Vasquez, P. C., Boykin, D. W., and Baumstark, A. L.**, ^{17}O NMR spectroscopy (natural abundance) of heterocycles: anhydrides, *Magn. Reson. Chem.*, 24, 409, 1986.

52. **Bruice, T. C. and Pandit, U. K.**, The effect of geminal substitution ring size and rotamer distribution on the intramolecular nucleophilic catalysis of the hydrolysis of monphenyl esters of dibasic acids and the solvolysis of the intermediate anhydrides, *J. Am. Chem. Soc.*, 82, 5858, 1960.

53. **Kabalka, G. W., Reed, T. J., and Kunda, S. A.**, Synthesis of oxygen-17 labeled alcohols via organoborane reactions, *Synth. Commun.*, 13, 737, 1983.

54. **Turro, N. J., Chow, M.-F., and Rigaudy, J.**, Mechanism of thermolysis of endoperoxides of aromatic compounds. Activation parameters, magnetic field, and magnetic isotope effects, *J. Am. Chem. Soc.*, 103, 7218, 1981.

55. **Baumstark, A. L., Vasquez, P. C., and Balakrishnan, P.**, ^{17}O-Enriched α-azohydroperoxides: ^{17}O NMR spectroscopy, ^{17}O-labeling reagents, *Tetrahedron Lett.*, 26, 2051, 1985.

56. **Kobayashi, M., Minato, H., and Ogi, Y.**, Scrambling of oxygen in the diester produced from labeled diacyl peroxide-olefin reaction, *Bull. Chem. Soc. Jpn.*, 43, 1158, 1970.

57. **Christ, H. A., Diehl, P., Schneider, H. R., and Dahn, H.**, Chemische Verschiebungen in der Kern-magnetischen Resonanz von ^{17}O in Organischen Verbindungen, *Helv. Chim. Acta*, 99, 865, 1961.

58. **Kabalka, G. W., Delgado, M. C., Kunda, U. S., and Kunda, S. A.**, Synthesis of carbon-13 labeled aldehydes, carboxylic acids, and alcohols via organoborane chemistry, *J. Org. Chem.*, 49, 174, 1984.

59. **Turro, N. J., Cheng, C.-C., Wan, P., Chung, C.-J., and Mahler, W.**, Magnetic isotope effects in the photolysis of dibenzyl ketone on porous silica ^{13}C and ^{17}O enrichments, *J. Phys. Chem.*, 89, 1567, 1985.

60. **Buchachenko, A. L., Yasina, L. L., Makhov, S. F., Maltsev, V. I., Fedorav, A. V., and Galimov, E. M.**, Magnetic isotope effect and oxygen enrichment by oxygen-17 during the oxidation of polypropylene, *Dokl. Akad. Nauk SSSR*, 260, 1143, 1981.

61. **Buchachenko, A. L., Belyakov, V. A., and Maltesu, V. I.**, Magnetic isotope effect and enrichment of oxygen-17 in chain oxidation reactions, I. Theory, *Izv. Akad. Nauk SSSR, Ser. Khim.*, p. 1016, 1982.

62. **Belyakov, V. A., Maltsev, V. I., and Buchachenko, A. L.**, Magnetic isotope effect and enrichment of oxygen-17 in chain oxidation reactions. II. kinetics, *Izv. Akad. Nauk SSSR, Ser. Khim*, p. 1022, 1982.

63. **Marling, J.**, Isotope separation of oxygen-17, oxygen-18, carbon-13, and deuterium by ion laser induced formaldehyde photopredissociation, *J. Chem. Phys.*, 66, 4200, 1977.

Chapter 3

APPLICATIONS OF 17O NMR SPECTROSCOPY TO STRUCTURAL PROBLEMS IN ORGANIC CHEMISTRY: TORSION ANGLE RELATIONSHIPS

David W. Boykin* and Alfons L. Baumstark*

TABLE OF CONTENTS

I. INTRODUCTION AND SCOPE

Until recently, the application of ^{17}O nuclear magnetic resonance (NMR) spectroscopy to organic chemistry as a method for structure and conformation elucidation as well as a probe for assessing electronic distribution had been under-utilized significantly. This was the case in spite of the fact that it was known that well-resolved spectra for a number of important classes of organic compounds could be obtained.[1-3] The chemical shift range for oxygen-containing organic molecules is large and, consequently, makes oxygen an attractive nucleus for studying structure and bonding in organic chemistry.

The relative paucity of ^{17}O NMR studies in organic chemistry can be attributed, in part, to the low natural abundance (0.037%) and the quadrupolar properties of the ^{17}O nucleus.[1] The difficulties, broad lines and low S/N, encountered earlier have been largely eliminated by use of high field spectrometers and by making measurements at higher temperatures at relatively low concentrations of solute in low viscosity solvents.[1b] Enrichment by exchange or synthesis by use of $H_2^{17}O$ or other ^{17}O sources has dramatically improved the ease of obtaining ^{17}O spectra (see Chapter 2).

It has been shown that ^{17}O chemical shifts are much more sensitive to structural variation than those of ^{13}C and ^{15}N.[2] However, in addition to providing insights into electronic distribution, much other important information is available from ^{17}O NMR data. For example, recent work from our laboratories has shown that ^{17}O NMR spectroscopy is a powerful method for detection of steric effects on molecular structure in organic systems. Our steric effect studies are divided into two categories. This chapter deals with one category: systems in which steric interactions are characterized by rotation of functional groups around single bonds to relieve van der Waals interactions. The following chapter focuses on rigid systems in which steric interactions are partially accommodated by bond angle and bond length distortions. Intramolecular repulsive van der Waals interactions play a dominant role in determining the influence of steric interactions on the ^{17}O chemical shifts in the systems described in Chapter 4. Recent work by Chesnut has provided excellent insight into the influence of intramolecular van der Waals interactions on chemical shifts for several nucleii.[4,5]

This chapter describes the relationships developed between ^{17}O chemical shifts and the torsion angle between oxygen-containing functional groups, primarily carbonyl type oxygens, and aromatic rings in a number of systems in which van der Waals interactions may be relieved by rotation around a single bond (Figure 1). Isolated examples of the importance of steric effects on the ^{17}O NMR chemical shifts had appeared earlier; however, quantitative relationships between ^{17}O chemical shifts and torsion angles were only recently explored in our laboratories. In general, the systems described in this chapter exhibited torsion angles ranging from near zero to near 90°. Most of the systems examined seem to have minimized repulsive van der Waals interactions (*vide infra*) by torsion angle rotation, and, consequently, the changes in chemical shifts with torsion angle can be explained by a change in electron density on oxygen. Hence, a variation in the Q term of the Karplus-Pople equation[6] for the paramagnetic contribution to chemical shift (*vide infra*) results. Doubtlessly, torsion angle variation also affects the ΔE^{-1} term of the Karplus-Pople expression.

There are three possible consequences for the ^{17}O chemical shift of a functional group when torsion angle rotation occurs: deshielding, shielding, and no change. The deshielding case is expected when the rotated conformation leads to greater double bond character or reduced electron density on oxygen (an isolated carbonyl). The shielding case corresponds to the situation in which the carbonyl assumes more single bond character or increased electron density on oxygen. The circumstances in which no change in chemical shift occurs with rotation could arise when one delocalizing system is replaced by another (cross-conjugation). This chapter describes examples of each of the three theoretical possibilities.

^{17}O NMR chemical shifts depend[1-3] on the paramagnetic (deshielding) $\sigma_o{}^p$ and the dia-

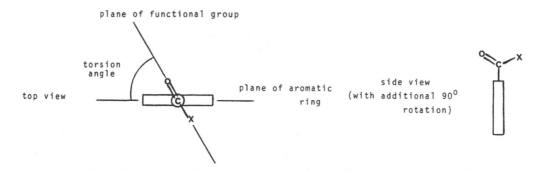

FIGURE 1. Perspective drawings which illustrate the torsion angle between the planes of an aromatic ring and carbonyl-containing functional group.

magnetic (shielding) σ_o^d screening constants (Equation 1). ^{17}O NMR chemical shifts are considered to be largely dependent upon the paramagnetic term.[1]

$$\delta_o = \sigma_o^p + \sigma_o^d \tag{1}$$

The Karplus-Pople equation[6] for the paramagnetic contribution for an oxygen nucleus (O) bound to some other nucleus (X) is given as Equation 2,

$$\sigma_o^p = \frac{-e^2h^2}{2m^2c^2\Delta E} (r^{-3})_{2po} [\Sigma Q_{ox}] \tag{2}$$

where E is the "average excitation energy", frequently approximated by the wavelength of the first maximum in the electronic spectra, $(r^{-3})_{2po}$ is the inverse of the mean volume of 2p orbitals on oxygen, and Q_{ox} is the charge density bond order matrix. All of these terms appear interrelated; however, a number[1,2] of empirical correlations between ^{17}O chemical shifts and other data taken to represent various terms in the Karplus-Pople expression have been reported For example, a correlation between chemical shift, δ, and ΔE^{-1} (measured by electronic spectra), has been reported[7] for C=O and N=O compounds, and for aliphatic ethers. A $\delta/\Delta E^{-1}$ correlation has been reported[8] where ionization potentials were used to approximate ΔE^{-1}. A correlation between bond order and ^{17}O chemical shift for chromium-oxygen compounds has been reported,[9] whereas local π-electron density was correlated with the ^{17}O chemical shift of acetophenones.[10] For a detailed discussion of ^{17}O chemical shift theory, the reader is referred to Chapter 1.

II. TORSION ANGLE EFFECTS

A. AROMATIC NITRO COMPOUNDS

In pioneering work, Christ and Diehl made the observation that the ^{17}O NMR chemical shift of o-nitrotoluene was downfield from that of p-nitrotoluene and attributed this deshielding shift to steric inhibition of resonance.[11] Other qualitative examples of steric inhibition of conjugation on the ^{17}O chemical shifts for acetophenones[12,13] and benzaldehydes[13] bearing a variety of *ortho*-substituents were noted. However, these earlier reports did not develop quantitative relationships or address the underlying causes of the shifts. It was reported that the nitro group signal for 3-methyl-4-nitropyridine-N-oxide **1** was 20 ppm downfield of that for 4-nitropyridine N-oxide **2**, and the shift differences were also explained in terms of steric inhibition of resonance.[14] This observation and the previous ones led to an expanded study of hindered aromatic nitro compounds in an attempt to quantify and understand the origin of these deshielding effects.[15]

NO$_2$ 573 ppm

NO$_2$ 592 ppm, CH$_3$

414 ppm

403 ppm

1

2

The [17]O chemical shift data for a series of sterically crowded aromatic nitro compounds (**3 to 19**) determined in acetonitrile at 75°C are shown in Table 1.[15] Also included in the table are chemical shift values determined under these conditions for unhindered nitro aromatic compounds used for reference. Comparison of the chemical shifts of nitrobenzene (**3**), 1-nitronaphthalene (**4**), and 9-nitroanthracene (**5**), revealed a substantial deshielding trend (Figure 2). Increasing nitro group-aromatic ring orbital overlap with increasing size might be expected to result in increasing single bond character of the nitro function and, thus, should be reflected by a shielding trend. The opposite was observed with the chemical shift trend corresponding to increasing nitrogen-oxygen double bond character of the respective nitro groups, which was explained in terms of increasing rotation of the nitro group from the plane of the aromatic ring. The *peri* interactions in **4** and **5** were expected to produce significant rotation of the nitro groups from the plane of the rings, and the X-ray data for **3** and **5** supported the prediction. The torsion angle that describes the orientation of the nitro group with respect to the plane of the aromatic ring was reported to be 0° for **3**[16] and 85° for **5**.[17]

TABLE 1
[17]O Chemical Shift Data (ppm) and Torsion Angles (deg) for Aromatic Nitro Compounds[15]

Compound no.	Name	Chemical shift (ppm)	X-ray torsion angle (deg.)
3	Nitrobenzene	575	0
4	1-Nitronaphthalene	605	
5	9-Nitroanthracene	637	85
6	2-Nitronaphthalene	575	
7	1,5-Dinitronaphthalene	612	49
8	1,3-Dinitronaphthalene	609, 578	
9	1,8-Dinitronaphthalene	599	43
10	*o*-Nitrotoluene	602	
11	*p*-Nitrotoluene	572	
12	2,4-Dimethylnitrobenzene	597	
13	2,3-Dimethylnitrobenzene	612	
14	2,6-Dimethylnitrobenzene	629	
15	2,4,6-Trimethylnitrobenzene	628	66
16	2,4,6-Tri-*tert*-butylnitrobenzene	657	
17	*p*-Dinitrobenzene	584	
18	*m*-Dinitrobenzene	579	
19	*o*-Dinitrobenzene	609	

In contrast to the chemical shift datum for **4**, the value of its isomer, 2-nitronaphthalene (**6**), devoid of *peri* interactions, was 575 ppm, unchanged from that of nitrobenzene. The chemical shift values for the dinitronaphthalenes 1,5-dinitro- (**7**), 1,3-dinitro- (**8**), and 1,8-dinitronaphthalene (**9**) also reflected the expected steric consequences. The value for **7** was in reasonable agreement with that of **4** when the electronic influences of the two nitro groups in **7** on each other were considered. X-ray data showed that the nitro groups of **7** were

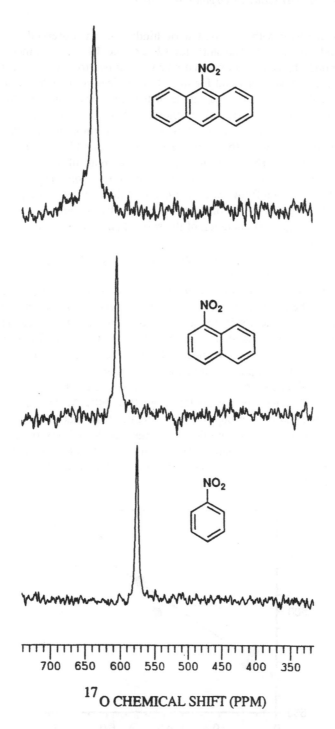

FIGURE 2. ^{17}O NMR spectra for nitrobenzene(3), 1-nitronaphthal-
ene(4) and 9-nitroanthracene(5).

rotated by 49° from the plane of the aromatic ring;[18] the downfield shift of **7** and **4** was thus consistent with increased nitrogen-oxygen double bond character. The two chemical shift values for **8** were in good agreement with the values of the corresponding nitro groups of **4** and **6**. The chemical shift value for **9** was shielded by 6 ppm in comparison to **4**. The X-ray analysis of **9** showed that the nitro groups in **9** were rotated from the plane of the aromatic ring by 43°.[19]

Increased deshielding with increased steric hindrance was observed for the *ortho*-substituted compounds in Table 1. The shifts for **12, 13,** and **14** reflected increasing nitrogen-oxygen double bond character, as expected when the nitro group is rotated from the plane of the aromatic ring. A similar trend for the ^{15}N chemical shift of the nitro group for **13** and **14** was reported; however, the magnitude of the shifts was substantially smaller.[20]

The ^{17}O chemical shift for 2,4,6-tri-*tert*-butylnitrobenzene (**16**) exhibited a large down-field shift, giving a value of 657 ppm. A reasonable assumption was that the electronic effects of the alkyl substituents for **15** and **16** were comparable; hence, the substantially greater downfield shift for **16** was attributed to one or more of several factors: increased average torsion angle of the nitro group from the plane of the aromatic ring for **16,** additional compressional effects arising from direct interaction of the *tert*-butyl groups with the oxygen atoms of the nitro group, distortion of the benzene ring from planarity, a possible result of extensive steric crowding, and alteration of the solvent structure near the nitro group by the large *t*-butyl groups.

The data for the nitro aromatic compounds studied suggested a quantitative relationship between the ^{17}O chemical shift of the nitro function and the torsional angle between the aromatic ring and their nitro groups. The value of the torsion angle (from X-ray data) was plotted vs. the ^{17}O chemical shift. Figure 3 shows the good correlation obtained between torsion angle and ^{17}O chemical shift for seven nitro aromatic compounds. The relationship for aromatic nitro groups is: torsion angle $= (0.76 \pm 0.14)\delta + 574$; $r = 0.987$, 95% confidence limits for error in slope. It appears that the solid state conformational preference is carried over to the solution phase and that the torsion angle observed in the solid state is a remarkably good estimate of the average solution phase angle. The symmetry of this system only allows dihedral angles (torsion angles) from 0 to 90°. At zero degrees, the charge density of the conjugated nitro group oxygen is greater than that of the 90° rotamer (non-conjugated nitro group). Simply considering charge density effects on chemical shift (Q term), the nonconjugated nitro group is expected to be deshielded relative to a conjugated one. Thus, the linear relationship is a consequence of change in charge density over this range of torsion angles.

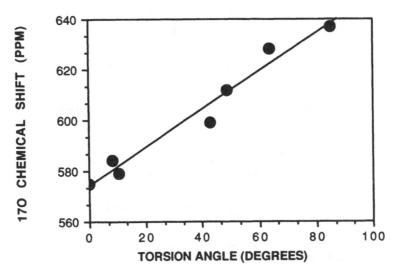

FIGURE 3.   Plot of torsion angles (X-ray) between aromatic rings and nitro groups vs. ^{17}O chemical shift (ppm) data.

B. HETEROAROMATIC NITRO COMPOUNDS

The ^{17}O NMR data-torsion angle approach was applied to a limited number of heteroaromatic nitro compounds.[21] The ^{17}O chemical shift value for 6-nitroquinoline, (**20**), was essentially the same as that of nitrobenzene (575 ppm) and indicated that the fusion of the heteroaromatic ring had only a small influence on the nitro resonance. The ^{17}O chemical shifts of 5-nitroquinoline, (**21**), and 5-nitroisoquinoline, (**22**), only showed small differences from the value for 1-nitronaphthalene (605 ppm). The chemical shift of **21** was deshielded by 25 ppm compared to its 6 isomer **20**; the difference was similar to that noted for the shifts of 1- and 2-nitronaphthalene. These results suggested similar torsion angles between the nitro group and the ring for both the carbocycles and heterocycles.

Since the results for **20** and **21** indicated only small influences arising from heteroaromatic ring fusion, the data from 8-nitroquinoline, (**23**), was used to explore the lone-pair/nitro group interaction. The ^{17}O chemical shift for 8-nitroquinoline, (**23**), was 24 ppm downfield from that of **21** and 51 ppm downfield from the value for nitrobenzene. On the assumption that the relationship developed for carbocyclic nitro compounds was applicable to 8-nitro-quinoline, the torsion angle of interest for **23** is predicted to be approximately 69°. Consequently, it appears that the ring nitrogen lone-pair/nitro group interaction is substantially greater than a nitro group/*peri* hydrogen interaction.

C. KETONES AND ALDEHYDES

The influence of electronic factors on the ^{17}O chemical shifts of aryl-carbonyl compounds have been extensively studied.[10,12,13] Sardella and Stothers[12] were the first to note the influence of steric factors on ^{17}O chemical shifts of aryl ketones. These workers found that the ^{17}O chemical shifts of *ortho*-substituted acetophenones were deshielded relative to unhindered derivatives and suggested that steric inhibition of resonance was the origin of these shifts. Several examples of steric inhibition of conjugation on the ^{17}O chemical shifts for aceto-phenones and benzaldehydes bearing *ortho*-substituents of widely varying electronic properties were reported by the Fiat group.[13] However, no quantitative relationships were determined in these early reports.

1. Aryl ketones

The ^{17}O chemical shifts of a number of sterically hindered aryl ketones (**24 to 36**) in acetonitrile solution at 75°C were reported[22,23] (Table 2). Qualitatively, the data showed that as the possibility for steric interactions between carbonyl groups and alkyl groups of the acetophenones increased, the carbonyl ^{17}O NMR signal was deshielded; compare aceto-phenone (**24**), *o*-methylacetophenone (**25**), and 2,4,6-trimethylacetophenone (**30**). A similar trend was observed for the ^{17}O chemical shifts for 1-acetylnaphthalene (**34**), 2-acetyl-na-phthalene (**35**), and 9-acetylanthracene (**36**).

TABLE 2
*17*O Chemical Shift Data (ppm) and Calculated Torsion Angles (deg) of Aryl Ketones[22,23]

Compound no.	Name	Chemical shift (ppm)	Torsion angle (deg.)
24	Acetophenone	552	1
25	2-Methylacetophenone	582	27
26	4-Methylacetophenone	546	1
27	2,4-Dimethylacetophenone	576	27
28	2,5-Dimethylacetophenone	582	26
29	3,4-Dimethylacetophenone	545	1
30	2,4,6-Trimethylacetophenone	601	57
31	2,4,5-Trimethylacetophenone	575	26
32	2,4,6-Tri-isopropylacetophenone	607	82
33	2,3,5,6-Tetramethylacetophenone	596	63
34	1-Acetylnaphthalene	585	38
35	2-Acetylnaphthalene	553	2
36	9-Acetylanthracene	613	69

As *peri*-hydrogen interactions became important, substantial downfield shifts were observed: for **35** the chemical shift was 553 ppm, for **34** with one *peri*-hydrogen adjacent to the carbonyl group, the signal appears at 585 ppm, and for **36** with two *peri*-hydrogens, the chemical shift was 613 ppm. As was observed for aromatic nitro compounds, the downfield shifts were consistent with steric inhibition of resonance; as the carbonyl group was rotated out of conjugation with the aromatic ring, greater double bond character was predicted for the carbon-oxygen bond. The earlier success with relating the *17*O chemical shifts of aromatic nitro compounds with their torsion angles led to the development of such a relationship for the aryl ketones. Most of the aryl ketones are liquids and, consequently, X-ray data are limited. Therefore, in order to estimate the torsion angles between the carbonyl group and the aromatic ring for the acetophenones (Table 2), a molecular mechanics approach (MM2) was employed. A plot of the *17*O chemical shift data vs. the estimated torsion angles is shown in Figure 4. The *17*O data were plotted directly as obtained since very large shifts were noted, and no attempt was made to correct for the relatively small electronic effects of the alkyl groups. Figure 4 shows a reasonable correlation for this data. The relationship found for the aryl ketones was: torsion angle = $(0.84 \pm 0.14) \delta + 533$; $r = 0.979$, 95% confidence limits for error in slope.

Tables 3 and 4 contain *17*O NMR data for a series of phenyl alkyl ketones (PhCOR) and methyl alkyl ketones (CH_3COR).[24] In these systems, smaller changes in chemical shift were noted, and it was necessary to correct for electronic effects. These results were used to provide insight into the influence of the combination of electronic and torsion angle effects on the chemical shifts of these systems. The data for the aliphatic ketones were used to estimate the contribution from electronic effects to the carbonyl chemical shift. The chemical shifts of the phenylketones were considered to be influenced by both electronic effects and torsion angle rotation. Deviations from shifts predicted based on electronic factors were attributed to torsion angle variation, and the magnitude of the torsion angle was estimated by employing the slope determined previously[23,24] from torsion angle-*17*O chemical shift relationships for electronically similar aryl ketones.

A shielding trend was noted for the phenyl alkyl ketones **37** to **40**, which was considered to arise from the inductive effect of the alkyl group and/or any gamma or higher effects, rather than torsion angle rotation. Molecular mechanics calculations for **37** to **40** predict little change in the torsion angles. In contrast to the results for the simple alkyl ketones,[1]

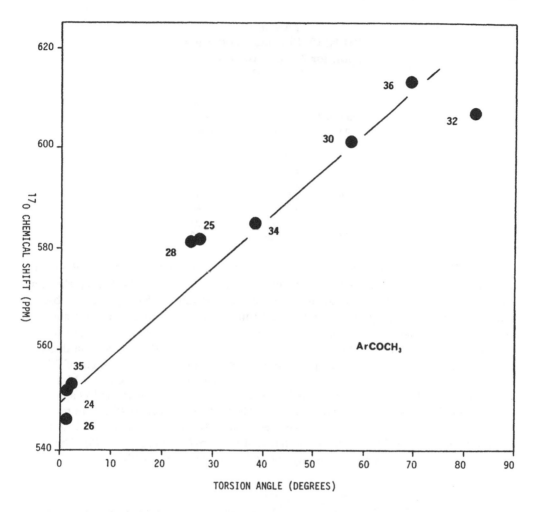

FIGURE 4.   Plot of calculated torsion angles (MM2) between the carbonyl group and the aromatic ring for aryl ketones vs. ¹⁷O chemical shift data.

TABLE 3
¹⁷O NMR Chemical Shift Data
(ppm) for Aryl Alkyl Ketones
(PhCOR)[24]

Compound no.	R	Chemical Shift (ppm)
37	Me	552
38	E*t*	540
39	*n*-Pr	543
40	*i*-Pr	535
41	*t*-Bu	565
42	Cyclopropyl	495
43	Cyclobutyl	530
44	Cyclopentyl	529
45	Cyclohexyl	538
46	Ph	552

TABLE 4
^{17}O NMR Chemical Shift Data
(ppm) for Methyl Alkyl Ketones
(CH_3COR)[24]

Compound no.	R	Chemical shift (ppm)
47	Me	571
48	Et	558
49	n-Pr	563
50	i-Pr	557
51	t-Bu	560
52	Cyclopropyl	521
53	Cyclohexyl	560

the ^{17}O NMR datum for *tert*-butyl phenyl ketone (41) showed significant deshielding and was assumed to arise from torsion angle rotation, which overrides the alkyl group shielding effect. It was predicted that 41 should have a chemical shift value of 541 ppm in the absence of torsion angle changes by using the chemical shift differences between acetone and pinacolone and 2-butanone and pinacolone as controls. Thus, the shift arising from torsion angle variation was estimated to be 24 ppm. Dividing the slope (0.84 δ/angle in degrees) from previous ketone results[23] by 24 ppm led to the prediction of a torsion angle of 29° for 41. A dipole moment approach had estimated the torsion angle value to be 63°.[25] Molecular mechanics calculations yield a value of 34°, in reasonable agreement with the ^{17}O approach.

The ^{17}O NMR spectra for several cycloalkyl phenyl ketones, (42 to 45), were reported. The cyclobutyl and cyclopentyl phenyl ketones 43 and 44 exhibit similar chemical shifts, which were slightly shielded compared with that of isopropyl phenyl ketone (40). The result for the cyclohexyl compound 47 was slightly deshielded compared with 40.

In comparison with 37 to 40, the ^{17}O signal for phenyl cyclopropyl ketone (42) was substantially shielded, appearing at 495 ppm. This value represents an upfield shift of 40 ppm when compared to the datum for its close structural analog, isopropyl phenyl ketone (40). This large shielding shift was interpreted as an indication of substantial overlap between the carbonyl group and the cyclopropane ring, which would be expected to increase the single bond character of the carbonyl group. A similar shielding result was found for methyl cyclopropyl ketone (52). The difference in the results[1] for methyl isopropyl ketone 50 and 52 is a 53 ppm shielding effect, which was greater than noted between the analogous phenyl alkyl ketones 40 and 42. The smaller shielding shift for phenyl cyclopropyl ketone was attributed to the competition for overlap between phenyl carbonyl and cyclopropyl carbonyl in the cross-conjugated system. The preferred conformation for maximum overlap of a cyclopropane ring and a conjugated π system has intrigued chemists for years.[26-28] It appears that ^{17}O NMR methodology should serve as a sensitive probe to test this point and explore the angular dependence.

The ^{17}O NMR spectral data for 1-indanone (54), 1-tetralone (55), and 1-benzosuberone (56) were reported and the chemical shifts showed significant sensitivity to ring size. The effect of fusion of a benzene ring on the cyclic ketones resulted in an upfield shift compared to the acyclic counterparts.

Using ^{17}O chemical shift values for cyclopentanone (543 ppm), cyclohexanone (558 ppm) and cycloheptanone (566 ppm),[1] the electronic effects of the different rings were deduced as outlined above. The difference values were used to estimate the torsion angles by assuming the torsion angle for 1-indanone (54) to be zero. Using the ^{17}O NMR approach outlined above for alkyl phenyl ketones yielded a torsion angle for 55 of 8° and for 56 of 39°. Molecular mechanics calculations gave torsion angle values of 0°, 2°, and 34° for 54, 55, and 56, respectively, which were in reasonable agreement with the ^{17}O results. Earlier UV studies had predicted the torsion angle for 54 and 55 to be zero and that for 56 to be 37°.[29] The ^{17}O method was in good agreement with the other approaches and thus provided a convenient method for the estimation of solution phase structure for cyclic aryl ketones.

2. Aldehydes

The ^{17}O chemical shifts of hindered, electronically similar aromatic aldehydes (57 to 63) are listed in Table 5.[24] The general trend of increasing chemical shift with increasing steric hindrance, similar to that noted above for other functional groups, was detected (30 ppm range).

Methods used previously to estimate aldehyde torsion angles had suggested rotation of the carbonyl groups from the aromatic plane in hindered systems analogous to the ones discussed here.[29] Fiat and co-workers had qualitatively noted the differences between ^{17}O chemical shifts for o- and p-substituted benzaldehydes.[13] As a means of estimating aldehyde torsion angles, the slope obtained[22,23] for the aryl ketone chemical shift-torsion angle relationship

TABLE 5
^{17}O Chemical Shift Data (ppm) and Estimated Torsion Angles (deg) for Aromatic Aldehydes (ArCHO)[24]

Compound no.	Ar	Chemical shift (ppm)	Estimated torsion angle (deg.)[a]
57	Ph	564	0[b]
58	4-MeC$_6$H$_4$	557	0[b]
59	2-MeC$_6$H$_4$	575	21
60	2,4,6-Me$_3$C$_6$H$_2$	585	33
61	1-Naphthyl	575	13
62	2-Naphthyl	564	0
63	9-Anthryl	597	39

[a] ±4°; see text.
[b] Assumed value.

(0.84 δ/angle in degrees) was used. In addition, a 0° torsion angle for benzaldehyde (57) and 4-methylbenzaldehyde (58) was assumed. The chemical shift difference between 57 or 58, as appropriate, and the remaining aldehydes divided by the ketone data slope[23] provided the estimated torsion angles for 59 to 63 listed in Table 5. The torsion angle estimated for 59 by the ^{17}O approach (21°) is identical with that obtained from UV data in earlier work.[30] The value obtained for 60 (33°) was in reasonable agreement with the literature value of 28° for a close analog, 2,6-dimethybenzaldehyde.[30] However, the result for 60 (33°) was rather different from the literature value[30] (22°) obtained from UV data. The torsion angle for anthracene-9-carboxaldehyde (63) estimated by the ^{17}O method (39°) was higher than that found in the solid state (27°).[31] Molecular mechanics calculations for the aromatic aldehydes (57 to 62) failed to show a torsion angle rotation; however, a rotation of 44° was predicted for 63. The calculations for 57 to 62 predicted a bending of the aldehyde-aromatic ring C-C bond.

These results demonstrate that, in homologous series of compounds for which the deshielding shifts are relatively small, the contribution to ^{17}O chemical shifts from steric factors can be deduced upon correction for electronic effects. Frequently, steric hindrance leads to torsion angle rotation, which can be estimated by using the previously developed ^{17}O chemical shift correlations. This ^{17}O NMR method allows the facile estimation of the solution phase structure for a variety of carbonyl compounds.

3. 1,2-Diketones

In a recent study, the inter-carbonyl dihedral angle in a series of 1,2-diketones was related to the ^{17}O chemical shift.[32] The data for the 1,2-diketones (64 to 70) are shown in Table 6. Cerfontain and co-workers found for the 1,2-diketones that as the dihedral angle increased toward 90°, corresponding to a reduction of overlap of the two carbonyl systems, the chemical shift moved upfield and that the chemical shift reversed direction on going from a dihedral angle of 90° to 180°. Thus, an approximate cosine relationship was observed. The direction of the chemical shift change for the diketones as a function of dihedral angle (0° to 90°) is in the opposite direction of that noted for aryl carbonyl compounds discussed above. The 1,2-diketone chemical shifts parallel the MNDO calculated π-electron density at oxygen (Q_o^π) for glyoxal used as a model; π electron density at oxygen increases as the overlap between the two π systems decreases. The mean excitation energy, ΔE, for these diketones also varies with dihedral angle, and a linear relationship was found between the ^{17}O chemical shift and the product $\Delta E \cdot Q_o^\pi$. The upfield shift observed was attributed to an expansion of the 2p orbitals, resulting from the increase in Q_o^π, which was expected to cause a reduction in the r^{-3} term of the Pople-Karplus equation and result in a reduction in the magnitude of the paramagnetic term (σ_N^p). The change in ΔE also affects σ_N^p in the same direction.

TABLE 6
^{17}O Chemical Shift Data (ppm) for 1,2-Diketones[32]

Compound no.	Name	Chemical shift (ppm)
64	3,3,5,5-Tetramethylcyclopentane-1,2-dione	573
65	Homoadamantane-4,5-dione	572
66	3,3,6,6-Tetramethylcyclohexane-1,2-dione	564
67	3,3,7,7-Tetramethylcycloheptane-1,2-dione	557
68	2,2,5,5-Tetramethylhexane-3,4-dione	558
69	Butane-2,3-dione	569
70	Cyclododecane-1,2-dione	576

The difference in direction of chemical shift change on rotation of a carbonyl group from coplanarity in the diketones and the aryl ketones deserves comment. One interpretation is that for coplanar aryl carbonyl compounds the electron density on the carbonyl oxygen is greater (larger r^{-3}) because of participation of aryl ring electron density. As the dihedral angle increases in the aryl ketone system, the carbonyl group becomes more like an isolated carbonyl and assumes more double bond character (less charge density on oxygen). For the 1,2-diketone, the planar conformation more closely resembles an isolated carbonyl. This explanation can be alternatively expressed in a valence-bond representation by noting that the canonical form A is favored in the planar conformer of the diketone and in the rotated conformer of the aryl ketone, whereas form B is favored in the rotated conformer of the diketone and the planar conformation of the aryl ketone.

Another example of an upfield shift of a carbonyl group signal caused by rotation from conjugation is described below in the section on *N*-arylacetamides.

D. AROMATIC CARBOXYLIC ACIDS, ESTERS AND AMIDES

1. Carboxylic Acids

The ^{17}O chemical shifts for representative aliphatic[33] and aromatic carboxylic acids[34] have been reported to be sensitive to electronic effects of substituents. Table 7 contains the ^{17}O chemical shift values for a series of hindered but electronically similar aromatic carboxylic acids.[23] As previously noted for other carboxylic acids,[33] only one ^{17}O signal for this functional group is detected. The equivalence of the two oxygens has been attributed to fast proton exchange[33] in dimeric or higher aggregates. Qualitatively, it is apparent (Figure 5) that as steric hindrance to coplanarity of the carboxylic function and the aromatic ring was increased, the magnitude of the carboxyl chemical shift increased. The direction of the shift was consistent with the rotation of the functional group from the plane of the aromatic ring, which would be expected to increase its overall double-bond character. Calculated torsion angles (MM2) for these compounds are also included in Table 7. An excellent correlation

TABLE 7
^{17}O Chemical Shift Data (ppm) and Torsion Angles (deg) for Aromatic Carboxylic Acids[23]

Compound no.	Ar-CO₂H	Chemical Shift[a]	Torsion Angle (deg)[b]
71	Ph	250.5	2
72	4-MeC₆H₄	249	2
73	2-MeC₆H₄	265	29
74	2,3-Me₂C₆H₃	269	29
75	2,6-Me₂C₆H₃	280	55
76	2,4,6-Me₃C₆H₂	280	55
77	1-Naphthyl	267	35
78	2-Naphthyl	251.5	2
79	9-Anthryl	287	67

[a] Both oxygens.
[b] Calculated by molecular mechanics method (MM2).

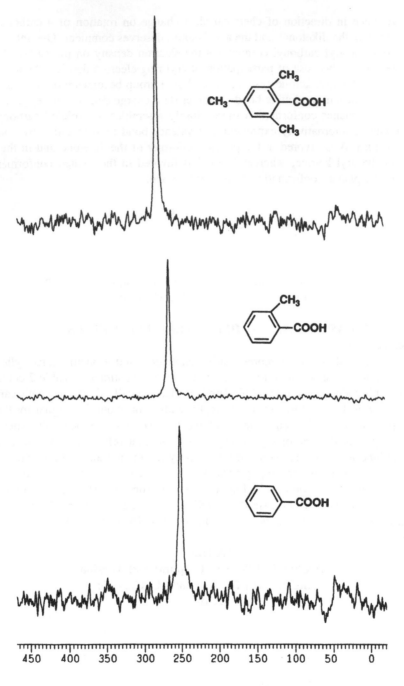

FIGURE 5.   ¹⁷O NMR spectra for enriched benzoic acid(**71**), 2-methylbenzoic acid(**73**), and 2,4,6-dimethylbenzoic acid(**76**).

between estimated torsion angle and ¹⁷O chemical shift is reflected in Figure 6. Torsion angle data appear in the literature for two compounds included in the study. The torsion angle estimated by the MM2 method for **75** (55°) was in good agreement with the X-ray value[35-37] (51.5°) and the value (50.7°) estimated by ¹³C NMR methods.[38] The calculated value for **74** (29°) did not agree with the X-ray data[35-37] (10°) but was reasonably consistent

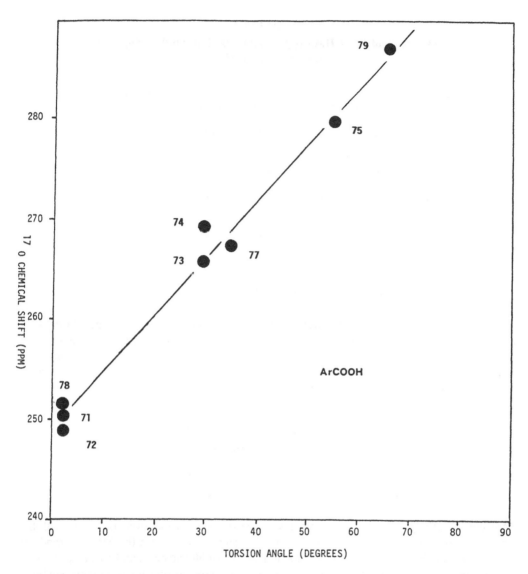

FIGURE 6. Plot of calculated torsion angles (MM2) between the carbonyl group and the aromatic ring for aryl carboxylic acids vs. ^{17}O chemical shift data.

with the value obtained from a ^{13}C NMR approach (25°).[38] It seems likely that the conformation in solution was different from that of the solid state for **74**. Results from linear regression analysis of the calculated torsion angles (MM2) and ^{17}O chemical shift data for the acids gave an excellent correlation (r = 0.994).

2. Esters

The ^{17}O chemical shift data for the variously hindered aromatic esters are listed in Table 8.[23] Earlier, the ^{17}O chemical shifts of esters have been found to be affected by both substituents attached to the carbonyl carbon[34,39] and to the single-bonded oxygen.[40-42] The ^{17}O chemical shifts of aromatic esters have been shown to be sensitive to electronic effects of substituents.[34,39] Variation of the size of the alkyl group attached to single-bonded oxygen significantly influences the chemical shift of the single-bonded oxygen, and, although less pronounced, a readily detected effect is observed on the carbonyl oxygen resonance.[42] In the series listed in Table 8, chemical shift differences arising from electronic effects were small and could be ignored, with little effect on the correlation.

TABLE 8

^{17}O Chemical Shift Data (ppm) and Torsion Angles (deg) for Aromatic Esters[23]

Compound no.	Ar-CO$_2$Me	Chemical shifts		Torsion angle (deg)[a]
		δ(C=O)	δ (–O–)	
80	Ph	340	128	2
81	4-MeC$_6$H$_4$	339[b]	127[b]	2
82	2-MeC$_6$H$_4$	359	138.5	29
83	2,3-Me$_2$C$_6$H$_3$	363	141	29
84	2,6-Me$_2$C$_6$H$_3$	377	150	54
85	2,4,6-Me$_3$C$_6$H$_2$	376	149	54
86	1-Naphthyl	361	139	33
87	2-Naphthyl	341	129	2
88	9-Anthryl	385	154	67
89	2,4,6-t-Bu$_3$C$_6$H$_2$	392	162	76

[a] Calculated by molecular mechanics method (MM2).
[b] From Reference 34; taken in acetone at 40°C.

The signals for both the carbonyl oxygen and the single-bonded oxygen both were deshielded substantially in compounds which have greater steric interactions (Figure 7). These results were consistent with previous findings; rotation of the functional group from the plane of the aromatic ring was expected to increase the double-bond character of both oxygen atoms and result in deshielding of both oxygen signals.

Molecular mechanics (MM2) calculations were carried out on the aromatic esters with the OCH$_3$ group in the *s*-Z conformation. The torsion angle values from the MM2 method are also included in Table 8. The calculated torsion angles for **80** and **85** (2 and 54°, respectively) are in reasonable agreement with the torsion angles predicted by dipole moment studies[43] (0° and 47°). A plot of the calculated torsion angles vs. the chemical shift for both the double- and single-bonded oxygen is shown in Figure 8; excellent correlations were obtained for data corresponding to both oxygen signals. The slope from data analysis of the single-bonded oxygen results is approximately one-half that of the slope for the carbonyl oxygen data. Their average value was approximately the value noted for similar treatment of the data for carboxylic acids, which is in agreement with the concept that the carboxylic acid data are a consequence of proton exchange between the oxygens.

3. Amides

Extensive studies on simple amides and peptides have appeared, and the role of solvent, electronic factors and hydrogen bonding on their chemical shifts has been reported.[44-48] Limited studies on aromatic amides have appeared. The ^{17}O chemical shift data reported for electronically similar hindered aromatic amides **90** to **99** are listed in Table 9.[23] It is clear from the data that the chemical shift range for the amides is compressed compared to that of the ketone and ester carbonyl data described above. The reduction in range was attributed to the fact that for the least hindered amide, benzamide (**90**), the amide-aromatic ring torsion

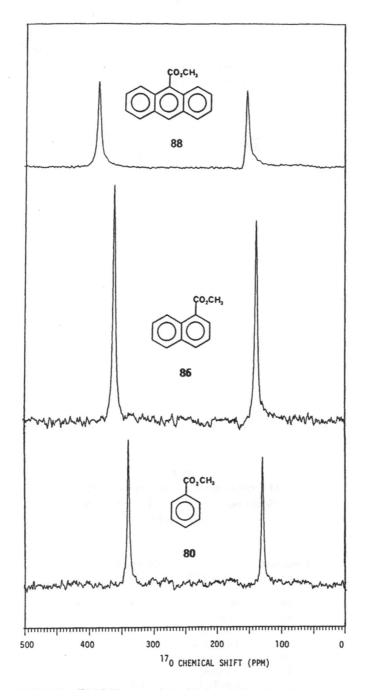

FIGURE 7. ¹⁷O NMR spectra for enriched methyl benzoate(**80**), methyl 1-naphthoate(**86**), and methyl 9-anthroate(**88**).

angle was not zero, as in the case for acetophenone and methyl benzoate, but was 26°, as determined by X-ray analysis (MM2 value 28°).[49] The quality of the correlation for the simple amides (Figure 9) was not as good as that obtained for ketones and esters. A linear relationship is shown in Figure 9, even though the data for the simple amides (ArCONH$_2$) appeared to exhibit scatter or curvature (which may be attributable to hydrogen-bonding interactions).

In order to avoid the probable complications arising from hydrogen-bonding, a limited

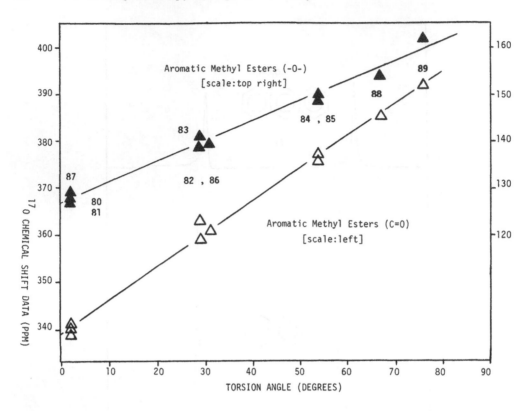

FIGURE 8.   Plot of calculated torsion angles (MM2) between the carboxylate group and the aromatic ring for aryl esters vs. ^{17}O chemical shift data.

TABLE 9
^{17}O Chemical Shift Data (ppm) and Calculated Torsion Angles (deg) for Aromatic Amides[23]

Compound no.	Ar-CONH$_2$	Chemical shift	Torsion angle (deg)[a]
90	Ph	329	28
91	4-MeC$_6$H$_4$	327	28
92	2-MeC$_6$H$_4$	350	44
93	2,6-Me$_2$C$_6$H$_3$	353	60
94	1-Naphthyl	359	51
95	2-Naphthyl	331	27
96	9-Anthryl	365	71
	Ar-CONMe$_2$		
97	Ph	348	62
98	1-Naphthyl	352	71
99	9-Anthryl	357	78

[a]   Calculated by molecular mechanics method (MM2).

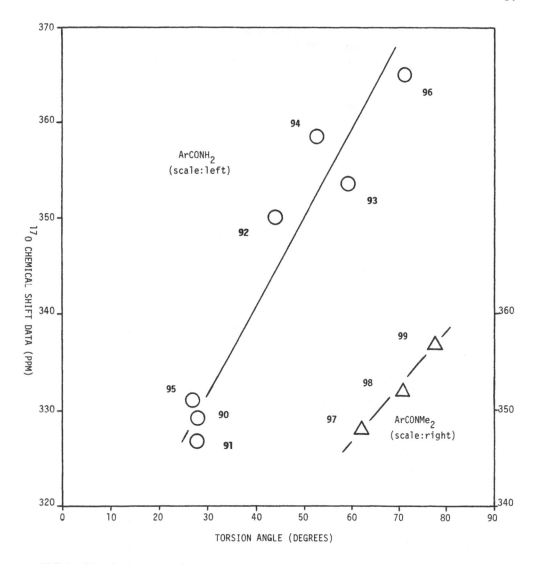

FIGURE 9. Plot of calculated torsion angles (MM2) between the carboxamide group and the aromatic ring for aromatic amides vs. ^{17}O chemical shift data.

number of *N,N*-dimethylcarboxamides (**97** to **99**) (Table 9) were examined. The added steric interactions arising from the *N,N*-dimethyl group substantially contributed to the torsion angle in this system as noted in Table 9. The predicted (MM2) torsion angle for *N,N*-dimethylbenzamide (**97**) was 62°. Because of the additional steric hindrance due to *N*-substitution, the range of torsion angles was substantially reduced and the chemical shift range was expected to be correspondingly smaller. Although the number of compounds studied was limited, it was apparent that the slope of the line for the *N,N*-dimethylamide data was reduced from that of the corresponding unsubstituted amides.

4. N-Arylacetamides

In a study of *N*-arylacetamides (Table 10), it was found that introduction of steric hindrance caused shielding of their ^{17}O chemical shifts.[50] This result is exactly the opposite of that obtained for the other hindered aryl carbonyl systems discussed, for which steric hindrance caused deshielding shifts (in agreement with the results found for 1,2-diketones). Molecular mechanics calculations for the *N*-arylacetamides showed a large change of torsion

TABLE 10
^{17}O Chemical Shift Data (ppm) and Torsion
Angles (deg) for Substituted Acetanilides[50]

Compound no.	ArNHCOMe	Torsion Angle (deg)[a]	$\delta(C=O)$
100	Ph	21	355.3
101	4-MeC$_6$H$_4$		352.0
102	2-MeC$_6$H$_4$	34	349.0
103	2,6-Me$_2$C$_6$H$_3$	58	343.0
104	1-Naphthyl	38	350.5
105	2-Naphthyl	19	358.1

[a] Molecular mechanics calculated angle.

angle with substituents (Table 10) which paralleled their ^{17}O chemical shift values. A quantitative torsion angle-^{17}O chemical shift relationship was also found (Figure 10) for the *N*-arylacetamide results.

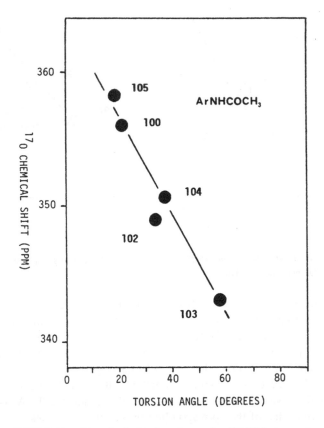

FIGURE 10.   Plot of calculated torsion angles (MM2) between the amide functional group and the aryl ring for *N*-arylacetamides vs. ^{17}O chemical shift data.

In the *N*-arylacetamide system, unlike the other aryl carbonyl systems, increasing steric hindrance caused a decrease in chemical shift of the carbonyl ^{17}O signal. This upfield shift was explained in terms of the varying importance of the canonical forms **106A** and **106B** as a function of substituents *ortho* to the NHAc group.

106A   106B

In the absence of a sterically large *ortho* substituent, contribution from **106A** appears important. Note the similarity in structure between **106A** and that of the 1,2-diketones. Introduction of *ortho* substituents would be expected to reduce the overlap between the amide function and the aromatic ring because of torsion effects to reduce van der Waals interactions. A decrease in the contribution of **106A** and, thereby, an increase in the importance of **106B** was expected to result from such an interaction. As the importance of the contribution from **106B** increased, the observed chemical shift of the amide carbonyl was expected to decrease. This upfield shift is consistent with a charge density argument since as the torsion angle increases, the structure approaches that of an isolated amide function (cf. *N*-methylacetamide), the chemical shift of which is upfield of that of conjugated *N*-arylacetamidcs. The results and the explanation for the change of chemical shift for the *N*-arylacetamides are analogous to those noted above for the 1,2-diketones.

The chemical shift value for *N*-methyl-*N*-phenylacetamide (**107**), 350 ppm, provided additional support for the argument developed to explain the upfield shifts observed for the hindered *N*-arylacetamides. Molecular mechanics calculations predicted a large torsion angle (62°) for **107**. Increased steric hindrance by introduction of an *N*-methyl group was expected to cause an increased torsion angle and thereby increase contributions from **106B**; consistent with this argument, the ^{17}O chemical shift of **107** was upfield of **100** by 6 ppm.

5. Aryl Acetates

A companion series of aryl acetates (**108** to **112**) was studied (Table 11), and only small changes with increasing steric hindrance were noted in their ^{17}O chemical shifts.[50] The relative insensitivity of chemical shift was explained, in part, by the fact that the calculated (MM2) torsion angles were large and only modest changes in torsion angle were predicted with changes in steric hindrance. Interpretation of results for the aryl acetates is doubtlessly further complicated by the competition between overlap of the aromatic ring orbitals and one or the other of the lone-pair orbitals on the single-bonded oxygen and maintenance of overlap of the carbonyl orbitals and the other lone pair orbital on the single-bonded oxygen. The interplay of these two independent overlap networks seems to combine to hold the electron density on the carbonyl oxygen essentially constant as steric hindrance is changed.

TABLE 11
^{17}O Chemical Shift Data (ppm) and Calculated
Torsion Angles (deg) for Substituted Phenylacetates[50]

Compound no.	ArOCOMe	Torsion angle (deg)[a]	δ(C=O)	δ(-O-)
108	Ph	64	370.0	201.3
109	2-MeC$_6$H$_4$	68	371.0	199.6
110	2,6-Me$_2$C$_6$H$_3$	73	371.0	196.6
111	1-Naphthyl	76	371.6	195.3
112	2-Naphthyl	66	371.3	201.3

[a] Molecular mechanics calculated angle (±2°).

A modest change of chemical shift of the single bond oxygen (Table 11) with increasing hindrance was noted. Thus, the aryl acetates appear to be an example of the case in which no change in ^{17}O chemical shift should be noted on change of torsion angle.

E. MISCELLANEOUS SYSTEMS
1. *N*-Phenyl Phthalimides
In a study of steric interactions in hindered *N*-substituted imides (**113** to **115**), a torsion angle relationship was found between the two ring systems of *N*-aryl phthalimides[51] and the carbonyl oxygen chemical shifts. The *N*-aryl phthalimide series differs from those previously described in that, in this case, the carbonyl functional group is locked in a specific geometry and the conjugated aryl ring must rotate to relieve van der Waals interactions.

The ^{17}O NMR data for substituted *N*-aryl phthalimides showed little effect attributable to electronic factors; however, van der Waals interactions were clearly an important factor. Molecular mechanics calculations for **113** to **115** predicted dihedral angles (torsion angles) for the aryl ring with the phthalimide system of 50°, 64° and 75°, respectively. The X-ray structure of **113** showed a dihedral angle of 56° in reasonable agreement with the calculations. A plot of ^{17}O chemical shift vs. calculated dihedral angle for **113** to **115** is shown in Figure 11. A reasonable correlation is seen which suggests that the ^{17}O NMR shifts for the *N*-aryl phthalimides resulted primarily from reduction of van der Waals repulsions by torsion angle rotation (loss of overlap) of the *N*-aryl group. Interestingly, since the ^{17}O data still showed some sensitivity to electronic effects, it was concluded that minimization of van der Waals repulsion was not complete. This suggested that some conjugation between the two systems was retained at the expense of complete minimization of van der Waals interactions. The ^{17}O results were consistent with the IR[52] and ^{13}C NMR[53] results; however, they do not agree with dipole moment studies[54-56] which had suggested that the two rings of *N*-aryl phthalimides are co-planar.

2. Anhydrides
In an extensive study of several different series of anhydrides,[57] it was found that the chemical shifts for both types of anhydride oxygens were sensitive to structural changes as illustrated by **116, 117** and **118**.

The single-bonded oxygen was deshielded in the cyclic structures **117** and **118** compared to the one in acetic anhydride **116**, indicative of greater π overlap with the carbonyl system in the cyclic structures. Conversely, the chemical shift data for the carbonyl oxygens were shielded in the cyclic systems in comparison with acetic anhydride, which presumably reflects greater single bond character for these carbonyl groups. It was apparent from the benzoic

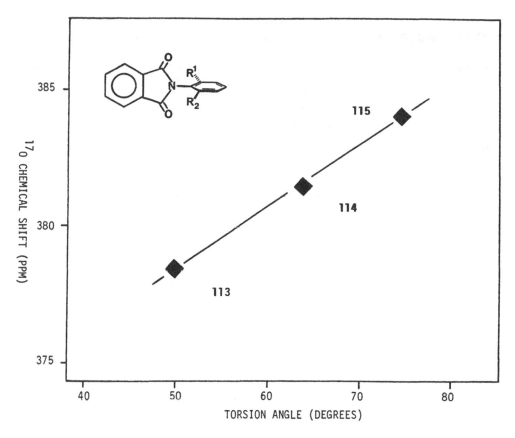

FIGURE 11. Plot of calculated torsion angles (MM2) between the N-aryl group and the phthalimide ring vs. ^{17}O chemical shift data.

anhydride data (398 and 242 ppm) that delocalization affects both types of oxygens in the same manner: both are shielded compared to acetic anhydride. The trend in chemical shifts for **116** to **118** parallels their O=C–O–C dihedral angle. The planar systems **117** and **118** are shielded relative to the rotated acetic anhydride **116**; this system provides yet another case analogous to the 1,2-diketones and the N-arylacetamides. It is clear then that torsion angle relationships are important in these systems as well.

3. Cyclic Peroxides

An extensive ^{17}O NMR study of several series of cyclic peroxides was reported by Salomon et al.[58] The chemical shifts of the cyclic peroxides were shown to be insensitive to solvent changes. Chemical shift data for these compounds, **119** to **128**, are listed in Table 12. The chemical shift range for the cyclic peroxides was rather large, from 232 to 318 ppm. Unlike alkyl ethers,[59] no correlation between ^{17}O chemical shifts and ionization potentials, relatable to the Q term of the Karplus-Pople expression, was noted[58] for the cyclic peroxides. Clearly, the peroxide case is more complex since two ionization potentials (η_o- and σ_o-) are associated with the lone pair of electrons of oxygen. No straightforward relationship between atomic charge and ^{17}O chemical shifts for these systems was apparent; consequently, some other factor was sought to explain the chemical shift differences noted. It was suggested that conformational changes contributed to the variation in ^{17}O chemical shifts. The C-O-O-C dihedral angles for the compounds in Table 12 vary from 0° to approximately 90°. No quantitative relationship between dihedral angle and ^{17}O chemical shift was found. However, it can be seen, very generally, that as the dihedral angle increases,

TABLE 12
¹⁷O Chemical Shift Data (ppm) and Dihedral Angle (deg) for Cyclic Peroxides[58]

Compound no.	Compound	Chemical shift	Dihedral angle
119		310	0
120		303	0
121		283	0
122		250	21
123		263	32
124		280	50
125		259	51
126		247	54
127		232	65
128		254	88

the chemical shift decreases. The large chemical shift of **119** to **121** was attributed to the interaction of neighboring lone-pair electrons arising from their co-planar arrangement. This interaction is presumably another example of the repulsive van der Waals interactions resulting in an increase in the r^{-3} term which would account for the downfield position of the oxygen signal relative to those of the other cyclic peroxides. It seems likely that the consequences of the changes in the C-O-O-C dihedral angle play a role in determining the chemical shifts of cyclic peroxides. However, the chemical shift values for these systems doubtlessly result from the interaction of several factors which cannot be readily sorted out at this time.

III. SUMMARY

The magnitude of ¹⁷O chemical shift changes with torsion angle can be very large for a variety of functional groups — up to 60 ppm or more — and thereby provides a sensitive and direct method of assessment of the consequences of steric interactions. It is clear from the studies reported to date that ¹⁷O NMR spectroscopy is an extremely sensitive and valuable method for assessing solution phase geometries for molecules containing any one of a number of different functional groups.[3] Table 13 summarizes the quantitative relationships between ¹⁷O chemical shifts and torsion angles which have been developed. These results, along with

TABLE 13
^{17}O Chemical Shift Data Correlated with Calculated Torsion Angles (MM2)[23]

Series	Slope (δ/angle deg)[a]	r[b]	Intercept	η[c]
ketones (C=O)	0.84 ± 0.14	0.979	553 ± 6	10
esters (C=O)	0.70 ± 0.04[a]	0.997	339 ± 2	10
esters (–O–)	0.43 ± 0.03	0.992	127 ± 2	10
acids	0.56 ± 0.05	0.994	249 ± 2	9
amides (C=O)	0.84 ± 0.05	0.942	308 ± 16	7
N,N-dimethylamides (C=O)	0.6[d]	0.991	313[d]	3
nitro compounds (NO₂)	0.76 ± 0.14	0.987	574 ± 16	7

[a] Error limits shown are 95% confidence limits.
[b] Correlation coefficient.
[c] Number of data points.
[d] Caution limited number of data points.

^{17}O chemical shift measurements, should allow for the rapid estimation of torsion angles for a wide variety of compounds.

The three possible consequences of ^{17}O chemical shift for torsion angle rotation have been observed. It has been found that torsion angle rotation can produce either downfield or upfield chemical shifts, and in one case no change in chemical shift on torsion angle rotation was noted. It was suggested that the direction of the shift depends upon the charge density change at oxygen upon rotation of the functional group. Previously, for systems under discussion here it was suggested that repulsive van der Waals interactions were minimized by rotation of the functional group from the plane of conjugation. The MM2 estimated total steric energies of a number of different compounds do not correlate with ^{17}O chemical shift. In an effort to clarify the nature of the steric interaction, we have estimated "local" van der Waals energies by a molecular mechanics approach. On comparing the difference in the calculated van der Waals energy for isomeric hindered and non-hindered ketones and esters, very small energy delta values were found. For example, for the naphthalene series α and β isomers were contrasted (for the ortho substituted compounds para isomers were used), for the 2,6-disubstituted systems their 3,5-disubstituted isomers were used, and for the 9-anthryl compounds the 2-anthryl isomers were used. The result suggests that for these systems "local" repulsive van der Waals interactions have been essentially minimized by torsion angle rotation. Furthermore, it was noted above that the ^{17}O chemical shift data for the same set of isomeric hindered and non-hindered ketones, esters, and N-arylacetamides exhibited large differences in their chemical shift values. A plot of the estimated "local" van der Waals energies for a wide range of structurally different aryl ketones, esters, and N-aryl-acetamides vs. the delta values in their chemical shifts shows an extremely large variation in chemical shift, all at essentially zero "local" van der Waals steric energies (Figure 12). Despite the fact that there is considerable error in this approximation of "local" van der Waals energies, this result provides strong support for the absence of repulsive steric interaction in these series. Consequently, the chemical shift changes with torsion angle should be attributed, in large part, to changes in charge density at the oxygen atom arising, indirectly, from minimization of van der Waals interactions. Charges in ΔE^{-1} with torsion angle change are also doubtlessly important since a linear relationship between the product $\Delta E^{-1} \cdot Q$ and ^{17}O chemical shifts for 1,2-diketones was found. The results from these conformationally rotatable systems indicate a distinctly different origin for the shift differences than that observed for rigid systems[3,51,60,61] (Chapter 4), where deshielding shifts are shown to correlate directly with repulsive van der Waals interactions.

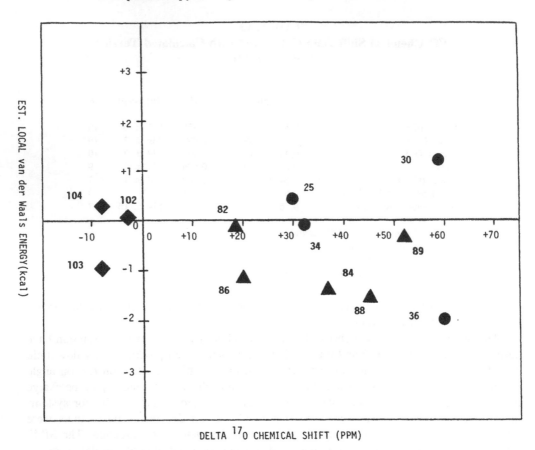

FIGURE 12.   Plot of delta values for MM2 calculated van der Waals energies vs. delta values for ¹⁷O chemical shifts for selected carbonyl compounds.

IV. EXPERIMENTAL CONSIDERATIONS

A number of the problems earlier associated with acquiring ¹⁷O NMR data have been minimized. High field instruments (greater than 50 MHz) for ¹⁷O (corresponding to 400 MHz for ¹H) have minimized base line-roll difficulties associated with acoustic ringing. Selection of low viscosity solvents, especially the avoidance of solvents such as chloroform, which are capable of hydrogen bonding, helps narrow linewidths. Acquisition of data at elevated temperature also contributes to reduction of linewidth.

The use of moderately enriched samples dramatically reduces the acquisition time required to obtain reasonable S/N values (see Chapter 2 for enrichment methods). A description of data acquisition and manipulation is contained in Chapter 4.

ACKNOWLEDGMENTS

Acknowledgment is made to the Donors of the Petroleum Research Fund, administered by the American Chemical Society, for partial support of this research, to NSF (CHE-8506665), and to the Georgia State University Research Fund. A.L.B. was a fellow of the Camille and Henry Dreyfus Foundation, 1981-86. Instrumentation used in these studies was supported in part by a NSF Equipment Grant (CHE 8409599).

REFERENCES

1. **Kintzinger, J.-P.**, Oxygen NMR characteristic parameters, in *NMR-17. Oxygen-17 and Silicon-29,* Diehl, P., Fluck, E., and Kosfeld, R., Eds., Springer-Verlag, New York, 1981, 1.

1a. **Kintzinger, J.-P.**, Oxygen-17 NMR, in *NMR of Newly Accessible Nuclei,* Vol. 2, Laszlo, P., Ed., Academic Press, New York, 1983, 79.

2. **Klemperer, W. G.**, in *The Multinuclear Approach to NMR Spectroscopy,* Proc. NATO Adv. Study Inst. on The Multinuclear Approach to NMR Spectroscopy, Lambert, J. B. and Riddel, F. G., Eds., D. Reidel, Dordrecht, Holland, 1983, 245.

3. **Boykin, D. W. and Baumstark, A. L.**, 17-O NMR spectroscopy: assessment of steric pertubation of structure of organic compounds, *Tetrahedron,* 45, 3613, 1989.

4. **Li, S. and Chesnut, D. B.**, Intramolecular van der Waals interactions and 13-C chemical shifts: substituent effects in some cyclic and bicyclic systems, *Magn. Reson. Chem.,* 24, 96, 1986.

5. **Li, S. and Chesnut, D. B.**, Intramolecular van der Waals interactions and chemical shifts: a model for γ and δ-effects, *Magn. Reson. Chem.,* 23, 625, 1985.

6. **Karplus, M. and Pople, J. A.**, Theory of carbon NMR chemical shifts in conjugated molecues, *J. Chem. Phys.,* 38, 2803, 1963.

7. **Figgas, B. N., Kidd, R. G., and Nyholm, R. S.**, Oxygen-17 nuclear magnetic resonance of inorganic compounds, *Proc. Chem. Soc., London,* A269, 469, 1962.

8. **Delseth, C. and Kintzinger, J.-P.**, ^{17}O NMR Aliphatic aldehydes and ketones, additivity of substituent effects and correlation with ^{13}C NMR, *Helv. Chim. Acta,* 59, 466, 1976.

9. **Kidd, R. G.**, Relationship between oxygen-17 nuclear magnetic resonance chemical shifts and π-bonding to oxygen in the dichromate ion, *Can. J. Chem.,* 36, 170, 1967.

10. **Brownlee, R. T. C., Sadek, M., and Craik, D. J.**, Ab initio MO calculations and 17-O NMR at natural abundance of *p*-substituted acetophenones, *Org. Magn. Reson.,* 21, 616, 1983.

11. **Christ, H. A. and Diehl, P.**, Lösungsmitteleinflüssen und spinkopplungen in der kernmagnetischen resonanz von ^{17}O *Helv. Phys. Acta,* 36, 170, 1963.

12. **Sardella, D. J. and Stothers, J. B.**, Nuclear magnetic resonance studies. XVI. Oxygen-17 shieldings of some substituted acetophenones, *Can. J. Chem.,* 47, 3089, 1969.

13. **St. Amour, T. E., Burger, M. I., Valentine, B. and Fiat, D.**, 17-O NMR studies of substituent and hydrogen-bonding effects in substituted acetophenones and benzaldehydes, *J. Am. Chem. Soc.,* 103, 1128, 1981.

14. **Boykin, D. W., Balakrishnan, P., and Baumstark, A. L.**, 17-O NMR spectroscopy of heterocycles. Steric effects for N-oxides, *Magn. Reson. Chem.,* 23, 695, 1985.

15. **Balakrishnan, P. and Boykin, D. W.**, Relationship of aromatic nitro group torsion angles to 17-O chemical shifts, *J. Org. Chem.,* 50, 3661, 1985.

16. **Trotter, J.**, The crystal structure of nitrobenzene at $-30°C$, *Acta Crystallogr.,* 12, 884, 1959.

17. **Trotter, J.**, The crystal structure of some anthracene derivatives. V. 9-Nitroanthracene, *Acta Crystallogr.,* 12, 237, 1959.

18. **Wyckoff, R. W. G.**, Ed., *Crystal Structures,* Vol. 1.(Part 2), Interscience, New York, 1971, 402.

19. **Akopyan, Z. A., Kitaigorodski, A. I., and Struchkov, Y. I.**, Steric hindrance and molecular conformation (XII). Crystal and molecular structure of 1,8-dinitron aphthalene, (transl.), *J. Struct. Chem.,* 6, 690, 1965.

20. **Lichter, R. I. and Levy, G. C.**, *Nitrogen-15 Nuclear Magnetic Resonance Spectroscopy,* John S. Wiley & Sons, New York, 86, 1979.

21. **Balakrishnan, P. and Boykin, D. W.**, Natural abundance 17-O NMR spectroscopy of heteroaromatic nitro compounds, *J. Heterocycl. Chem.,* 23, 191, 1986.

22. **Oakley, M. G. and Boykin, D. W.**, Relationship of torsion angle to 17-O NMR data for aryl ketones, *J. Chem. Soc. Chem. Commun.,* p. 439, 1986.

23. **Baumstark, A. L., Balakrishnan, P., Dotrong, M., McCloskey, C. J., Oakley, M. G., and Boykin, D. W.**, 17-O NMR spectroscopy: torsion angle relationships in aryl carboxylic esters, acids, and amides, *J. Am. Chem. Soc.,* 109, 1059, 1987.

24. **Boykin, D. W., Balakrishnan, P., and Baumstark, A. L.**, 17-O NMR studies in aryl-alkyl ketones and aromatic aldehydes, *Magn. Reson. Chem.,* 25, 248, 1987.

25. **Pinkus, A. G. and Custard, H. C., Jr.**, Dipole moments of alkyl mesityl ketones and some aliphatic and phenyl analogs, *J. Phys. Chem.,* 74, 1042, 1970.

26. **Ferguson, L. N.**, *Organic Molecular Structure,* Willard Grant Press, Boston, 1975, chap. 5.

27. **McQuillin, F. J. and Baird, M. S.**, *Alicyclic Chemistry,* 2nd ed., Cambridge University Press, Cambridge 1983, chap. 1.

28. **Greenberg, A. and Stevenson, T. A.**, Structures and energies of substituted strained organic molecules, in *Molecular Structure and Energetics,* vol. 3, Liebman, J. F. and Greenberg, A., Eds., VCH, Deerfield Beach, 1986, 193.

29. **Jaffe, H. H. and Orchin, M.**, Theory and Applications of Ultraviolet Spectroscopy, John S. Wiley & Sons, New York, 1962, chap. 15.

30. **Braude, E. A. and Sondheimer, F.**, Studies in light absorption. XI. Substituted benzaldehydes, acetophenones and related compounds. The effects of steric conformation on the electronic spectra of conjugated systems, *J. Chem. Soc.*, p. 3754, 1955.

31. **Trotter, J.**, The crystal structures of some anthracene derivatives. VI. 9-Anthraldehyde, *Acta Crystallogr.*, 12, 922, 1959.

32. **Cerfontain, H., Kruk, C., Rexwinkel, R., and Stunnenberg, F.**, Determination of intercarbonyl dihedral angle of 1,2-diketones by 17-O NMR, *Can. J. Chem.*, 65, 2234, 1987.

33. **Delseth, C., Nguyen, T. T.-T., and Kintzinger, J.-P.**, Oxygen-17 and carbon-13 NMR. Chemical shifts of unsaturated carbonyl compounds and aryl derivatives, *Helv. Chim. Acta*, 63, 498, 1980.

34. **Balakrishnan, P., Baumstark, A. L., and Boykin, D. W.**, 17-O NMR spectroscopy: Effect of substituents on chemical shifts for *p*-substituted benzoic acids, methyl benzoates, cinnamic acids, and methyl cinnamates, *Org. Magn. Reson.*, 22, 753, 1986.

35. **Anca, R., Martinez-Carrera, S., and Garcia-Blanco, S.**, The crystal structure of 2,6-dimethylbenzoic acid, *Acta Crystallogr.*, 23, 1010, 1967.

36. **Cano, F. H., Martinez-Carrera, S., and Garcia-Blanco, S.**, The crystal structure of 2,4,6,-trimethyl-benzoic acid, *Acta Crystallogr. Sect. B*, 26, 659, 1970.

37. **Smith, P., Florencio, F., and Garcia-Blanco, S.**, The crystal structure of 2,3-dimethylbenzoic acid, *Acta Crystallogr. Sect. B*, 27, 2255, 1971.

38. **Guilleme, J., Diez, E. D., and Bermejo, F. J.**, Aryl carbon chemical shifts of Methylbenzoic Acids and Methylbenzoate Anions, *Magn. Reson. Chem.*, 23, 442, 1985.

39. **Liepin'sh, E. E., Zitsmane, J. A., Ignatovich, L. M., Lukevitis, E., Guvanova, L. I., and Voronkov, M. G.**, 17-O NMR chemical shifts of trifluorosilylmethyl *p*-substituted benzoates and their carbon analogs, *Zh. Obshch. Khim.*, 53, 1789, 1982.

40. **Sugawara, T., Kawada, Y., and Iwamura, H.**, Oxygen-17 NMR chemical shifts of alcohols, ethers and esters, *Chem. Lett.*, 12, 1371, 1978.

41. **Orsini, F. and Ricca, G. S.**, Oxygen-17 NMR chemical shifts of esters, *Org. Magn. Reson.*, 22, 653, 1984.

42. **Monti, D., Orsini, F., and Ricca, G. S.**, Oxygen-17 NMR spectroscopy. Effect of substituents on chemical shifts for substituted benzoic acids, phenylacetic and methyl benzoates, *Spectrosc. Lett.*, 19, 91, 1986.

43. **Pinkus, A. G. and Lin, E. Y.**, Electric dipole moment studies of carboxylic ester conformations: alkyl acetates, benzoates, 2,4,6-trimethyl-benzoates and 2,3,5,6-tetramethylbenzoates, *J. Mol. Struct.*, 24, 9, 1975.

44. **Burgar, M. I., St. Amour, T. E., and Fiat, D.**, 17-O and 14-N NMR studies of amide systems, *J. Phys. Chem.*, 85, 502, 1981.

45. **Canet, D., Goulin-Ginet, C., and Marchal, J. P.**, Accurate determination of parameters for 17-O in natural abundance by fourier transform NMR, *J. Magn. Reson.*, 22, 537, 1976.

46. **Ruostesuo, P., Hakkinen, A.-M., and Peltola, K.**, Carbon-13, nitrogen-15 and oxygen-17 NMR chemical shifts of 1-ethyl-2-pyrrolidione in some solvents, *Spectrochim. Acta*, 41A, 739, 1985.

47. **Steinschneider, A. and Fiat, D.**, Carbonyl 17-O NMR of amino acid and peptide carboxamide and methyl ester derivatives, *Int. J. Pep. Protein Res.*, 23, 591, 1984.

48. **Valentine, B., Steinschneider, A., Dhawan, D., Burgar, M. I., St. Amour, T., and Fiat, D.**, Oxygen-17 NMR of peptides, *Int. J. Pep. Protein Res.*, 25, 56, 1985.

49. **Penfold, P. R. and White, J. C. B.**, *Acta Crystallogr.*, 12, 30, 1959.

50. **Boykin, D. W., Deadwyler, G. H., and Baumstark, A. L.**, 17-O NMR studies on substituted *N*-arylacetamides and aryl acetates: torsion angle and electronic effects, *Magn. Reson. Chem.*, 26, 19, 1988.

51. **Baumstark, A. L., Dotrong, M., Oakley, M. G., Stark, R., and Boykin, D. W.**, 17-O NMR study of steric interactions in hindered N-substituted imides, *J. Org. Chem.*, 52, 3640, 1987.

52. **Matsuo, T.**, Carbonyl absorption bands in the infrared spectra of some cyclic imides with a five-membered ring, *Bull. Chem. Soc. Jpn.*, 37, 1844, 1964.

53. **Khadim, M. A. and Colebrook, L. D.**, Carbon-13 delta shifts and steric interaction in *N*-aryl-1-isoindolinones and isoindoline-1,2-diones, *Magn. Reson. Chem.*, 23, 259, 1985.

54. **Lumbroso, H. and Dabbard, R.**, Effets inductif et mesomere dans les molecules organiques substituees. VII.-Phtalimides substitutes, *Bull. Soc. Chim. Fr.*, p. 749, 1959.

55. **Ascoria, A., Barassin, J., and Lumbroso, H.**, Effets inductif et mesomere dans les molecules organiques substituees. XIII.-Sur les moments electriques de divers derives substitues du phenyl-N succinimide et du phenyl-N phtalimide, *Bull. Soc. Chim. Fr.*, p. 2509, 1963.

56. **Barassin, J.**, Molecular configuration of some derivatives of pyridine and *N*-phenylsuccinimide, *J. Liebigs Ann.*, 8, 656, 1963.

57. **Vasquez, P. C., Boykin, D. W., and Baumstark, A. L.**, 17-O NMR spectroscopy (natural abundance) of heterocycles: anhydrides, *Magn. Reson. Chem.*, 24, 409, 1986.

58. **Zagorski, M. G., Allen, D. S., Salomon, R. G., Clennan, E. L., Heah, P. C., and L'Esperance, R. P.**, Oxygen-17 nuclear magnetic resonance chemical shifts of dialkyl peroxides: large conformational effects, *J. Org. Chem.*, 50, 4484, 1985.

59. **Beraldin, M.-T., Vauthier, E., and Fliszar, S.**, Charge distributions and chemical effects. XXVI. Relationships between nuclear magnetic resonance shifts and atomic charges for ^{17}O nuclei in ethers and carbonyl compounds, *Can. J. Chem.*, 60, 106, 1982.

60. **Baumstark, A. L., Balakrishnan, P., and Boykin, D. W.**, 17-O NMR spectroscopy as a probe of steric hindrance in phthalic anhydrides and phthalides, *Tetrahedron Lett.*, p. 3079, 1986.

61. **Boykin, D. W., Baumstark, A. L., Kayser, M. M., and Soucy, C. M.**, 17-O NMR spectroscopic study of substituted phthalic anhydrides and phthalides, *Can. J. Chem.*, 65, 1214, 1987.

Chapter 4

APPLICATIONS OF ^{17}O NMR SPECTROSCOPY TO STRUCTURAL PROBLEMS IN RIGID, PLANAR ORGANIC MOLECULES

Alfons L. Baumstark* and David W. Boykin*

TABLE OF CONTENTS

I. INTRODUCTION AND SCOPE

^{17}O nuclear magnetic resonance (NMR) spectroscopy is rapidly developing into an important method for examining a wide variety of structural problems[1,2,3] and may provide new insights into the understanding of chemical reactivity in hindered systems (*vide infra*). A number of ^{17}O NMR studies have focused on electronic factors,[4] while others have investigated conformational effects.[5] Recent work has shown that quantitative relationships can be formulated between ^{17}O NMR data and torsion angles for aromatic nitro compounds,[6] acetophenones,[7] aryl ketones,[8] and aromatic carboxylic acids and derivatives.[9] The structural information relating to torsion angle relationships accessible by ^{17}O NMR methodology has been discussed in detail in the preceding chapter.

This chapter will concentrate on ^{17}O NMR studies of rigid, planar systems, work that was ongoing in our laboratories concurrent with our torsion angle studies. In particular, the effect of steric interactions on the ^{17}O NMR chemical shift data of *N*-oxides and carbonyl functional groups will be emphasized. The ^{17}O NMR data for these sterically hindered systems have been shown to correlate (*vide infra*) with in-plane bond angle distortions. The nontorsional deshielding effects are thought to be indicative of repulsive van der Waals interactions. Thus, these deshielding effects are fundamentally different from those in which torsion angle changes occur.

Heteroatom NMR chemical shifts are described by diamagnetic and paramagnetic expressions. ^{17}O chemical shifts[10] are dominated by paramagnetic effects which are often formulated as the Karplus-Pople equation:[1-3,11]

$$\sigma_o^p = \frac{-e^2h^2}{2m^2c^2\Delta E}\,(r^{-3})_{2p_o}\,[Q_{ox}]$$

Recently, Chesnut has reported[12] correlations between the local van der Waals steric energy and the chemical shift of a resonant nucleus (^{13}C, ^{15}N, and ^{31}P) within several classes of saturated aliphatic phosphines, amines, hydrocarbons, and monofunctional alcohols. Chesnut suggested[12] that the repulsive van der Waals interactions were associated with deshielding effects in the observed chemical shifts. A theory was developed that suggested, in part, that the net deshielding effect was the result of a contraction of the orbitals on the resonant nucleus, due to repulsive van der Waals interactions, leading to a corresponding increase in the r^{-3} term of the paramagnetic expression of the Karplus-Pople equation. This concept can be applied successfully to our ^{17}O studies (*vide infra*).

II. NONTORSIONAL EFFECTS: *N*-OXIDES

A. BACKGROUND

Previously, little was known about steric effects on ^{17}O chemical shifts; only studies on saturated carbocyclic and heterocyclic systems[5,13] and *ortho*-substituted benzaldehydes and acetophenones[4a,14] had appeared. The steric effect on chemical shifts observed for benzaldehydes and acetophenones was explained in terms of steric inhibition of resonance.[4a,14] No studies had probed the effect of steric hindrance on ^{17}O NMR data in rigid, planar systems. Pyridine *N*-oxides and related heterocyclic *N*-oxides appeared to be ideal systems for the study of steric interactions since rotation of the functional group out of the plane of an aromatic ring was not possible. In addition, the ^{17}O chemical shifts of the *N*-oxide group for pyridine *N*-oxides had been found to be remarkably sensitive to electronic and solvent effects.[15,16] The sensitivity of the *N*-oxide group to structural changes[15] made it an attractive group for studying the effect of changes steric environments on ^{17}O chemical shifts.

B. PYRIDINE *N*-OXIDES

To assess the effect of changing steric environment on the *N*-oxide ^{17}O chemical shift, a series of 2-substituted pyridine *N*-oxides were chosen for study.[17] The ^{17}O chemical shift data for 2- and 4-substituted pyridine *N*-oxides (**1** to **10**) in acetonitrile at 75°C, are listed in Table 1. It can be seen from the data of Table 1 that the signals for 4-alkyl substituted compounds were shielded when compared with that of pyridine *N*-oxide,[15] while comparison of the chemical shift data of the 2-alkylpyridine *N*-oxides with their 4-alkyl isomers showed that the 2-alkyl isomers were deshielded[17] relative to the 4-alkyl isomers. The chemical shift data for the 4-substituted compounds are readily explained in terms of normal electronic effects. Representative spectra for pyridine *N*-oxide and 2-*t*-butylpyridine *N*-oxide are shown in Figure 1. Assuming that electronic effects of alkyl groups in the 2- and 4-positions are comparable and that solvent disruption is small, the deshielding observed with the 2-alkyl isomers must be attributed to compressional effects. A compressional shift of 6 to 14 ppm was found for the 2-alkyl groups ranging from methyl to isopropyl. Introduction of a 2-*tert*-butyl group resulted in a significant increase in deshielding (23 ppm), presumably the result of greater steric interactions and/or a consequence of additional disruption of solvent structure around the polar *N*-oxide function. A slight diminution of the deshielding effect (see Δ values in Table 1) with increasing size of the 2-alkyl group (except for the 2-*tert*-butyl group) was noted.[17] The ^{17}O chemical shift of 2,6-dimethylpyridine *N*-oxide (**10**) was observed at 350 ppm. Assuming only electronic effects of the two methyl groups to be operative, a chemical shift of 323 ppm would be predicted. Based on this assumption, a compressional effect of 27 ppm was obtained for the 2,6-dimethyl compound, an effect comparable to that noted for the 2-*tert*-butyl group.

TABLE 1
^{17}O Chemical Shift Data (± 1 ppm) for
Substituted Pyridine *N*-Oxides[17] in
Acetonitrile at 75°C

Compound no.	R	δ (ppm)	Δ ($\delta_{2R} - \delta_{4R}$)
1	H	349	—
2	2-Me	350	14
3	4-Me	336	
4	2-Et	346	10
5	4-Et	336	
6	2-*n*-Pr	342	6[a]
7	2-*i*-Pr	342	6[a]
8	2-*t*-Bu	361	23
9	4-*t*-Bu	338	
10	2,6-Di-Me	350	27

[a] δ_{4R} assumed to 336 ppm.

In order to determine if compressional effects were observable for a less polar function group, structurally similar *t*-butylanisoles were chosen for study.[17] The ^{17}O chemical shifts for *o*- and *p-tert*-butylanisoles **11a** and **b** were obtained in acetonitrile at 75°C (49 and 44 ppm, respectively). The chemical shift data for the anisoles were near those reported in chloroform for a series of *para*-substituted anisoles.[18] The ^{17}O chemical shift of *o-tert*-butylanisole **11a** was 5 ppm downfield from its analog *p-tert*-butylanisole **11b** (see Scheme 1). This shift was in the same direction (deshielding) as observed for similarly substituted *N*-oxides; however, the magnitude of the shift is less by a factor of approximately 5. It is clear from this data that the *N*-oxide oxygen in pyridine *N*-oxides is more sensitive to steric effects than is the anisole oxygen. This may indicate that more than one factor is operative in determining the value of the chemical shift. For the polar *N*-oxide function, some of the deshielding may be due to solvent disruption.

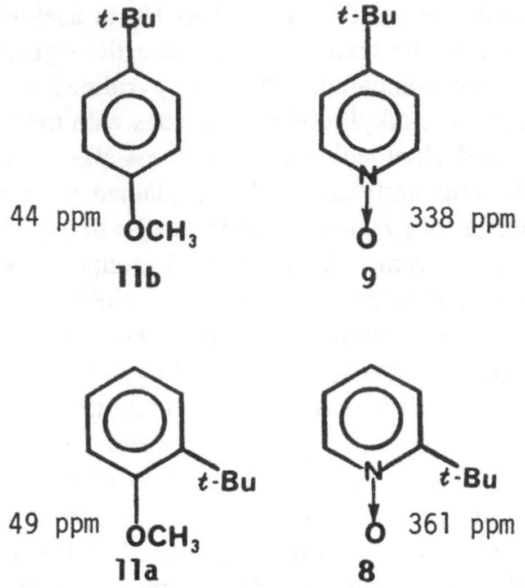

SCHEME 1.

To test for a reciprocal steric effect of the *N*-oxide function on the ^{13}C chemical shifts of the carbon of the *tert*-butyl group, the ^{13}C spectra were recorded[17] for 2- and 4-*tert*-butylpyridine *N*-oxides in CDCl$_3$. The differences in chemical shifts of the quaternary carbon are for the 2- and 4-*tert*-butyl compounds 1.8 ppm and for the methyl carbons -3.3 ppm. For closely related structural analogs, 2- and 4-*tert*-butylanisoles, the delta values were 0.8 and -1.8 pm for the quaternary and methyl carbons, respectively.[19] The magnitude of compressional effects on the ^{17}O resonance for the *N*-oxide function was greater by a factor of seven than that noted for the methyl carbon of the *tert*-butyl group. In saturated heterocyclic systems the carbon and oxygen compressional shifts have been found to differ by a factor of approximately three.[5b,13]

C. BENZOQUINOLINE *N*-OXIDES

In order to test for other examples of compressional effects on the ^{17}O chemical shifts of *N*-oxides, the ^{17}O spectra of quinoline *N*-oxide (12), benzo[f]quinoline *N*-oxide (13a) and benzo[h]quinoline *N*-oxide (13b) were measured[17] under conditions similar to those for the pyridine *N*-oxides.

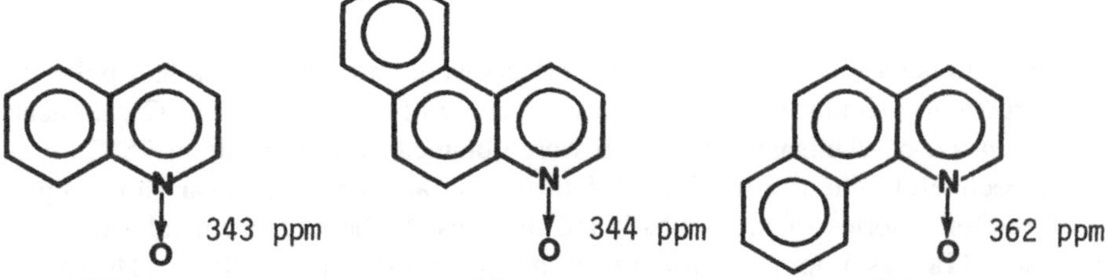

A small shielding effect (6 ppm) on benzene ring fusion was observed on the ^{17}O chemical shift of the *N*-oxide function (compare pyridine *N*-oxide 1 and quinoline *N*-oxide 12). It

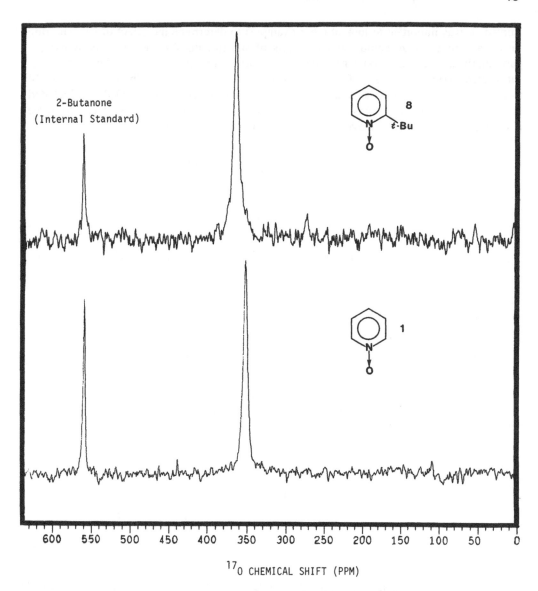

FIGURE 1. ^{17}O NMR spectra of pyridine N-oxide (**1**) and 2-*t*-butylpyridine *N*-oxide (**8**) in acetonitrile at 75° (0.5% 2-butanone as internal standard).

was previously observed that fusion of a benzene ring to quinoline and related carbonyl systems resulted in a shielding effect on the carbonyl oxygen ^{17}O chemical shift by approximately 30 ppm.[20] In contrast, when the ^{17}O chemical shift values for the two benzo-quinoline N-oxides **13a** and **13b** were compared, significant deshielding (18 ppm) was noted for **13b**. The location of the fused benzene ring on **13b** can be expected to give rise to a large steric interaction between the N-oxide function and the peri-like C-H, which could account for the deshielding of the N-oxide function.

D. QUINOLINE N-OXIDES

The above studies showed,[17] for a number of 2-substituted pyridine N-oxides and for three benzopyridine N-oxides, that compressional effects play a significant role in determining ^{17}O chemical shifts. Because of the limited number of heterocyclic aromatic N-oxides studied, it was important to look at other examples to determine the effect of benzene ring fusion and to compare compressional effects of substituents for the quinoline *N*-oxides, particularly at the 8-position where peri interactions are involved, with those of the 2-substituted pyridine *N*-oxides. To this end several isomeric methyl quinoline *N*-oxides, **14** to **17**, were examined and the data shown below;[21] isoquinoline *N*-oxide (**18**) was included for reference. Figure 2 contains representative natural abundance ^{17}O NMR spectra for three *N*-oxides that illustrate the magnitude of the compressional effects.

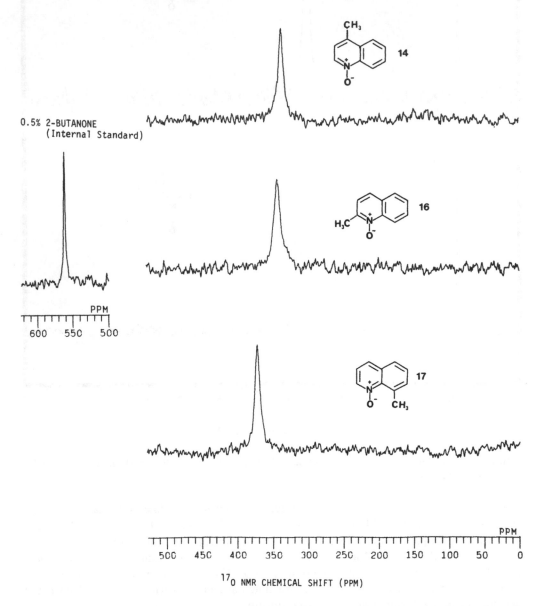

FIGURE 2. ^{17}O NMR spectra (natural abundance) of 4-methylquinoline (**14**), 2-methylquinoline (**16**), and 8-methylquinoline (**17**) in acetonitrile at 75°. (Insert: 2-butanone as internal standard.)

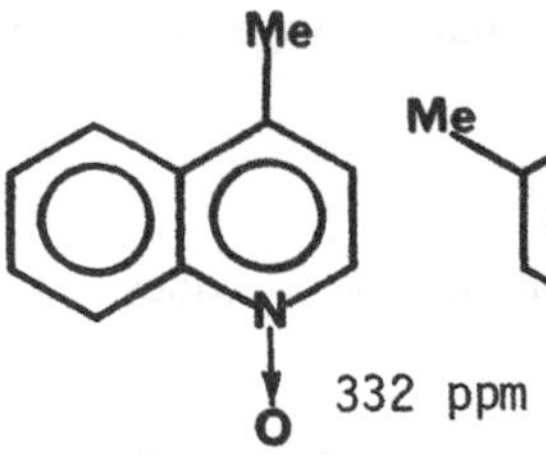

14   332 ppm   15   341 ppm   16   341 ppm

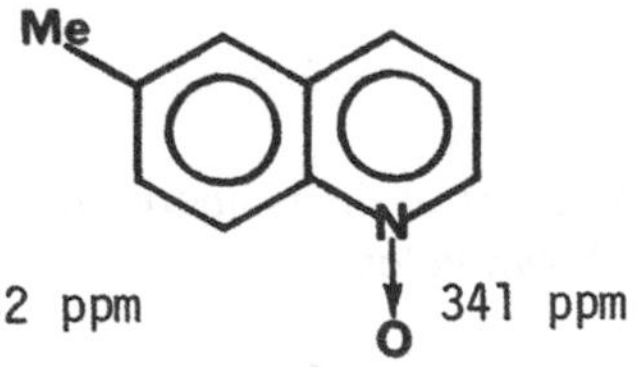

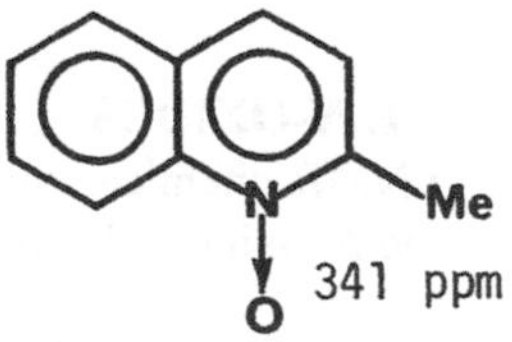

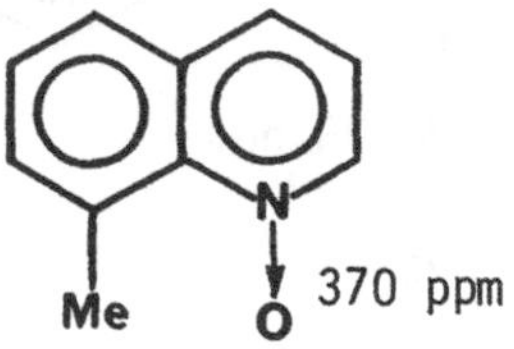

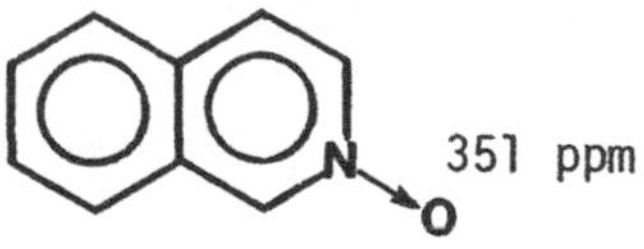

17   370 ppm   18   351 ppm

It was noted that the chemical shifts for **14** and **16** were shielded by approximately 5 ± 1 ppm compared with those of the appropriate pyridine *N*-oxides.[15,17] This shielding effect was consistent with those observed previously for the *N*-oxides of quinoline and two benzoquinolines and, thus, appears to be a general effect of benzene ring fusion. The chemical shift of **16** was deshielded by 9 ppm compared to its 4-isomer **14**, which was similar to the chemical shift differences observed for the 2- and 4-methylpyridine *N*-oxides.[17] The electronic effect of the methyl at position-6 was greatly reduced in magnitude compared to that at position-4. Overall, electronic effects of alkyl groups in the heterocyclic ring were comparable to those noted[17] for pyridine *N*-oxides; however, the effects of those substituents in the fused benzene ring appeared to be more complex. As expected, the 8-substituted compound **17** was deshielded by 29 ppm compared to its electronically equivalent isomer **15**. This chemical shift difference for the 8-methyl group is equivalent to that observed for a 2-*t*-butyl group in the pyridine system.

8-Hydroxyquinoline *N*-oxide, **19**, represents an interesting case.[21]

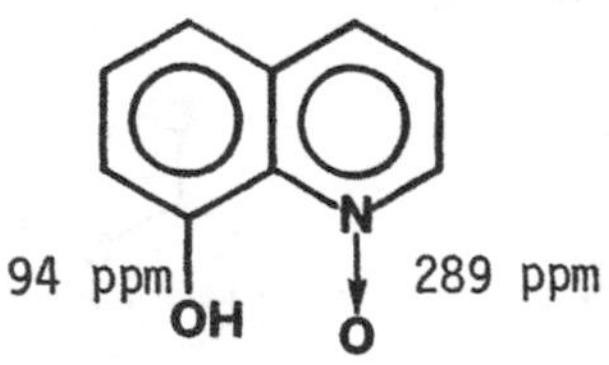

94 ppm   289 ppm

19

The ^{17}O NMR signal for **19** at 289 ppm (*N*-oxide) was substantially shielded from that of quinoline *N*-oxide **12**, at 343 ppm. Shielding, attributed to intramolecular hydrogen-bonding, has been noted for the carbonyl oxygens of 5,8-dihydroxynaphthoquinone (see Chapter 5).[20] However, the magnitude of the shielding for **19**, exclusively attributable to hydrogen-bonding, is difficult to assess. Compressional effects of the 8-hydroxy group upon the *N*-oxide oxygen would be expected to be deshielding; consequently, the difference of 54 ppm for the chemical shifts of **12** and **19** should be regarded as the minimum effect due to hydrogen-bonding. Note that the hydroxy oxygen (94 ppm) participating in the hydrogen bond was deshielded relative to the signal for phenol (79 ppm)[22] and related phenolic compounds (87 ppm).[20] The analysis of the effects of structural variation on single bonded oxygen atoms is not well-documented. The effects appear to be larger than earlier thought.[23] It also seems probable that the oxygen of the donor for intramolecular hydrogen-bonding experiences deshielding.[20,24]

E. DIAZENE *N*-OXIDES

The ¹⁷O NMR chemical shift values of the isomeric diazine *N*-oxides, **20** to **22**, and of benzopyrazine *N*-oxide, **23**, are shown below.[21]

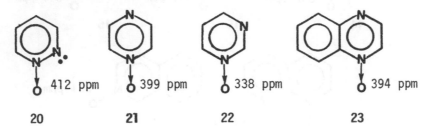

412 ppm	399 ppm	338 ppm	394 ppm
20	**21**	**22**	**23**

When compared with that of pyridine *N*-oxide **1**, the order of ¹⁷O chemical shifts was **20**>**21**>**1**>**22**, with the larger value corresponding to the greater double bond character for the NO bond. This order of double bond character was in reasonable agreement with that suggested by Paudler and Jovanovic[25] based upon ¹⁵N chemical shifts. The only difference in the bond order estimates based on ¹⁵N data and the ¹⁷O results was that the former predicted that **1** and **22** should exhibit comparable NO bond orders. The ¹⁷O results indicated that back donation was somewhat greater for pyridine *N*-oxide **1** than for pyrimidine *N*-oxide **22**. The ¹⁷O chemical shift of **20** was found downfield from that of **21** by 13 ppm. However, this downfield shift difference may not be solely attributable to differences in back donation of the two isomers. It is possible that compressional effects of the lone pair of electrons on the adjacent nitrogen are contributing to the deshielding observed for **20**. The ¹⁷O chemical shift of the benzodiazine **23** was shielded by 5 ppm compared to its diazine parent **21**. This represents another example of the consistent shielding effect of benzene ring fusion *ortho* to the *N*-oxide function.

To assess the effect of compressional effects (deshielding) on the ¹⁷O chemical shifts in the diazine series, ¹⁷O NMR spectra of 4-methylpyrimidine 1-oxide, **24**, and 4-methyl-pyrimidine 3-oxide, **25**, were obtained.[21]

These two *N*-oxides were chosen because their structure proof[26] was difficult and had shown an earlier assignment to be incorrect. Since the electronic effects of the methyl group in isomers **24** and **25** are equivalent, any difference in chemical shift of the two can be attributed to compressional effects. Based upon pyridine *N*-oxide data[15,17], the electronic effect of the methyl group should be shielding for the signal of **24** and **25** by 13 ppm. The 1-oxide, **24**, which should be devoid of compressional effects, exhibited a chemical shift of 324 ppm shielded by 14 ppm compared to **22** in agreement with predictions based upon the pyridine system. The chemical shift of **25** should experience the same degree of shielding (13 ppm) arising from electronic effects as for **24**, but its signal should also be deshielded by 14 ppm arising from compressional effects (based upon pyridine results), resulting in essentially no chemical shift difference from its parent **22**. The chemical shift observed for **25** was 335 ppm in good agreement with the above prediction. Thus the application of the ^{17}O NMR methodology would have resulted in correct structural assignment.

F. SUMMARY

Significant compressional effects[17,21] have been shown for 2-substituted pyridine *N*-oxides, for benzo[h]quinoline, for various substituted quinolines, and for electronically similar *N*-oxides of diazines. These deshielding effects are large enough in many cases to use safely the ^{17}O chemical shifts to distinguish between isomers, and this approach should be applicable to other systems. It should also be noted that these ^{17}O NMR studies, carried out at natural abundance on 0.5 *M* solutions in acetonitrile at 75°, usually required 6 to 8 h of instrument time (2 to 3 times greater than that for many other functional groups). This time requirement, although inconvenient, should not deter one from using this methodology.

The origin of these deshielding effects is speculative. The greater sensitivity to compressional effects of the *N*-oxide oxygen compared with that of the anisoles can be attributed to the *N*-oxide double bond character and/or difference in solvation between the polar and neutral molecules. Steric (deshielding) effects comparable to those of the *N*-oxide system are seen in carbonyl systems (*vide infra*). These results suggest that steric interactions may be greater when substantial π-interactions are involved. Possibly involved in the sensitivity difference may be the influence of alteration of the solvent structure around the polar *N*-oxide function on introduction of large hydrocarbon groups in close proximity. In contrast to the results for compounds, for which steric inhibition of resonance can be used[27] to explain the shift differences, the data for the *N*-oxides, while of comparable magnitude, are due to a different mode of interaction. Paudler and Jovanovic[25] have explained the ^{15}N NMR data and IR data for other sterically hindered *N*-oxides by steric inhibition of back donation through alteration of coplanarity of the NO group and the ring. It is not clear that this type of ring-NO group deformation is required to explain these data. As we have shown for planar carbonyl systems (*vide infra*), in-plane bond deformation (*N*-oxide) due to repulsive van der Waals effects[12] could also explain the ^{17}O deshielding effects. The effect of solvent structure changes, if any, on the ^{17}O NMR chemical shifts of these polar functional groups is an additional factor to consider that may further complicate the interpretation.

III. IN-PLANE DEFORMATIONS: CARBONYLS

A. PHTHALIC ANHYDRIDES AND PHTHALIDES

The significant deshielding effects on the ^{17}O chemical shift data for sterically hindered heteroaromatic *N*-oxides,[17,21] where torsional effects are limited, are useful for identifying certain types of isomeric compounds. However, the identification of the origin of the steric factors that affect ^{17}O NMR data is unclear at present but should be understood to significantly increase insight into molecular structure by use of ^{17}O NMR methodology. To begin to sort out nontorsional, steric factors, it was essential to study a system which had a less polar functional group with a well-defined, planar geometry and which showed a large chemical shift range. Certain cyclic anhydrides seemed to satisfy these criteria,[28] and, in addition, in some cases the two carbonyls had been reported to exhibit differential reactivity.[29] The phthalic anhydride system was chosen for study since it appeared that large steric interactions were possible. Indeed, the ^{17}O NMR data for a series of 3-substituted phthalic anhydrides (**26**) and corresponding phthalides (**27** to **28**) clearly showed[30,31] that the chemical shifts are sensitive to nontorsional bond angle deformations.

26a-j   27a-i   28a-g

The ^{17}O NMR data (natural abundance) for phthalic anhydride (**26a**), 3-methyl-phthalic anhydride (**26b**), 3-*t*-butylphthalic anhydride (**26c**), phthalide (**27a**), 7-methyl-phthalide (**27b**), 7-*t*-butylphthalide (**27c**), and 4-*t*-butylphthalide (**28a**) were obtained[30] at 75°C in acetonitrile to check the feasibility of the approach and to assign the chemical shift data. The results are summarized in Table 2.[9,10] For both **26b** and **26c**, two well-defined ^{17}O signals for the sterically different carbonyl groups, separated by 11 and 29 ppm, respectively, were observed. The single bond oxygen for all the anhydrides appeared at 263 ± 1 ppm. A representative spectrum of 3-methylphthalic anhydride is shown in Figure 3. The ^{17}O data for the two isomeric *t*-butyl phthalides (**27c**, **28a**) allowed the assignment of the downfield (deshielded) carbonyl signal in the substituted anhydrides to the carbonyl adjacent to the substituent. Note that the "carbonyl" signal for **27c** is downfield of that of **28a** by 27 ppm. The downfield shifts noted for **26b**, **26c**, **27b**, and **27c** cannot be due to electronic effects. Previous work on the benzoate system had shown that the electronic effects of alkyl groups on the carbonyl resonance were modestly shielding (2 ppm).[32a] Hence, the chemical shift differences between **26a** and its methyl and *t*-butyl analogs (**26b** and **26c**) would be roughly 2 ppm larger if a correction for electronic effects had been made. The deshielding effect in both series (**26b**, **27b**) for an *ortho*-methyl group was 9 to 12 ppm while a similarly located *t*-butyl group produced a 22 to 27 ppm shift. Interestingly, the ^{13}C NMR chemical shifts of the two carbonyl carbons in **26b**, **c** were essentially identical (within 1 ppm). Thus, the ^{17}O NMR methodology provided new insights into the effect of steric hindrance on the carbonyl function.

TABLE 2
^{17}O NMR Data (±1 ppm) for 3-Substituted Phthalic Anhydrides 26a-c and Phthalides (27a-c, 28a) in Acetonitrile at 75°[30]

	Anhydrides				Lactones		
R	Compound no.	δ (C=O)	δ (–O–)	R	Compound no.	δ (C=O)	δ (–O–)
H	26a	374	263	H	27a	320	170
Me	26b	383	264	7-Me	27b	332	170
		372					
t-Bu	26c	396	262	7-*t*-Bu	27c	346	168
		367					
				4-*t*-Bu	28a	319	173

As a check of the ^{17}O chemical shift assignments in series **26**, qualitative shift reagent studies were carried out[32b] on **26a** and **26c** in CDCl$_3$. The results for the carbonyl signals for **26c** showed that the deshielded signal was relatively insensitive to the shift reagent [Eu(FOD)$_3$].[5c,32c] The other carbonyl signal showed a sensitivity similar to that observed with the parent anhydride **26a**. Thus, the results of the shift reagent study were consistent with the chemical shift assignments. The results are summarized in Figure 4.

In an effort to gain insight into the influence of substituents of varying electronic character on the properties and reactivities of the two carbonyl functions of 3-substituted phthalic

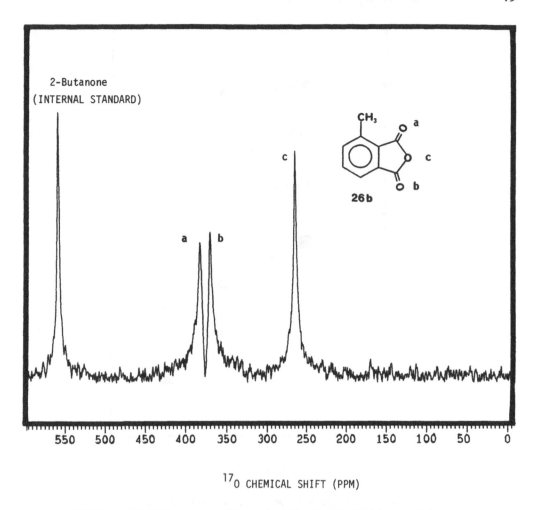

FIGURE 3.   ^{17}O NMR spectrum of 3-methylphthalic anhydride (26b) in acetonitrile at 75°.

anhydrides, the [17]O chemical shifts of additional series of phthalic anhydrides (26d to j) and related phthalides (27d to i and 28b to g) were studied.[31] [17]O chemical shift data for the 3-substituted phthalic anhydrides (26b to k), 7-substituted phthalides (27d to i) and 4-substituted phthalides (28b to g) in acetonitrile at 75°C, measured at natural abundance, are given in Table 3 and Table 4. All the substituted anhydrides showed two carbonyl signals in the 370 to 395 ppm region and one signal near 263 ppm.

The assignment of the two carbonyl signals for the remaining anhydrides of series 26 was made by using the results from the corresponding phthalides 27 and 28 (Tables 2 to 4) as discussed above.[30] It is clear from the data for the phthalide series 27 that the effect of a substituent *ortho* to the planar carbonyl was deshielding, regardless of the electronic character of the substituent. Consequently, the assignments for the two carbonyl signals of the anhydrides shown in Tables 2 to 3 are made by analogy with the corresponding phthalides. With the exception of the 3-*t*-butyl and the 3-nitro compounds in series 26, the effect of substituents was a uniform deshielding value of 9 ± 1 ppm. This, in part, is presumably a result of repulsive van der Waals interactions.[12] The single bonded oxygen signal for all 3-substituted anhydrides, 26, was insensitive to substituents with all chemical shift values at 263 ± 1 ppm. The ^{17}O NMR signal for the carbonyl group *meta* to the substituent was relatively insensitive to the 3-substituent ranging in value from 367 to 377 ppm. The chemical shift of the carbonyl *ortho* to the substituent for the 3-methoxy compound 26e was deshielded

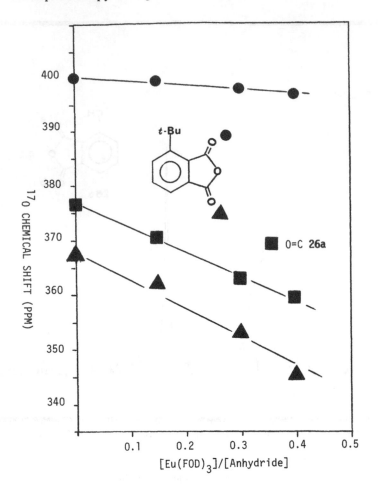

FIGURE 4.   Shift reagent studies on phthalic anhydride (**26a**) and 3-*t*-bu-tylphthalic anhydride (**26c**) in CDCl₃ at 55°C.

TABLE 3
¹⁷O NMR Data (± 1 ppm) for 3-Substituted Phthalic
Anhydrides 26d-j in Acetonitrile at 75°C³¹

Compound no.	R	δ (C=O)₁	δ (C=O)₂	δ (–O–)	δ (R)
26d	3-C₂H₅O	370	384	264	91
26e	3-CH₃O	371	383	264	65
26f	3-F	376	385	264	
26g	3-Cl	375	386	264	
26h	3-Br	374	385	265	
26i	3-I	373	382	263	
26j	3-NO₂	377	395	262	

relative to the parent **26g** by 9 ppm, and the shift data for the 3-nitro compound **26j** was deshielded by 22 ppm relative to the parent. Thus, the carbonyl groups *ortho* to the 3-substituents were deshielded relative to the parent compound by all substituents regardless of the electronic character of the groups. The deshielding effect seemed to be the result of steric interactions of the 3-substituent and the lone pairs on the carbonyl group.

The chemical shift data for the carbonyl group of the 7-substituted phthalides (**27**) were

TABLE 4
^{17}O NMR Data (± 1 ppm) for 7- and 4-Substituted Phthalides (27d-i, 28b-g) in Acetonitrile at 75°C[31]

| | 7-Substituted | | | | 4-Substituted | | | |
R	Compound no.	δ (C=O)	δ (–O–)	δ (R)	Compound no.	δ (C=O)	δ (–O–)	δ (R)
CH₃O	27d	333	168	59	28b	323	170	49
F	27e	334	171		28c	325	169	
Cl	27f	335	170		28d	327	170	
Br	27g	334	170		28e	327	171	
I	27h	330	171		28f	327	171	
NO₂	27i	337	172	616	28g	325	174	582

all deshielded by 10 to 26 ppm compared to that of the parent molecule **26a**, which exhibited a carbonyl signal at 320 ppm. In series **27**, electron-donating or electron-withdrawing substituents resulted in deshielding as observed in series **26**. The resonances for the lactone (dicoordinate) oxygen of series **27** were essentially invariant, all appearing at 170 ± 2 ppm. The chemical shift values for the carbonyl oxygen for series **28** were only slightly affected by substituent, appearing at 323 ± 4 ppm, as were those of the single bonded (–O–) oxygen which appeared at 171 ± 3 ppm and could be explained as normal electronic effects.

The molecules in series **26** and **27** were expected to be planar; however, since the magnitude and direction of these shift differences in series **26** and **27** were similar to those attributed to torsional variations,[6-9] this possibility was evaluated and could be ruled out. Results[27] for sterically hindered ketones showed that a torsional rotation of the carbonyl of at least 20° would be required to yield the deshielding of 25 ppm observed for the chemical shifts in **26c** and **27c**. Molecular mechanics calculations[33] (MM2) for **26** to **28** predicted[30] that all the ring systems were planar. In addition, the calculations indicated that no significant changes in bond lengths were anticipated within each series of compounds. However, the calculations did predict substantial in-plane distortions of the bond angles in both rings in close proximity to the R group at the juncture of the two rings. Representative results for the alkyl-substituted anhydrides and phthalides are shown in Table 5. To obtain independent corroboration of the predicted geometry, the crystal structure (Figure 5) of the most distorted anhydride, 3-t-butylphthalic anhydride, **26c**, was obtained. Analysis of a single crystal showed that all the atoms for **26c** with the exception of t-butyl methyl groups, were completely planar.

The trends discernible in Table 5 showed that for compounds with the larger R groups, the angles represented, particularly by entries 1, 2, 3 and 4, were larger and those represented by entries 5, 7 and 13 were smaller. The carbonyl group *peri*-like to the R-group showed significant bond angle variation with R-group size, whereas the relatively unhindered carbonyl group was essentially unaffected. Presumably the in-plane distortions observed reflect (partial) minimization of van der Waals interactions. The trends noted in Table 5 paralleled the carbonyl ^{17}O chemical shift data. The predicted distortions for the phthalides **27a** to **27c** also paralleled those observed for **26a** to **26c**, whereas little distortion of the carbonyl angles for **28a** was estimated. The carbonyl ^{17}O data for the phthalides reflected this trend (Tables 2 to 4). For all five compounds no distortion of the bond angle involving the single bonded (–O–) oxygen (entry 6) was predicted by the calculations; the ^{17}O data for O_9 is essentially constant in each series (Tables 2 to 4). Comparison of the MM2 data and the X-ray results for **26c** showed they were in reasonable agreement. The two methods were in poorest agreement for the angles represented by entries 3, 6 and 9; MM2 calculations underestimated the bond angles although the trends were consistent with ^{17}O data.

It is of interest to note that there is no appreciable difference in the data for the compounds

TABLE 5
Calculated (MM2) Bond Angles for Anhydrides 26a-c and
Phthalides 27a-c, 28a[30]

Entry	Type	26a	26b	26c		27a	27b	27c	28a
				Angle degrees			Angle degrees		
1	C_2C_3R	121	122	125	(124.7)[a]	121	122	125	121
2	$C_3C_2C_{10}$	129	130	133	(132.9)	128	129	131	126
3	$C_2C_{10}O_{11}$	126	126	128	(133.3)	125	126	128	125
4	$C_2C_{10}O_9$	109	110	111	(107.4)	111	112	112	111
5	$O_{11}C_{10}O_9$	123	123	120	(119.3)	123	122	119	123
6	$C_7O_9C_{10}$	103	103	103	(111.3)	107	107	107	107
7	$C_1C_2C_{10}$	108	107	105	(105.7)	108	107	105	109
8	$C_2C_1C_7$	108	108	109	(109.6)	109	110	111	107
9	$C_1C_7O_8$	126	126	126	(132.0)	—	—	—	—
10	$C_1C_7O_9$	109	109	109	(107.0)	102	102	101	103
11	$O_8C_7O_9$	123	123	123	(121.0)	—	—	—	—
12	$C_2O_3C_4$	116	115	113	(113.7)	116	115	113	116
13	$C_6C_1C_7$	129	128	126	(126.8)	130	129	125	131
14	C_1C_6R''	—	—	—	—	120	120	121	123

[a] X-ray data.

26f to 26i even though the substituents (halogen) appear quite different sterically. Molecular mechanics calculations (MM2) carried out on this series (26f to 26i) showed[31] similarities in plane distorations to those found for 26b. Apparently, the increase in halogen size is offset by the lengthening halogen-carbon bond distance resulting in the effective steric interactions remaining roughly constant. A possible explanation for the relatively invariant deshielding effects noted for ^{17}O chemical shifts for the anhydrides is that the repulsive van der Waals interactions between substituents and the *ortho* carbonyl group are roughly constant. The two substituents which lead to the large downfield shifts, 3-*t*-butyl 26c (22 ppm) and 3-nitro 26j (21 ppm), exhibit opposite electronic effects — the former a moderate electron donor, the latter strongly electron withdrawing. In the case of 26c, the large effective size caused larger in-plane distortions and gave rise to increased repulsive van der Waals interactions.[12] The large downfield shift for 26j was a composite of the electronic deshielding effect of the nitro group as well as the influence of repulsive van der Waals forces.

Due to steric congestion, the nitro group of 3-nitrophthalic anhydride 26j is forced to rotate from the plane, and a reciprocal effect from this interaction is noted on the nitro group ^{17}O signal. On comparing the ^{17}O shifts of the nitro signal for 3- and 4-nitrophthalic anhydride, some information regarding torsion angle rotation for the 3-nitro group can be deduced. Note that the nitro groups of both isomers are electronically equivalent. Thus, the difference in chemical shifts (NO_2 group) for the two isomers (27 ppm) can be attributed to torsion angle rotation, which minimizes the lone pair repulsion of the carbonyl oxygen. Using the slope (0.76 δ/angle°) for the nitro group ^{17}O chemical shift-torsion angle relationship previously reported,[6,9] the 3-nitro group can be estimated to be rotated 35° from the plane of the aromatic ring. This is in contrast to the steric effect on the carbonyl which appears to be only in-plane distortions in nature.

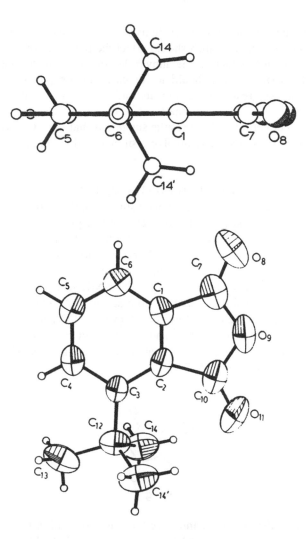

FIGURE 5.   X-ray structure of 3-*t*-butylphthalic anhydride (**26c**).
See Table 5 for data on selected bond angles.

The effect of substituents on the ^{17}O chemical shifts for the 3-substituted phthalic anhydrides (**26**) and 7-substituted phthalides (**27**) were similar, both experiencing a comparable magnitude of deshielding in relationship to the respective parent molecules.[30,31] The magnitude of the deshielding effect for series **27** was surprisingly constant for all the groups except the large *t*-butyl group and to a lesser extent the nitro group. The relative consistency in chemical shift for the substituents of varying size and electronic effects can be explained in terms of competition between simple electronic effects and van der Waals interactions (in-plane). Series **28** has a *meta* relationship between the carbonyl group and the substituent. (The small ^{17}O chemical shift differences can be explained by normal electronic effects.) An interesting conclusion to be made from these results for all the 3-substituted compounds listed in Table 2 to 3 is that the ^{17}O chemical shift values suggest that reactivity difference[29,34] for the two carbonyl groups of these compounds cannot be explained by simple electronic effects alone.

Differential reactivity of carbonyl functions in similar systems has been reported, and explanations for the differences have included steric blocking of the attacking reagents as well as the influence of electronic effects of the substituents.[29,34] The present study clearly showed that steric interactions in these systems were not limited to steric blocking but also included molecular distortions, which should affect the reactivity of the carbonyl functions. An interpretation of the ^{17}O data for the anhydrides suggests greater double bond character for the hindered carbonyl which should lower its reduction potential, whereas the other carbonyl oxygen shows greater charge density and, thus, should be more likely to form a complex with Lewis acids. This combination of effects must be considered in explanations of reactivity data. Furthermore, these results suggested that the ^{17}O NMR data reported[17,21] for sterically hindered N-oxides may involve a similar deshielding mechanism indicative of analogous deformations.

B. PHTHALIMIDES, IMIDES AND PHTHALAMIDES

^{17}O NMR data on a series of hindered 3-substituted phthalic anhydrides and corresponding phthalides have been shown[30,31] to correlate with in-plane bond angle distortions. The nontorsional ^{17}O chemical shift effects observed for the anhydrides, thought to be indicative of repulsive van der Waals interactions, could be used to rationalize the regiospecificity of reduction reactions. Sterically hindered imide systems also show regiospecificity[35] in reductions and, thus, should be investigated by ^{17}O NMR techniques. ^{17}O NMR data for a series of N-substituted phthalimides **29a-d** and a series of N-substituted succinimides **30a-d** and maleimides **31a-d** presented below show[36] that the ^{17}O NMR chemical shift data can provide additional insights into structure and reactivity in relation to steric phenomena.

^{17}O NMR data were obtained[36] (natural abundance) for a series of N-substituted phthalimides (**29a-g**), in acetonitrile at 75°C.

29a-g

The data for the phthalimides are summarized in Table 6. With the exception of the parent compound **29a**, the signal for the carbonyl oxygens was deshielded as the size of the N-substituent increased, despite similar electronic effects (shielding) of the alkyl groups. For example, the substitution of an N-isopropyl group for an N-methyl group yielded a 9 ppm downfield shift while the similar effect of the N-t-butyl group was 20 ppm (compare **29b** to **d**). In the unsymmetrical compound **29f** separate signals for both carbonyl oxygens were detected. Since electronic effects are again negligible, the large difference (deshielding) observed was indicative of significant repulsive van der Waals interactions. The magnitude of this shift (28 ppm) suggested in-plane distortions caused by the partial relief of the steric interactions of the substituent on the aromatic ring with the carbonyl oxygen similar to that observed[30] in the analogous phthalic anhydride system (~25 ppm). The results for the doubly hindered compound **29g** showed that the ring substituent deshielding effect and that due to the N-substituent were additive (*vide infra*).

The large deshielding effects noted in **29b** to **d** were surprising. The ^{17}O chemical shift data for N-substituted succinimides (**30a-d**) and maleimides (**31a-d**) were examined[36] to test the generality of this finding. The ^{17}O NMR chemical shift data for the N-substituted succinimides and maleimides showed deshielding effects of similar magnitude to those for the phthalimides. In addition, ^{17}O NMR data for the analogous N-substituted phthalamides (**32a-c**) showed deshielding effects with large N-substituents in agreement with the above three imide systems. The data are summarized in Table 7. The signals for compounds with

TABLE 6
^{17}O Chemical Shift Data[36] (± 1 ppm) for
Substituted Phthalimides 29 in
Acetonitrile at 75°C

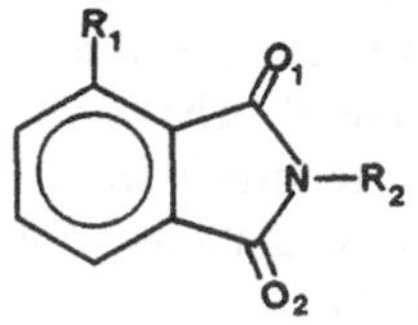

Compound no.	R_1	R_2	δ (C=O)$_1$	δ (C=O)$_2$
29a	H	H	379.0	379.0
29b	H	Me	374.0	374.0
29c	H	i-Pr	383.0	383.0
29d	H	t-Bu	394.0	394.0
29e	H	Ph	378.3	378.3
29f	t-Bu	H	407.3	370.6
29g	t-Bu	t-Bu	423.3	385.3

TABLE 7
^{17}O NMR Chemical Shift Data (± 1 ppm) for N-Substituted Malimides,
Succinimides, and Phthalamides in Acetonitrile at 75°C[36]

Succinimides			Malimides			Phthalamides		
Compound no.	N-R	δ (C=O)	Compound no.	N-R	δ (C=O)	Compound no.	N-R	δ (C=O)
30a	H	373.5	31a	H	411	32a	H	282
30b	Me	371	31b	Me	407	32b	Me	281
30c	t-Bu	392	31c	t-Bu	426	32c	t-Bu	300
30d	Ph	376	31d	Ph	412			

N-t-butyl groups were deshielded 20 ± 1 ppm relative to those for the N-methyl compounds for all four cases. The signals for the N-phenyl compounds were deshielded by 5 ± 1 ppm relative to those for the N-methyl compounds. The chemical shift data for the parent compounds (R=H) of each group (**29a, 30a, 31a, 32a**) are complicated by the presence of a hydrogen-bonding component. Simple hydrogen bonding to a carbonyl group should clearly result in an upfield shift of the ^{17}O signal.[4a,20,37,38] The effect of N-H donation to another system on the carbonyl of the donor imide is not as clear.[38,39] The overall effect of differential hydrogen bonding is complex.[38] Thus, the chemical shift differences between these compounds and those for the N-substituted compounds were difficult to interpret.

30a–d

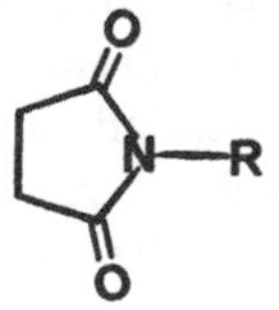

31a–d

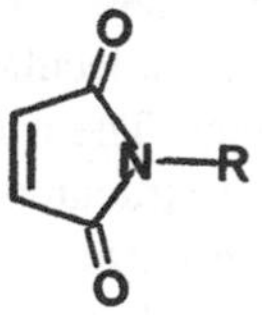

32a–c

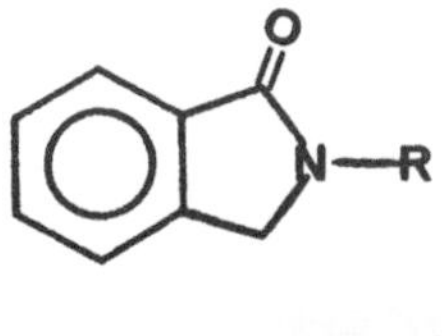

Molecular mechanics (MM2) calculations[33] were carried out[36] on **29a** to **d**; see Table 8 for selected bond angles (entries 1 to 9). The calculations predicted in-plane angle distortions of the planar molecules with the larger *N*-substituents. The bond angle of the carbonyl group opened toward the ring (entries 1 and 2), and the internal bond angle of the imide (entry 4) diminished as the *N*-substituent increased in size. In contrast to the 3-substituted anhydride system, the distortion for the *N*-substituted imides resulted in a symmetrical opening of the carbonyl angles (entries 1 and 2, Table 8). The X-ray structures of phthalimide[40] **29a** and *N-t*-butylphthalimide[36] **29d** have been reported. Both structures were found to be planar in agreement with molecular mechanics calculations. Unfortunately, the X-ray results showed that the crystal structures for **29a** and **29d** were not symmetrical around an axis through the nitrogen bisecting the molecule, making quantitative comparisons with the molecular mechanics calculations difficult. Despite this, the bond angle distortions noted were in qualitative agreement with those predicted by the calculations (Table 8).

TABLE 8

Molecular Mechanics Calculated Bond Angles ($\pm 1°$) and X-ray Data for Phthalimides 29a, d, g[36]

Entry	Angle	29a		29d		29g	
		MM2	(X-ray)[a]	MM2	(X-ray)[b]	MM2	(X-ray)[c]
1	$O_1C_2N_3$	124°	(124.8°)	126°	(126.3°)	124°	(125.5°)
2	$C_{11}C_2N_3$	105°	(105.2°)	107°	(108.6°)	108°	(106.8°)
3	$C_{10}C_{11}C_2$	131°	(130.0°)	131°	(131.9°)	133°	(132.8°)
4	$C_2N_3C_4$	114°	(112.2°)	111°	(110.6°)	111°	(110.5°)
5	$N_3C_4O_5$	124°	(125.4°)	126°	(129.6°)	126°	(126.6°)
6	$N_3C_4C_6$	105°	(106.2°)	107°	(104.6°)	107°	(107.4°)
7	$C_4C_6C_7$	131°	(120.3°)	131°	(125.9°)	128°	(127.2°)
8	$C_2N_3R_2$	122°	—	122°	(122.1°)	122°	(122.6°)
9	$C_{11}C_{10}R_1$	—	—	—	—	126°	(125.2°)

[a] Values taken from Reference 40.
[b] Sigma values $\simeq$ 1.5.
[c] Sigma values $\simeq$ 0.5.

The [17]O signal for the double-hindered carbonyl of **29g** was deshielded by 50 ppm, which was much larger than any other effects seen. Moreover, the magnitudes of the deshielding effects on the carbonyl signals were consistent with expectations based upon the [17]O data for **29a**, **28b** and **29e**. The X-ray structure of **29d** was obtained[36] (Figure 6) for comparison with molecular mechanics calculations (Table 8) and again confirmed the molecule to be planar. The calculations predicted essentially identical values for the angles for $(C=O)_2$ in both structures (**29d** and **29g**), and the X-ray results were qualitatively in agreement; compare entries 1 and 5 for compounds **29a** and **29g**. The doubly hindered carbonyl, $(C=O)_1$, is being influenced in opposing directions by the two *t*-butyl groups

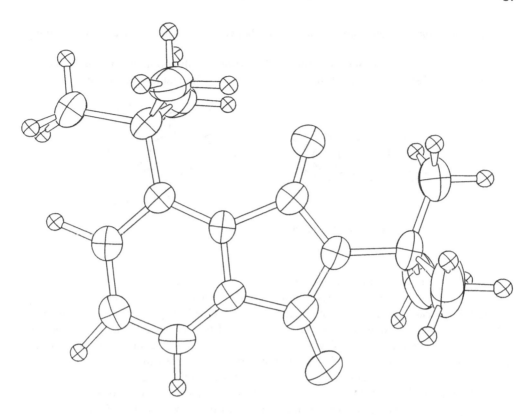

FIGURE 6.   X-ray structure of *N-t*-butyl-3-*t*-butylphthalimide (**29g**). See Table 8 for data on selected bond angles.

such that no unusual distortion is apparent in the structure. However, the ^{17}O results showed that this carbonyl, $(C=O)_1$, was subject to severe van der Waals interactions.

Interestingly, the ^{13}C NMR signals of sterically hindered imides were found[36] to be extremely insensitive to compression effects. For example, the two carbonyl signals for **29g** were within 0.5 ppm of one another, whereas the ^{17}O NMR data for the double-bonded oxygens attached to these carbons differed by 50 ppm.

The ^{17}O NMR results for imide systems gave interesting new insights into ground state structure. In certain systems (cf. **29g**) the ^{17}O method provided detailed information not accessible by other methods. The results for planar amides (**32a-c**) show that the *N*-substituent deshielding effects are not limited to imides. However, it is not clear that these effects will be predominant in conformationally mobile (acyclic) systems. The ^{17}O NMR results may provide insights into the regiospecificity of imide reductions.[35] For reductions in which electron transfers are rate determining, one would expect the carbonyl which shows the most deshielded ^{17}O chemical shift value to undergo reaction preferentially. For example, the regiospecific zinc reduction of a hindered imide[41] is consistent with the above expectation.

C. ANTHRAQUINONES

To further evaluate compressional effects on the ^{17}O NMR data of planar compounds, a series of anthraquinones was investigated.[42] This system allows one to look at stronger *peri*-like interactions. In the parent compound anthraquinone, the carbonyl and the α-position substituent are essentially parallel. This is in contrast to the anhydride and imide series in which the carbonyls are slightly tilted away from the 3-position. Thus the anthraquinone system should have greater steric interactions between the α-position and the carbonyl function.

The ^{17}O NMR data for anthraquinone **(33)**, 2-*t*-butylanthraquinone **(34)** and 1-*t*-buty-lanthraquinone **(35)** were obtained in acetonitrile at 75°. The results are shown:

33 R = H 524 ppm

34 R = t-Bu 522 ppm br.

35

The chemical shift data for **33** and **34** were found to be essentially identical, as expected based on electronics effects. However, the ^{17}O NMR spectrum of **35** showed two broad signals for the carbonyl groups. One signal was deshielded by greater than 50 ppm from that of **33** or **34**. This downfield signal was assigned to the sterically hindered carbonyl based on the previously discussed work. The second signal at 522 ppm was normal. The deshielding effect observed for **35** was much larger than that seen for the previous systems and indicates that greater steric interactions were present in this case.

Molecular mechanics (MM2) calculations[33] were carried out[42] on **33** to **35**. The anthra-quinone system was predicted to be planar in these cases. However, for **35**, large in-plane distortions of the carbonyl and *t*-butyl group angles were calculated. An X-ray analysis of **35** (Figure 7) yielded an excellent correlation with the molecular mechanics calculations to confirm the conclusion. Work is in progress on disubstituted anthraquinones to look at the ^{17}O NMR effects on the doubly hindered carbonyl.

D. SUMMARY

In conclusion, ^{17}O NMR compressional effects on the carbonyl systems provide a direct method to assess steric hindrance in planar systems. The ^{17}O NMR deshielding effects paralleled[30,31,36,42] the relative degree of in-plane distortions in all the planar systems inves-tigated. The in-plane distortions are reflective of repulsive van der Waals interactions (that have been only partially relieved). Plots of total van der Waals energy (MM2 calculation)[33] vs. [17]O chemical shift show a fair correlation with increasing van der Waals interactions (Fig. 8), whereas total steric energy does not correlate.[43] This relationship must be corrected to reflect the local van der Waals contributions which, as Chesnut has shown[12], should be responsible for deshielding effects. The difference in van der Waals energy could be estimated by comparison of total van der Waals energies of the sterically hindered (deshielded) com-pound with that for an isomer. Only compounds in which no conformational problems were present were chosen. Thus, 4-substituted phthalic anhydrides and 6-substituted phthalides were taken as models for series **26** and **27**. No reasonable model was available for the *N*-substituted imides; however, the di-*t*-butyl compound **29g** could be modeled as above. The 2-*t*-butyl compound was taken for the model for **35**. The estimated local van der Waals energies for several series of planar compounds gave a reasonable correlation[43] (see Figure 9) with the change in [17]O chemical shift (relative to the parent compound in each series).

The above correlation of "local" van der Waals repulsion with the change in [17]O chemical shifts agrees with Chesnut's postulation[12] on the r^{-3} term of the Karplus-Pople equation. However, since the carbonyls are clearly distorted in these cases, the ΔE^{-1} and the charge density terms would also be expected to be affected.[12]

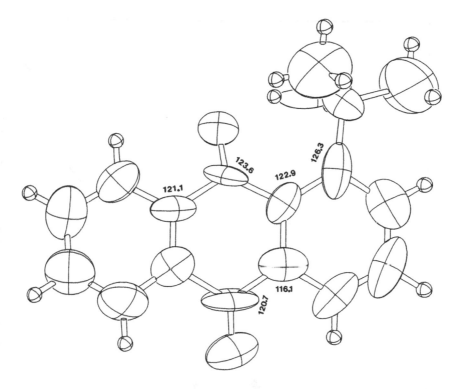

FIGURE 7.   X-ray structure of 1-*t*-butylanthraquinone. (Selected bond angle data included.)

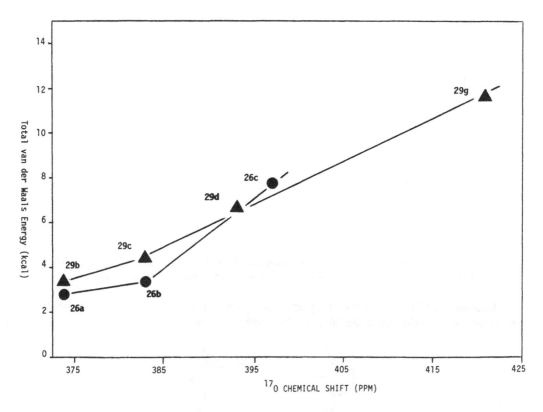

FIGURE 8.   Plot of total van der Waals energy (MM2) vs. ^{17}O chemical shift for selected 3-substituted phthalic anhydrides ● and *N*-substituted phthalimides ▲.

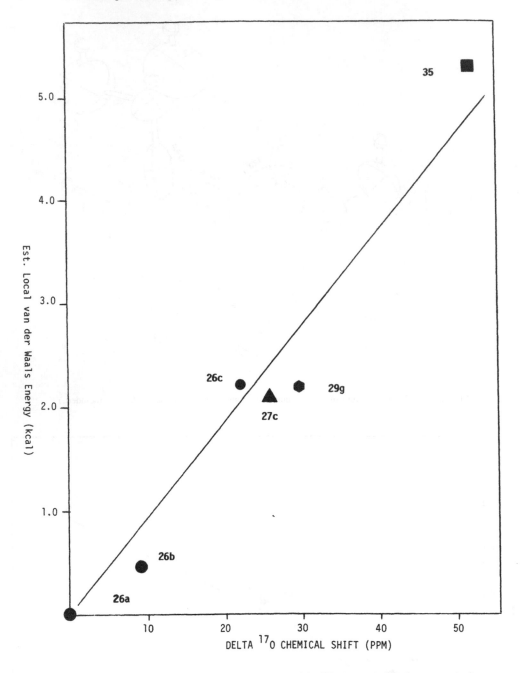

FIGURE 9.   Plot of estimated local van der Waals energy vs. delta ^{17}O chemical shift (δ compound - δ parent of series) for phthalic anhydrides ●, phthalides ▲, phthalimides ⬣ , and anthraquinones ■.

Caution must be exercised in assigning ''compression effects.'' For example, consider the case[28] of succinic anhydride **36** and 2,2-dimethylsuccinic anhydride **37**.

Compound **37** showed two carbonyl signals: 379 and 367 ppm while that for **36** was found at 376 ppm. It was tempting to assign the deshielded signal to the "hindered" carbonyl group. However, labeling experiments clearly showed[28] that the carbonyl next to the gem-dimethyl group was shielded by 9 ppm as would be expected based on results for conformationally mobile systems.[44] This indicates that there are no repulsive van der Waals interactions in **37** and that, in fact, attractive van der Waals interactions[12] may be dominating the chemical shift changes.

IV. REPRESENTATIVE EXPERIMENTAL CONDITIONS

The ^{17}O spectra were recorded on a JEOL GX-270 or on a Varian VXR-400 spectrometer equipped with a 10 mm broad-band probe. In general, spectra were acquired at natural abundance at 75°C in acetonitrile (Aldrich, anhydrous gold label under nitrogen) containing 1% of 2-butanone or acetone as an internal standard. The concentration of the carbonyl compounds usually employed in these experiments was 0.5 M. The signals were referenced to external deionized water at 75°C. The 2-butanone resonance (558 ± 1 ppm) or acetone (571 ± 1 ppm) was used as an internal check on the chemical shift measurements for these compounds. The instrumental settings for the GX-270 at 36.5 MHz were: spectral width 25 kHz, 2k data points, 90° pulse angle (28 μs pulse width), 200 μs acquisition delay, 40 ms acquisition time, and 40,000 to 100,000 scans. The instrumental settings for the VXR-400 at 54.22 MHz were: spectral width 35 kHz, 2 k data points, 90° pulse angle (40 μs pulse width), 100 to 200 μs acquisition delay, 29 ms acquisition time, and 20,000 to 40,000 scans. The spectra were recorded with sample spinning and without lock. The signal-to-noise ratio was improved by applying a 5 to 25 Hz exponential broadening factor to the FID prior to Fourier transformation. The data point resolution was improved to ± 0.1 ppm on the VXR-400 and ± 0.2 ppm on the GX-270 by zero filling up to 8 K data points. The reproducibility of the chemical shift data is estimated to be < ± 1.0.

ACKNOWLEDGMENTS

Acknowledgment is made to the Donors of the Petroleum Research Fund, administered by the American Chemical Society, for partial support of this research, to NSF (CHE-8506665), and to the Georgia State University Research Fund. A.L.B. was a fellow of the Camille and Henry Dreyfus Foundation, 1981 to 1986. Instrumentation used in these studies was supported in part by a NSF Equipment Grant (CHE 8409599).

REFERENCES

1. **Kintzinger, J.-P.**, Oxygen-17 NMR, in *NMR of Newly Accessible Nuclei*, Vol. 2, Laszlo, P., Ed., Academic Press, New York, 1983, 79.
2. **Klemperer, W. G.**, Application of 17-O NMR spectroscopy to structural problems, in *The Multinuclear Approach to NMR Spectroscopy*, Lambert, J. B. and Riddel, F. G., Eds., Reidel, Dordrecht, Holland, 1983, 245.
3. **Kintzinger, J.-P.**, Oxygen NMR characteristic parameters, in *NMR-17. Oxygen-17 and Silicon-29*, Diehl, P., Fluck, E., and Kosfeld, R., Eds., Springer-Verlag, New York, 1981, 1.
4. For representative papers see: (a) **St. Amour, T. E., Burger, M. I., Valentine, B., and Fiat, D.**, 17-O NMR studies of substituent and hydrogen-bonding effects in substituted acetophenones and benzaldehydes, *J. Am. Chem. Soc.*, 103, 1128, 1981; (b) **Fraser, R. R., Ragauskas, A. J., and Stothers, J. B.**, Nitrobenzene valence bond structures: evidence in support of 'through-resonance' from 17-O shieldings, *J. Am. Chem. Soc.*, 104, 6475, 1982; (c) **Lipkowitz, K. B.**, A reassessment of nitrobenzene valence bond structures, *J. Am. Chem. Soc.*, 104, 2647, 1982; (d) **Craik, D. J., Levy, G. C., and Brownlee,**

R. T. C., Substituent effects on 15-N and 17-O chemical shifts in nitrobenzenes: correlations with electron densities, *J. Org. Chem.*, 48, 1601, 1983; (e) **Brownlee, R. T. C., Sadek, M., and Craik, D. J.**, *Ab initio* MO calculations and 17-O NMR at natural abundance of *p*-substituted acetophenones, *Org. Magn. Reson.*, 21, 616, 1983.

5. For representative papers see: (a) **Manoharan, M. and Eliel, E. L.**, 17-O NMR spectra of tertiary alcohols, ethers, sulfoxides, and sulfones in the cyclohexyl and 5-substituted 1,3-dioxanyl series and related compounds, *Magn. Reson. Chem.*, 23, 225, 1984; (b) **Eliel, E. L., Liu, K.-T., and Chandrasekaran, S.**, 17-O NMR spectra of equatorial and axial hydrocyclohexanes and 5-hydroxy-1,3-dioxanes and their methyl ethers, *Org. Magn. Reson.*, 21, 179, 1983; (c) **Sammakia, T. H., Harris, D. L., and Evans, S. A., Jr.**, Oxygen-17 spectral investigation of 3-alkoxy-*trans*-3,4-disubstituent-thiolone 1,1-dioxides and related compounds, *Org. Magn. Reson.*, 22, 747, 1984; (d) **Dyer, J. C., Harris, D. L., and Evans, S. A., Jr.**, Oxygen-17 nuclear magnetic resonance spectroscopy of sulfoxides and sulfones. Alkyl substituent induced chemical shift effects, *J. Org. Chem.*, 47, 3660, 1982; (e) **Crandall, J. K., Centeno, M. A., and Borresen, S.**, Oxygen-17 nuclear magnetic resonance. 2. Cyclohexanones, *J. Org. Chem.*, 44, 1184, 1979.

6. **Balakrishnan, P. and Boykin, D. W.**, Relationship of aromatic nitro group torsion angles to 17-O chemical shifts, *J. Org. Chem.*, 50, 3661, 1985.

7. **Oakley, M. G. and Boykin, D. W.**, Relationship of torsion angle to 17-O NMR data for aryl ketones, *J. Chem. Soc. Chem. Commun.*, p. 439, 1986.

8. **Boykin, D. W., Balakrishnan, P., and Baumstark, A. L.**, 17-O NMR studies on aryl-alkyl ketones and aromatic aldehydes, *Magn. Reson. Chem.*, 25, 248, 1987.

9. (a) **Baumstark, A. L., Balakrishnan, P., Dotrong, M., McCloskey, C. J., Oakley, M. G., and Boykin, D. W.**, 17-O NMR spectroscopy: torsion angle relationships in aryl carboxylic esters, acids, and amides, *J. Am. Chem. Soc.*, 109, 1059, 1987; (b) **Boykin, D. W., Deadwyler, G. H., and Baumstark, A. L.**, 17-O NMR studies on substituted *N*-arylacetamides and aryl acetates: torsion angle and electronic effects, *Magn. Reson. Chem.*, 26, 19, 1988.

10. **Butler, L. G.**, The NMR parameters for oxygen-17, in *17-O NMR Spectroscopy in Organic Chemistry*, Boykin, D. W., Ed., CRC Press, Boca Raton, FL, 1990, chap. 1.

11. **Karplus, M. and Pople, J. A.**, Theory of carbon NMR chemical shifts in conjugated systems, *J. Chem. Phys.*, 38, 2803, 1963.

12. (a) **Li, S. and Chesnut, D. B.**, Intramolecular van der Waals interactions and 13-C chemical shifts: substituent effects in some cyclic and bicyclic systems, *Magn. Reson. Chem.*, 24, 96, 1986; (b) **Li, S. and Chesnut, D. B.**, Intramolecular van der Waals interactions and chemical shifts: a model for β and γ-effects, *Magn. Reson. Chem.*, 23, 625, 1985.

13. **Barbarella, G., Dembech, P., and Tugnoli, V.**, 13-C and 17-O Chemical shifts and the conformational analysis of mono- and di-methyl-substituted thiane 1-oxide and thiane 1,1-dioxide, *Org. Magn. Reson.*, 22, 402, 1984.

14. **Sardella, D. J. and Stothers, J. B.**, Nuclear magnetic resonance studies. XVI. Oxygen-17 shieldings of some substituted acetophenones, *Can. J. Chem.*, 47, 3089, 1969.

15. **Boykin, D. W., Baumstark, A. L., and Balakrishnan, P.**, 17-O NMR spectroscopy of 4-substituted pyridine *N*-oxides: substituent and solvent effects, *Magn. Reson. Chem.*, 23, 276, 1985.

16. Results consistent with those in CH_3CN (Ref. 15) were obtained in DMSO. See: Sawada, M., Takai, Y., Kimura, S., and Misumi, S., A 17-O NMR study. SCSs of 4-substituted pyridine 1-oxides in DMSO: importance of dual enhanced resonance contributions with pi-donor and pi-acceptor substituents, *Tetrahedron Lett.*, p. 3013, 1986.

17. **Boykin, D. W., Balakrishnan, P., and Baumstark, A. L.**, 17-O NMR spectroscopy of heterocycles. Steric effects for *N*-oxides, *Magn. Reson. Chem.*, 23, 695, 1985.

18. **Katoh, M., Sugawara, T., Kawada, Y., and Iwamura, H.**, 17-O Nuclear magnetic resonance studies, V. 17-O Shieldings of some substituted anisoles, *Bull. Chem. Soc. Jpn.*, 52, 3475, 1977.

19. **Berger, S.**, The *t*-butyl group as sensor group of the ortho effect, *Tetrahedron*, 32, 2451, 1976.

20. **Chandrasekaran, S., Wilson, W. D., and Boykin, D. W.**, 17-O NMR studies on polycyclic quinones, hydroxyquinones and related cyclic ketones: models for anthracycline intercalators, *Org. Magn. Reson.*, 22, 757, 1984.

21. **Boykin, D. W., Balakrishnan, P., and Baumstark, A. L.**, Natural abundance 17-O NMR spectroscopy of heterocyclic *N*-oxides and di-*N*-oxides. Structural effects, *J. Heterocycl. Chem.*, 22, 981, 1985.

22. **Sugawara, T., Kawada, Y., Katoh, M., and Iwamura, H.**, Oxygen-17 nuclear magnetic resonance. III. Oxygen atoms with a coordination number of two, *Bull. Chem. Soc. Jpn.*, 52, 3391, 1979.

23. **Wysocki, M. A., Jardon, P. W., Mains, G. J., Eisenbraun, E. J., and Boykin, D. W.**, Steric effects in the 17-O NMR spectroscopy of aromatic methyl ethers, *Magn. Reson. Chem.*, 25, 331, 1987.

24. **Jaccard, G. and Lauterwein, J.**, Intramolecular hydrogen bonds of the C=O . . . H—O type as studied by oxygen-17 NMR, *Helv. Chim. Acta*, 69, 1469, 1986.

25. **Paudler, W. W. and Jovanovic, M. V.**, Backdonation and interrelationships between 15-N, 13-C chemical shifts and infrared absorption frequencies in heterocyclic *N*-oxides, *Heterocycles*, 19, 93, 1982.

26. **Ogata, M., Watanabe, H., Tori, K., and Kano, H.**, 4-Methylpyrimidine *N*-oxides, *Tetrahedron Lett.*, p. 19, 1964.

27. For a detailed discussion see the preceding chapter: Boykin, D. W. and Baumstark, A. L., Applications of 17-O NMR spectroscopy to structural problems in organic chemistry: torsion angle relations, in *17-O NMR Spectroscopy in Organic Chemistry*, Boykin, D. W., Ed., CRC Press, Boca Raton, FL, 1990, chap. 3.

28. **Vasquez, P. C., Boykin, D. W., and Baumstark, A. L.**, 17-O NMR spectroscopy (natural abundance) of heterocycles: anhydrides, *Magn. Reson. Chem.*, 24, 409, 1986.

29. (a) **Swenton, J. S., Jackson, D. K., Manning, M. J., and Raynolds, P. W.**, Model studies for anthracyclinone synthesis. The chemistry of 1-lithio-3,3,6,6-tetramethoxycyclohexa-1,4-diene, an umpolung for quinone, *J. Am. Chem. Soc.*, 100, 6182, 1978; (b) **Kayser, M. M. and Eisenstein, O.**, Theoretical study of regioselectivity in nucleophilic addition to unsymmetrical cyclic anhydrides. Intrinsic reactivity and influence of the cation, *Can. J. Chem.*, 59, 2457, 1981; (c) **Makhlouf, M. A. and Rickborn, B.**, Regioselective reduction of anhydrides by L-selectride, *J. Org. Chem.*, 46, 4810, 1981; (d) **Newman, M. S. and Scheurer, P. G.**, The behavior of 3-chlorophthalic anhydride in Friedel-Crafts and Grignard condensations, *J. Am. Chem. Soc.*, 78, 5004, 1956; (e) **Allahdad, A. and Knight, D. W.**, An investigation of the Wittig reaction between a series of monosubstituted phthalic anhydrides and ethoxycarbonmethyl-lidenetriphenylphosphorane, *J. Chem. Soc. Perkin Trans. I*, p. 1855, 1982.

30. **Baumstark, A. L., Balakrishnan, P., and Boykin, D. W.**, 17-O NMR spectroscopy as a probe of steric hindrance in phthalic anhydrides and phthalides, *Tetrahedron Lett.*, p. 3079, 1986.

31. **Boykin, D. W., Baumstark, A. L., Kayser, M. M., and Soucy, C. M.**, 17-O NMR spectroscopic study of substituted phthalic anhydrides and phthalides, *Can. J. Chem.*, 65, 1214, 1987.

32. (a) **Balakrishnan, P., Baumstark, A. L., and Boykin, D. W.**, 17-O NMR spectroscopy: effect of substituents on chemical shifts for *p*-substituted benzoic acids, methyl benzoates, cinnamic acids, and methyl cinnamates, *Org. Magn. Reson.*, 22, 753, 1986; (b) **Boykin, D. W., Baumstark, A. L., and Subramanian, T. S.**, unpublished results; (c) see: **Peter, J. A., Nieuwenhuizen, M. S., and Raber, D. J.**, Analysis of multinuclear lanthamide-induced shifts: 1. investigations of some approximations in the procedures for separation of diamagnetic, contact, and pseudocontact shifts, *J. Magn. Reson.*, 65, 417, 1985.

33. (a) **Burkert, U. and Allinger, N. L.**, *Molecular Mechanics*, American Chemical Society, Washington, D.C., 1982; (b) **Osawa, E. and Musso, H.**, Applications of molecular mechanics calculations in organic chemistry, in *Topics in Stereochemistry*, Allinger, N. L., Eliel, E. L., and Wilen, S. H., Eds., John Wiley & Sons, New York, 1982, 117.

34. (a) **Kayser, M. M. and Morand, P.**, Regioselectivity of metal hydride reductions of unsymmetrically substituted cyclic anhydrides. Systems where steric hindrance along the preferred reaction path rationalization is not applicable, *Can. J. Chem.*, 58, 2484, 1980; (b) **McAlees, A. J., McCrindle, R., and Sneddon, D. W.**, Reduction of substituted phthalic anhydrides with sodium borohydride, *J. Chem. Soc. Perkin Trans I*, p. 2037, 1977; (c) **Braun, M.**, Regioselektive Synthese von Daunomycinon und τ-Rhodomycinon, *Tetrahedron Lett.*, p. 3871; 1980; (d) **Canonne, P., Lemay, G., and Belanger, D.**, Reaction of di(bromomagnesio) alkanes with unsymmetrically substituted cyclic anhydrides, *Tetrahedron Lett.*, p. 4167, 1980.

35. See: **Menger, F. M.**, Directionality of organic reactions in solution, *Tetrahedron*, 39, 1013, 1983.

36. **Baumstark, A. L., Dotrong, M., Oakley, M. G., Stark, R., and Boykin, D. W.**, 17-O NMR study of steric interactions in hindered *N*-substituted imides, *J. Org. Chem.*, 52, 3640, 1987.

37. **Reuben, J.**, Hydrogen-bonding effects on oxygen-17 chemical shifts, *J. Am. Chem. Soc.*, 91, 5725, 1969.

38. **Baumstark, A. L., Vasquez, P. C., and Balakrishnan, P.**, 17-O Enriched α-azohydroperoxides: 17-O NMR spectroscopy, 17-O labeling reagents, *Tetrahedron Lett.*, p. 2051, 1985.

39. (a) **Valentine, B., Steinschneider, A., Dhawan, D., Burgar, M. I., St. Amour, T., and Fiat, D.**, Oxygen-17 NMR of peptides, *Int. J. Pept. Protein Res.*, 25, 56, 1985; (b) **Burgar, M. I., St. Amour, T. E., and Fiat, D.**, 17-O and 15-N NMR studies of amide systems, *J. Phys. Chem.*, 85, 502, 1981.

40. **Matzat, E.**, Die Kristallstruktur des Phtalimids (Kladnoit), *Acta Crystallogr.*, B28, 415, 1972.

41. **Brewster, J. H. and Fusco, A. M.**, Steric effect in the reduction of *N*-methyl-1,2-naphthalimide with zinc, *J. Org. Chem.*, 28, 501, 1963.

42. **Baumstark, A. L., Dotrong, M., Stark, R. R., and Boykin, D. W.**, 17-O NMR spectroscopy: origin of deshielding effect in rigid, planer molecules, *Tetrahedron Lett.*, p. 2143, 1988.

43. **Baumstark, A. L. and Boykin, D. W.**, 17-O NMR spectroscopy: assessment of steric perturbation of structure in organic compounds, *Tetrahedron*, 45, 3613, 1989.

44. **Orsini, F. and Ricca, G. S.**, Oxygen-17 NMR chemical shifts of esters, *Org. Magn. Reson.*, 22, 653, 1984.

Chapter 5

^{17}O NMR SPECTROSCOPY: HYDROGEN-BONDING EFFECTS

Alfons L. Baumstark* and David W. Boykin*

TABLE OF CONTENTS

I. INTRODUCTION AND SCOPE

The hydrogen bond is fundamental to the structure of many organic systems. In recent years extensive reviews of hydrogen bonding have appeared.[1] Numerous physical methods have been employed to evaluate hydrogen bonding in a wide variety of molecules. For hydrogen-bonding systems involving the oxygen atom, [17]O nuclear magnetic resonance (NMR) spectroscopy provides the opportunity for direct observation of one or more of the participants in the hydrogen-bonding array.

^{17}O NMR spectroscopy is an important method[2] for assessment of numerous structural problems in organic chemistry.[3] Quantitative relationships between ^{17}O NMR chemical shifts and torsion angles for a variety of aromatic carbonyl systems have been developed[2,4-6] (see Chapter 3). In addition, a general correlation between repulsive van der Waals interactions and ^{17}O NMR data for rigid, planar aromatic carbonyl systems has been demonstrated[2,7] (see Chapter 4). The sensitivity of the carbonyl oxygen chemical shift to intermolecular hydrogen bond formation has been known since the initial report by Christ and Diehl[8] and the followup study of Reuben.[9] Those investigators noted that the chemical shift of acetone is shielded by 52 ppm on extrapolation to infinite dilution in water.[8,9] Intermolecular hydrogen-bonding effects have also been noted in the ^{17}O NMR data for N-oxides,[10] simple amides and peptides,[11] and carboxylic acids.[12] ^{17}O NMR intramolecular hydrogen-bonding studies have been reported for enols of β-dicarbonyl compounds,[13] aryl carbonyl compounds,[14] quinones,[15] N-oxides,[16] anisole derivatives,[17] amides,[18] and α-azohydroperoxides[19] (*vide infra*).

This chapter will review the effect of intramolecular and intermolecular hydrogen bonding on ^{17}O NMR chemical shifts including hydrogen-bonding solvent interactions from a historical viewpoint. The discussion is not intended to be encyclopedic; however, an attempt has been made to include the majority of papers in the area. A section of this chapter describes intramolecular hydrogen-bonding studies involving NH and OH proton donors on the ^{17}O NMR data for carbonyl groups that are currently ongoing in our laboratories.

II. INTRAMOLECULAR HYDROGEN-BONDING EFFECTS

A. HISTORICAL PERSPECTIVE

The earliest reports of ^{17}O NMR data on intramolecular, hydrogen-bonded compounds are contained in the classic survey works of Christ et al. in 1961 and 1963.[8] The ^{17}O NMR spectrum for the enol of acetylacetone, **1**, was found to show only one signal at 269 ppm. The datum was interpreted to be indicative of fast chemical exchange between the two symmetric enol forms (Scheme 1); hence, only the average of the two extreme oxygen environments could be seen.

1a   **1b**

SCHEME 1

Christ and Diehl were also the first to show that intramolecular hydrogen bonding in salicylaldehyde, *o*-hydroxybenzaldehyde, caused a large upfield shift (60 ppm relative to benzaldehyde) in the carbonyl signal similar to that observed for the solvent effect on [17]O data for acetone of hydrogen bonding with water.[8]

In 1967, Gorodetsky et al described a method using ^{17}O NMR spectroscopy to determine relative concentrations of the two enol tautomers of asymmetric β-diketones.[13] The ^{17}O NMR data of eight enol pairs **2** to **9** were reported (Table 1). In addition, the ^{17}O NMR data on tropolone **10** and the enol of 1,2-cyclohexadione **11** were reported. As noted for each enol pair, only one signal was observed for tropolone indicating that fast chemical exchange was taking place. On the other hand, two signals were observed for **11**.

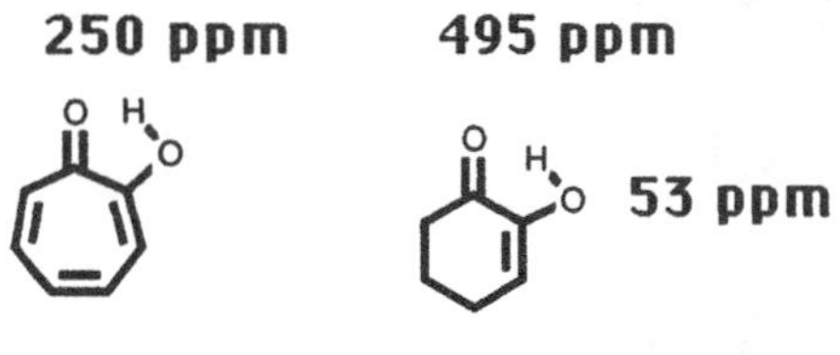

250 ppm   495 ppm

53 ppm

10   **11**

In 1979, Winter and Zeller[20] published the X-ray structure of the enol of benzoylacetone **2** and confirmed the earlier ^{17}O NMR data. In addition, the ^{17}O NMR signals were found to show no solvent (CHCl$_3$ vs. toluene) or temperature dependence (range 130°). The results were explained by a model in which the C and O atoms are fixed and the hydrogen (proton) tunnels between the two oxygen atoms.

An extensive ^{17}O NMR study of acetophenones and benzylaldehydes by Fiat and co-workers in 1981 showed the importance of intramolecular hydrogen-bonding effects.[14] The [17]O chemical shifts for *ortho* hydroxy substituted acetophenones, *ortho*-hydroxybenzaldehyde and *ortho*-aminoacetophenone were noted to be at higher field than expected based on the *para* isomers (Scheme 2).

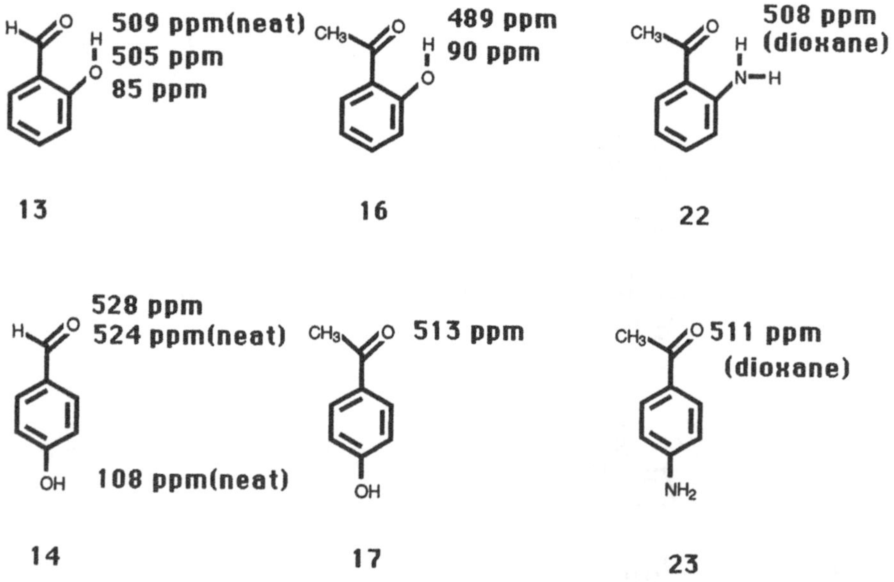

509 ppm(neat)   489 ppm   508 ppm
505 ppm   90 ppm   (dioxane)
85 ppm

13   **16**   **22**

528 ppm
524 ppm(neat)   513 ppm   511 ppm
                               (dioxane)

108 ppm(neat)

14   **17**   **23**

SCHEME 2

For the series of polysubstituted acetophenones, the substituted effects were thought to be small and superimposed on the dominating intramolecular hydrogen-bonding effect. The data are summarized in Table 2. It was concluded that, in the case of proton donating *ortho* substituents, the carbonyl oxygen prefers an orientation in close juxtaposition to the group.

TABLE 1
^{17}O NMR Data for Enols of β-Dicarbonyl Compounds

No.	Compound[a]	δ(O₁) ppm	δ(O₂) ppm	K[a]	Ref.
1		269 274	269 274	≅1	8 13
2		233 239	291 294	1.3	13 20
3		430	88	0.15	13
4		408	108	0.2	13
5		205	360	2.0	13
6		175	359	2.3	13
7		231	295	1.3	13
8		354	169	0.43	13
9		395	121	0.25	13

TABLE 2
^{17}O NMR Data on Selected Acetophenones and Benzaldehydes in Dioxane[14]

No.	Compound	δ(C=O) ppm	δ(R) ppm
12	Benzaldehyde	562	—
13	o-OH-Benzaldehyde	505	85
14	p-OH-Benzaldehyde	528	108
15	Acetophenone	550	—
16	o-OH-Acetophenone	488	90
17	p-OH-Acetophenone	513	N.R.[a]
18	2,4-Dihydroxyacetophenone	459	N.R.[a]
19	2,6-Dihydroxyacetophenone	476	N.R.[a]
20	2,3,4-Trihydroxyacetophenone	466	N.R.[a]
21	2,4,6-Trihydroxyacetophenone	441	N.R.[a]
22	o-NH$_2$-Acetophenone	508	—
23	p-NH$_2$-Acetophenone	511	—

[a] N.R. = not reported.

The relative strength of the hydrogen bond was thought to correlate with the observed trend in the ^{17}O NMR data.

In 1984, Lauterwein et al. reported an ^{17}O NMR study of the conformations of N-acetyl-L-proline in aqueous solution.[21a] Two sets of signals for the ^{17}O-enriched carboxy and amide groups were observed and assigned to the *cis/trans* conformers (Scheme 3). Surprisingly, both sets of signals were found to show parallel pH titration curves.

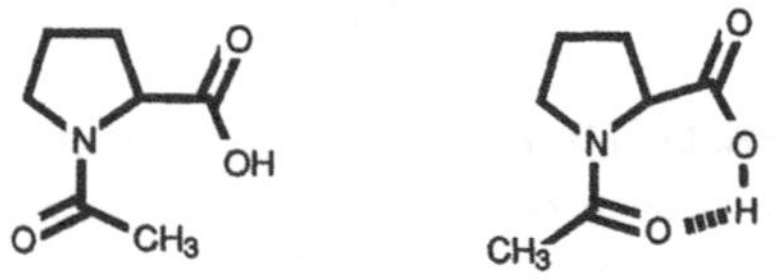

24

cis   **trans**

SCHEME 3

The data were interpreted to suggest the absence of an intramolecular hydrogen bond at low pH (in the "*trans*" conformer). The origin of the chemical shift differences was suggested to be an electric field effect of the amide bond. However, hydrogen-bonding effects from the solvent (water) are difficult to evaluate and would be expected to greatly complicate the situation. Thus, interpretation of the results in aqueous solution must be viewed with caution.

A more detailed account[21b] of the N-acetyl-L-proline results included a similar study on N-acetylsarcosine (AcMeN-CH$_2$-CO$_2$H). In addition to aqueous medium, ^{17}O NMR data were also reported for these compounds in organic solvents. The chemical shift difference for the *cis* and *trans* amide resonances (in organic solvents, at dilute concentrations) was interpreted to be due to intramolecular hydrogen bonding. The ^{17}O NMR data (at higher concentrations of solute) were interpreted to indicate aggregation and/or self-association. The probability of γ-turned structures in the various solvents was estimated for the ^{17}O data.

Boykin and co-workers in 1984 found that the ^{17}O NMR spectra of 2,5-dihydroxyben-zoquinone **25** and 5,8-dihydroxynaphthoquinone **26** showed only a single resonance for each compound (Scheme 4).[15a]

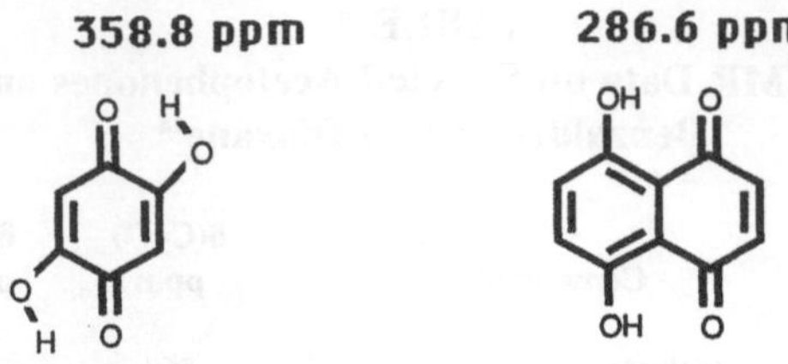

358.8 ppm   **286.6 ppm**

25   **26**

SCHEME 4

The data were interpreted to be indicative of rapid intramolecular proton exchange. On the other hand, 1,4-dihydroxyanthraquinone **27** was noted to show two ^{17}O NMR signals indicating that rapid exchange was not taking place in this case (Scheme 5).

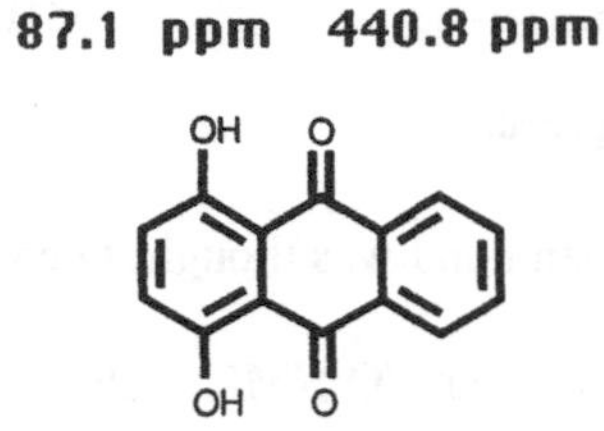

87.1 ppm   440.8 ppm

27

SCHEME 5

In addition, the ^{17}O NMR upfield shift of the carbonyl signal for the anthraquinone **27** was attributed to intramolecular hydrogen bonding.

Noting the data of Lambert[22] on ethyl-3-oxo-4-phenylbutanoate and diethyl-3-oxo-1,5-pentanedioate, Lapachev[23] reported the ^{17}O NMR data on the intramolecular hydrogen-bonded enols **28** to **30** of three β-keto esters (ethyl acetoacetate, ethyl benzoylacetate, and ethyl trifluoroacetoacetate, respectively). Unlike that for the unsymmetrical β-diketone cases,[13] only one enol is formed in the β-ketoester examples. In addition, the compounds were in slow exchange; signals were observed for both the keto and enol forms in two of the cases (Scheme 6). Equilibrium constants estimated from the ^{17}O NMR data were in good agreement with reported values.

δ(PPM)					δ(PPM)		
a	b	c		#	a_1	b_1	c_1
578	365	173	R=CH₃	2 8	124	296	165 (est.)
553	365	176	R=Ph	2 9	109	297	176
-	-	-	R=CF₃	3 0	96	315	170

SCHEME 6

In addition, Lapachav reported the ^{17}O NMR data for a series of phenols **31** to **34** with intramolecular hydrogen bonds (Scheme 7).[23,24]

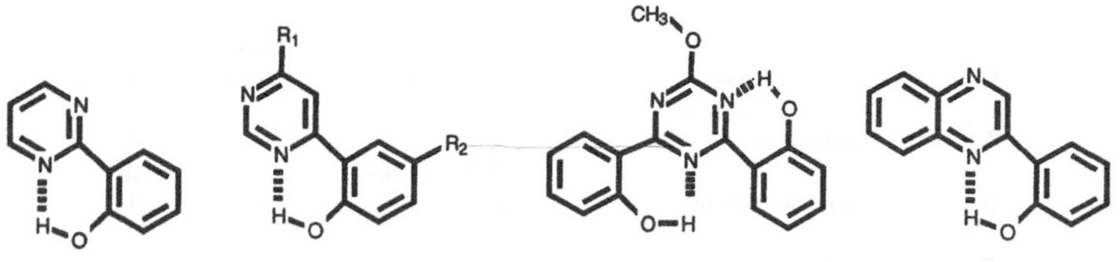

31	32a–e	33	34
96 ppm	$R_1=R_2=H$ 94 ppm	97 ppm	95 ppm
	$R_1=R_2=Br$ 95 ppm		
	$R_1=R_2=OCH_3$ 90 ppm		
	$R_1=Cl,R_2=H$ 95 ppm		
	$R_1=NEt_2,$ 97 ppm		
	$R_2=H$		

SCHEME 7

It was concluded that the chemical shifts (90 to 97 ppm) for these intramolecular hydrogen-bonded phenols were almost independent of the nature of the chelate's basic moiety. In addition, the ^{17}O chemical shift of the intramolecular hydrogen-bonded hydroxy group of 2-hydroxy-1-naphthaldehyde was also noted to be 95 ppm. Similarly, the ^{17}O chemical shift for the intramolecular hydrogen-bonded OH group in 8-hydroxyquinoline N-oxide (**35**) (see Chapter 9) was noted at 94 ppm by Boykin and co-workers.[16]

^{17}O NMR data on intramolecular hydrogen-bonded, acyclic α-azo hydroperoxides **36** have been reported (Scheme 8).

SCHEME 8

In this case, both five-membered and six-membered intramolecular hydrogen-bonded structures are possible.[19] The ^{17}O NMR data showed a large solvent dependence (benzene vs. acetonitrile vs. methanol), which was interpreted in terms of competitive hydrogen bonding with the solvent (see Chapter 9).

In a paper devoted solely to intramolecular hydrogen bonds of the C=O . . . HO type, Lauterwein reported ^{17}O NMR data on a series of quinones and ketones (Table 3)[15b] in chloroform solution. In addition, the results of Fiat[14] on o-hydroxyacetophenone and o-hydroxybenzaldehyde were confirmed. The ^{17}O NMR signals for the "peri" carbonyl of two α-hydroxy naphthoquinones were found to be displaced by $\sim$70 ppm from those of the parent compounds. These upfield shifts were attributed to intramolecular hydrogen bonding. The lack of a concentration dependence (no change in chemical shift between 0.1 and 0.005 M) showed that exclusively intramolecular hydrogen bonding was occurring in $CDCl_3$. Substituent and steric effects were thought to be one order of magnitude smaller than the

TABLE 3
17O NMR Data on Intramolecular Hydrogen-Bonded Carbonyl Compounds and Parent Compounds

No.	Compound	δ(C=O) ppm[a]	δ(OH) ppm	Ref.
37	Naphthoquinone	568.7	—	15b
		(580.7; Toluene)		(15a)
38	5-OH-Naphthoquinone	498.3	84.1	15b
		570.4		
39	2-Methylnaphthoquinone	558.2 br.	—	15b
40	5-OH-2-Methylnaphthoquinone	488.5	84.4	15b
		560.3		
15	Acetophenone	544.2	—	15b
		(550)		(14)
16	o-OH-Acetophenone	489.3	86.0	15b
		(488)	(90)	(14)
12	Benzaldehyde	557.4	—	15b
		(562)		(14)
13	o-OH-Benzaldehyde	504.3	(80.3)	15b
		(505)	(85)	(14)
14	1-Naphthaldehyde	571.8	—	15b
41	2-OH-1-Naphthaldehyde	472.0	96.4	15b
		(N.R.)	(95)	(23)
42	Fluorenone	510.1	—	15b
43	1-OH-Fluorenone	476.4	72.1	15b
44	Benzophenone	543.0	—	15b
45	2-OH-Benzophenone	485.0	84.4	15b
46	2,2'-Dihydroxybenzophenone	440.3	79.5	15b

[a]   Chemical shifts from Ref. 15b were obtained from CDCl₃ solutions at 40°; those from Reference 14 were from *p*-dioxane.

observed intramolecular hydrogen-bonding effects based on the ¹⁷O NMR data of the analogous *o*-methoxy substituted compounds. Torsion angle changes for the carbonyl groups were not included in the determination of ¹⁷O NMR hydrogen-bonding dependent shifts. In general, torsion angle and substituent effects must be considered in the analysis (*vide infra*) to obtain accurate estimation of hydrogen-bonding contributions. In this case torsion angle effects and electronic effects are opposing each other and apparently roughly cancel.

The ¹⁷O NMR chemical shift data for the carbonyls were found[15b] to show poor correlations (r~0.94) with both the ¹H and ¹⁷O NMR data for the OH groups. The ¹³C NMR data for the carbonyls was found not to show any functional relationship with ¹H or ¹⁷O NMR data. The chemical shifts of the intramolecular hydrogen-bonded OH groups were found to be downfield from those expected. The ¹⁷O NMR linewidths were found to decrease strongly on hydrogen bonding. The quadrupole coupling constants in α-hydroxy-1,4-naphthoquinone were determined. No correlation was observed between the variation in the ¹⁷O NMR quadrupole coupling constants and the ¹⁷O NMR chemical shift for intramolecular hydrogen-bonded carbonyls.

Recently, Lauterwein reported an ¹⁷O NMR study of intramolecular hydrogen bonding in *o*-anisic acid **47** and *o*-anisamide **48** in different solvents.[17] The ¹⁷O chemical shifts for the methoxy group were upfield in CDCl₃ compared to those in CD₃CN or CD₃OD (Scheme 9).[17]

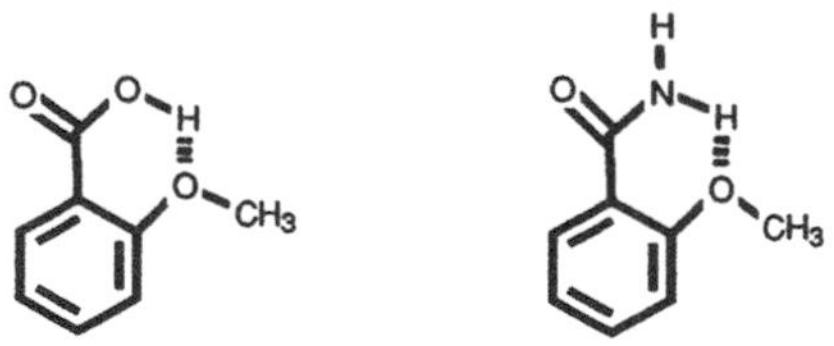

47   **48**

Solvent	δ(OMe) PPM	
CDCl$_3$	47.0	50.5
CD$_3$CN	51.6	52.1
CD$_3$OD	52.0	52.7

SCHEME 9

Based on ^{1}H data, the compounds were predicted to be 95% (intramolecular) hydrogen-bonded in CDCl$_3$ and 54% or less in the other solvents. The sensitivity of the methoxy group to hydrogen bonding was found to be much smaller than that of carbonyl groups. The ^{17}O data for *o*-anisic acid and *o*-anisamide were compared to those of the analogous methyl ester and *N,N*-dimethyl derivatives. Small deshielding effects were noted. *Ab initio* calculations were used to predict the geometry of the hydrogen bonds. The quantitative analysis did not consider torsion angle differences and must be viewed with caution. The authors concluded that variation in charge was the origin of the ^{17}O hydrogen-bonding shifts.

B. RECENT AND ONGOING STUDIES

Previous studies have firmly established ^{17}O NMR spectroscopy as a useful method for studying intramolecular hydrogen bonding. However, assessment of the influence of intramolecular hydrogen bonding on structure and the resulting ^{17}O chemical shift changes have not been fully elucidated. We have completed a study on a series of 2'- and 4'-amino and amido substituted acetophenones (**49** to **54**) and α-amido substituted acetophenones (**55** and **56**) which shows that the ^{17}O NMR data must be interpreted as a combination of torsion angle changes, electronic considerations and hydrogen-bonding shielding effect.[18]

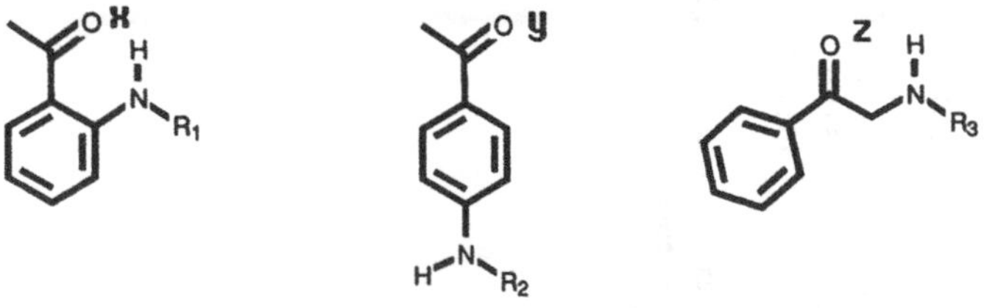

49,51,53   **50,52,54**   **55,56**

The ^{17}O NMR chemical shift data for 2'- and 4'-aminoacetophenone (**49** and **50**) 2'- and 4'-acetamidoacetophenone (**51** and **52**), 2'- and 4'-trifluoroacetamidoacetophenone (**53** and **54**), and α-acetamido- and α-trifluoroamidoacetophenone (**55** and **56**), obtained at natural abundance (0.5 *M*) in acetonitrile at 75°C, are summarized in Table 4. The ketone signals appeared between 519 and 551 ppm and those for the amido carbonyls appeared between 320 and 373 ppm as expected. For the *para* substituted compounds, the ketone and the

TABLE 4
¹⁷O Chemical Shift Data[18] (±1 ppm) for Compounds 49 to 56 in Acetonitrile[a] at 75° (and in CH_2Cl_2)[b]

Compound no.	R_1	δx	δR_1
49	H	524.1 (518.0)[b]	
51	COCH₃	535.4	373.3
53	COCF₃	531.0 (528.9)[b]	343.3 (341.5)[b]

Compound no.	R_2	δy	δR_2
50	H	518.7 (513.0)[b]	
52	COCH₃	539.5	366.3
54	COCF₃	551.0 (542)[b]	339.1 (338)[b,c]

Compound no.	R_3	δz	δR_3
55	COCH₃	532.6	337.1
56	COCF₃	530.0	319.6

a   Referenced to external water at 75°; 2-butanone as internal standard (558 ppm).
b   Referenced to external water at 35°; 2-butanone as internal standard (554 ppm).
c   Saturated solution.

amido carbonyl chemical shifts were as predicted based upon electronic effects reported in each series. However, the ^{17}O NMR chemical shifts for the ketone signals of the *ortho* isomers **51** and **53** were shielded relative to those of their *para* isomers in contrast to that of **49** relative to **50**. Since the electronic effect for the *para* substituent of compound **56** is essentially zero, the significant shielding (20 ppm) noted for the *ortho* isomer **53** is clearly the result of intramolecular hydrogen bonding. For the *N*-acetyl compounds, the ketone signal for the *ortho* compound **51** appears only slightly shielded (5 ppm) relative to **52** which could be interpreted to indicate a weak hydrogen bond. In contrast, the *ortho* amino compound **49** is deshielded relative to its *para* isomer **50**, misleadingly indicating a lack of hydrogen bonding (*vide infra*). The data for **49** and **50** were qualitatively in agreement with those in dioxane reported by Fiat.[14] To check for intermolecular hydrogen-bonding effects of solvent, the ^{17}O NMR spectra of the amino compounds **49** and **50** and the trifluoroacetoamido compounds **53** and **54** were taken in dichloromethane (Table 4). The net differences were essentially the same in both solvents. Compounds **55** and **56** were chosen to test for the effect of five-membered ring intramolecular hydrogen bonding on ^{17}O NMR data. The ketone chemical shifts for **55** and **56** were essentially identical and appear shielded relative to analogs indicative of hydrogen-bonding effects.

Molecular mechanics (MM2) calculations which include hydrogen bonding considerations were carried out on compounds **49** to **56**. The molecular mechanics calculations on these systems predict that the *ortho* compounds have significantly different geometries than their *para* isomers (Table 5). The calculated torsion angles for the functional groups for the *para* isomers **50, 52,** and **54** are essentially identical to those reported for the respective parent compounds.[6,25] Calculations for the *ortho* isomers **49, 51,** and **53**, which neglect hydrogen bonding, show larger torsion angle changes ($O_1C_2C_3C_4 \cong$ 35-31°) than those shown in Table 5, analogous to those reported for *ortho* alkyl groups.[6,25]

TABLE 5
MM2-Calculated[18] Hydrogen-Bonding Distances, Bond, and Torsion
Angles for 49, 51, 53

Compound no.	R	$H_6 \ldots O_1$(Å)	$O_1H_6N_5$	$O_1C_2C_3C_4$	$H_6N_5C_4C_3$	$O_1H_6N_5C_4$
49	H	1.97	122.3°	−12.8°	1.2°	−16.2°
51	$COCH_3$	1.95	120.1°	−16.5°	19.9°	−52.7°
53	$COCF_3$	1.97	117.3°	−12.1°	24.7°	−53.7°

Interestingly, the hydrogen-bonding distances are predicted to be essentially constant for **49, 51,** and **53**. The calculated bond angles for these six-ring, hydrogen-bonded systems are roughly 120°. Hydrogen bonding is predicted to reduce the torsion angle for the ketone carbonyls to between 12° to 17° from that expected based only on steric interactions. We have shown that torsion angle rotation of aryl ketones causes significant deshielding for the carbonyl signal (0.84 ppm/deg.)[6] It should be noted that earlier investigations had concluded that electronic contributions at the *ortho* position are substantially less than at the *para* position; this was due in part to model studies using *o*-methoxy groups,[14,15b] and the assumption of no conformational changes. At the time of those studies the importance of torsion angle changes was not appreciated. The result of torsion angle changes for the *ortho* analogs is to partially mask the electronic contribution because the effect on ^{17}O NMR chemical shift is in the opposite direction. Since for aryl carbonyls torsional rotation yields deshielding effects, while hydrogen bonding effects are shielding, the chemical shifts observed are a combination of these competing influences. The contribution of hydrogen bonding to the chemical shift can be estimated by subtracting the value for the *ortho* isomer from the sum of values of the torsion angle-rotation contribution and the chemical shift data for the *para* isomer (use of *para* isomer value corrects for electronic substituent effects). This approach yields shielding effects of 5, 18, and 30 ppm, respectively, for hydrogen bonding in **49, 51**, and **53**. This trend parallels the relative acidity of the N-Hs. The shielding value for compound **53** begins to approach that noted for phenolic analogs.[14] Casual inspection of the data for **49** and **50** might suggest that hydrogen bonding is unimportant in **49**; however, this method of analysis shows that hydrogen bonding contributes significantly to the structure of **49**. This approach assumes that the electronic effects of *ortho* and *para* substituents are identical; however, as a result of torsion angle differences, the electronic contribution for the *ortho* isomer may be somewhat reduced. If valid, this will lead to even larger shielding contributions for hydrogen bonding in these systems.

Molecular mechanics calculations for **55** and **56** showed identical hydrogen bond distances and angles (OHN) of 2.18 Å, and 105°. The amides in this case were predicted to be essentially planar unlike the aryl acetamides which were predicted to be slightly pyramidal. These results may explain, in part, the shielding seen for the amido carbonyl signals of **55** and **56** compared to that of the acetanilides.[25]

The apparent lack of competition of intermolecular hydrogen bonding to the solvent (acetonitrile) is intriguing. It is doubtful that the solvent is inert; more likely the 2′ and 4′ isomers are experiencing analogous interactions to yield the apparent insensitivity. Intramolecular hydrogen bonding of an amide NH to a methoxy group was deduced from a small solvent dependency; however, torsion angle dependence was not considered.[17] The ^{17}O NMR data seems sensitive to hydrogen bonding even in these systems which are conformationally mobile and appear to experience considerable changes in geometry. The data clearly show that torsion angles must be considered when analyzing hydrogen-bonding phenomena by ^{17}O NMR spectroscopy. Despite the mobility of this system, the ^{17}O NMR data indicate that larger shielding effects are seen with the more acidic NHs.

An ^{17}O NMR study of intramolecular hydrogen bonding of N-H to carbonyls in substituted fluorenones and anthraquinones[26] (Scheme 10, Tables 6 and 7) supports our previous conclusions.[18,27] In these cases torsion angle considerations have been minimized; the intramolecular hydrogen bond distances and angles would be expected to vary between the fluorenone and anthraquinone systems. Applying the previously described method of analysis

TABLE 6
^{17}O NMR Data[26] of Amino and Amido Substituted
Fluorenones in CH_3CN at 75°C[a]

No.	Compound	δ(C=O) ppm	δ(NHCO) ppm
42	Fluorenone	517.3	—
		(510.1)[a]	
57	1-NH$_2$-Fluorenone	486.3	—
58	3-NH$_2$-Fluorenone	483.3	—
59	1-NHCOCH$_3$-Fluorenone	500.8	378.1
60	3-NHCOCH$_3$-Fluorenone	507.1	371.0
61	1-NHCOCF$_3$-Fluorenone	499.5	348.7

[a]   Reference 15b in CDCl$_3$ at 40°C.

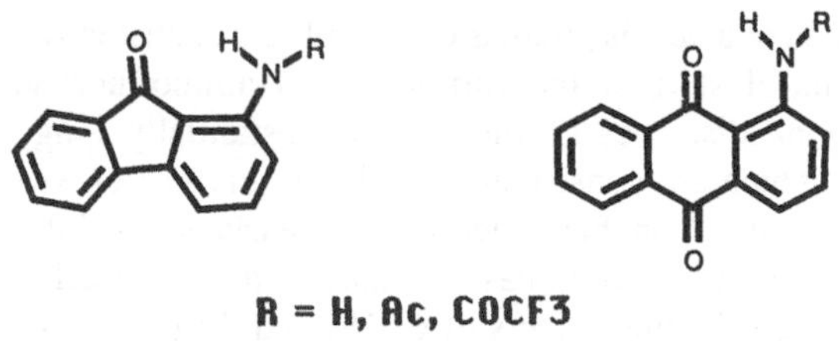

R = H, Ac, COCF3

SCHEME 10

TABLE 7
^{17}O NMR Data[26] of Enriched Amino and Amido Substituted Anthraquinones in CH_3CN and 75°C[a]

No.	Compound	$\delta(C=O)$ ppm
62	Anthraquinone	524
		(531)[b]
63	1-NH_2-Anthraquinone	485.5
		515
64	1-NHAc-Anthraquinone	496.5
		525.5
65	1-$NHCOCF_3$-Anthraquinone	494
		532

[a] The quinone carbonyls were enriched; hence, the -NHCO carbonyl signals for **64** and **65** were not detected.

[b] Reference 15a from toluene at 95°.

to the *N*-acetyl and *N*-trifluoromethyl fluorenones **59** and **61**, the upfield shifts of 4 and 17 ppm, respectively, are attributable to hydrogen bonding. These values are much less than deduced for the analogous acyclic examples (18 and 30 ppm). This result is consistent with MM2 calculations that predict the intramolecular hydrogen bonding distances are longer for the fluorenone cases (Table 8) than for those of the acetophenone and anthraquinone cases.

TABLE 8
MM2-Calculated[26] Hydrogen-Bonding Distances, Bond and Torsion Angles for Fluorenones (57, 59, 61) and Anthraquinones (63 to 65)

R	$H_6 \ldots O_1$(Å)	$O_1H_6N_5$	$C_2C_3C_4$	$O_1C_2C_3C_4$	$H_6N_5C_4C_3$	$O_1H_6N_5C_4$
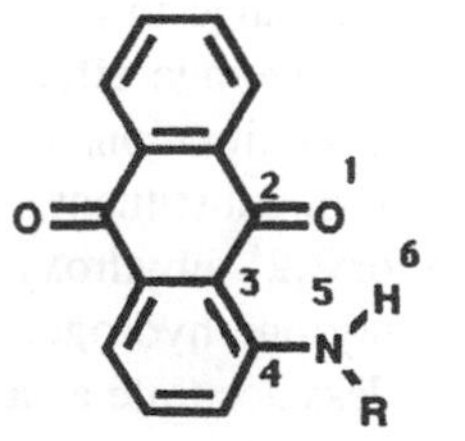 H	1.92	121.6°	120.7	0°	0°	0°
$COCH_3$	1.91	118.7°	120.9	0.9°	27.1°	−43.6°
$COCF_3$	1.93	115.8°	120.8	0.9°	30.7°	−48.0°
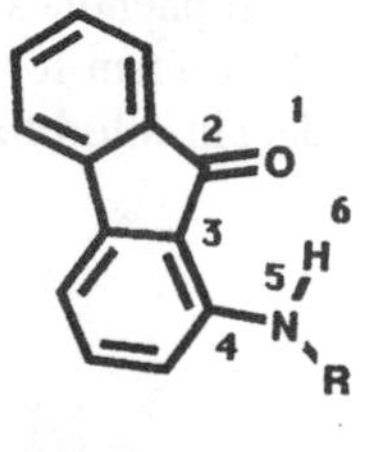 H	2.21	122.4°	127.7	0°	0°	0°
$COCH_3$	2.02	132.5°	127.6	0°	4.5°	−8.8°
$COCF_3$	2.07	127.1°	127.4	0°	18.1°	−30.9°

Interestingly, this method of analysis predicts essentially no contribution of hydrogen bonding to the ¹⁷O NMR chemical shift of the carbonyl of 1-aminofluorenone (**57**). The MM2 calculated hydrogen bond distances (Table 8) are substantially longer than those for the acetophenone systems which are consistent with the observed ¹⁷O NMR results.

The anthraquinone data,[26] on the other hand, clearly show intramolecular hydrogen bonding for both the amino and amido cases. Analysis as discussed above yields intramolecular hydrogen-bonding shielding effects of 6, 15, and 29 ppm, respectively, for **63, 64,** and **65**. The ¹⁷O NMR shielding effects attributable to hydrogen bonding are essentially identical with those of their acetophenone analogs, and the MM2 predicted hydrogen-bonded structures show similar geometric parameters (Table 8).

Our intramolecular hydrogen-bonding studies of amino and amido groups to the carbonyl moiety clearly show that the factors which determine ¹⁷O chemical shift can be deduced. The contribution of hydrogen bonding to chemical shift can be quantitated only after corrections for substituent effects (electronic effects) and torsional variation have been performed. Hydrogen-bond induced chemical shifts (Δδ) correlate reasonably well with the acidity of the N-H and seem to show a dependence on the hydrogen-bond distance. The angular dependence of the hydrogen bond on the ¹⁷O NMR data in systems such as these awaits further experimentation.

Applying the above approach to intramolecular hydrogen-bonded phenols (to carbonyls) allows the quantitation of the influence of hydrogen bonding on ¹⁷O NMR chemical shifts therein. The ¹⁷O NMR data for a series of *ortho*-hydroxy acetophenones in acetonitrile are listed in Table 9.[27] Interestingly, the advent of high resolution instruments allows the detection of proton-to-oxygen coupling in the phenol signal. For example, Figure 1 shows proton coupled and decoupled spectra of 2-hydroxyacetophenone. Coupling is expected under conditions where fast intramolecular proton transfer between the two oxygens occurs without proton spin exchange.[13] Other solvents, lower temperatures, and/or large broadening factors can obscure the coupling. The only previous example of similar coupling was deduced for the enol of 2-acetylcyclohexanone by line-shape analysis, and a J value of 77 Hz was noted by Gorodetsky.[13] In the present examples the observed coupling constants ranged from 75 to 90 Hz (Table 9).

Correcting the ¹⁷O NMR carbonyl chemical shifts for torsion angle rotation and electronic effects as described above for amino and amido systems yields the shielding contributions for hydrogen bonding (Δδ) shown in Table 9. For example, the hydrogen-bonding contribution to the ¹⁷O carbonyl chemical shift of **67** is estimated as follows: the parent system (acetophenone) chemical shift is 552, the electronic effect of a *para* methoxy group is −16 ppm, the electronic effect of the *ortho*-hydroxy group is taken as −21 ppm, and the torsion angle contribution is +3 ppm, which gives a predicted chemical shift of 518 ppm for **67**, assuming no effect of intramolecular hydrogen bonding. The observed chemical shift of 466 ppm is 52 ppm shielded from the calculated value. For all the systems, the range in contribution to chemical shift from hydrogen bonding is 43 to 67 ppm with the average effect near 55 ppm. Presumably, the range of Δδ is in part due to the error in approximations of torsion angle and to changes in the acidity of the phenol as a consequence of substituent.

Interestingly, the predicted ¹⁷O NMR hydrogen-bonding contribution for 2,2′-dihydroxy benzophenone **46** is 53 ppm which is consistent with the formation of only one hydrogen bond to the carbonyl group in acetonitrile. However, these Δδ values are less accurate as a result of the difficulty in estimation of the torsion angle changes in the benzophenones by molecular modeling methods. Previously, the ¹⁷O chemical shift data in chloroform for **46** were interpreted to show two hydrogen bonds to the carbonyl group.[15b] The data in chloroform

TABLE 9
^{17}O NMR Data[27] (±1 ppm) for Intramolecular Hydrogen-Bonded Phenols and Related Compounds in CH$_3$CN at 75°C[a]

No.	Compound	δ(C=O)	δ(OH)	J(OH)	Δδ[b]
16	o-OH-Acetophenone[a]	491	85.5[c]	86	43
17	p-OH-Acetophenone[a]	531	88		—
18	2,4-Dihydroxyacetophenone[a]	463	90, 94.5[c,d]		50
19	2,6-Dihydroxyacetophenone[a]	481	91		51
66	2-OH,6-OMe-Acetophenone	484	89[c]; [62.5 (OMe)]	87	58
67	2-OH,4-OMe-Acetophenone	466	91[c]; [67 (OMe)]	75	52
44	Benzophenone	552			
45	2-OH-Benzophenone[a]	492	85		51
		(486)[e]	(84)[e]		(36)[e]
46	2,2'-Dihydroxybenzophenone[a]	475	82		53
		(444)[e]	(80.5)[e]		(57)[e]
68	2-Acetyl-1-Naphthol	466	90[c]	80	67
69	1-Acetyl-2-Naphthol	512	93		60
70	Propiophenone	540	—		—
71	2-OH-Propiophenone	481	85[c]	91	45

[a] See Tables 3 and 4 for values in different solvents.
[b] Calculated ^{17}O NMR hydrogen bonding shielding effect; see text.
[c] Splitting of OH signal.
[d] Coupling was apparent; however, signal overlap prevented quantitation of J.
[e] In CDCl$_3$; see also Table 3.

are reproducible and substantially different from those obtained in acetonitrile. The Δδ values in this system suggest the formation of two hydrogen bonds in chloroform for **46**. Clearly, the ^{17}O NMR hydrogen-bonding-induced shift data (94 ppm) for the 9-carbonyl signal of 1,8-dihydroxyanthraquinone **72** in toluene at 75° are consistent with the formation of two intramolecular hydrogen bonds to the carbonyl group.[28] Also, this result is consistent with the X-ray structure of 1,8-dihydroxyanthraquinone.[29]

444 80.5 (CDCL3)
475 81.7 (CH3CN)

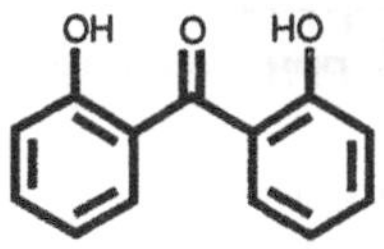

395 91.5 (toluene)

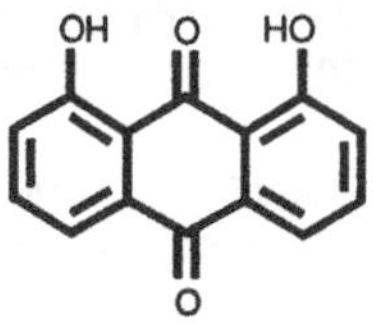

532

46

72

Application of corrections for substituent effects (electronic effects) and torsion angle rotation allows the determination of the contribution of ^{17}O NMR chemical shifts of carbonyl groups arising from intramolecular hydrogen bonding. Analysis of the geometry of the hydrogen-bonding array must also be performed since less than optimum geometry will also

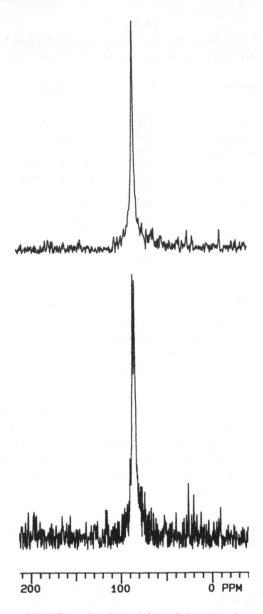

FIGURE 1.   Coupled and decoupled spectra of *o*-hydroxyacetophenone in acetonitrile at 75°C.[27]

dramatically reduce the effect. Large upfield shifts (ca. 55 ppm) are attributable to intramolecular hydrogen bond formation in acidic systems (phenols). Intramolecular hydrogen bonding involving the amido NH of trifluoroacetyl derivatives causes 30 ppm upfield shifts which are approximately twice as large an effect observed for acetyl NH's participation in intramolecular hydrogen bonding (18 ppm). The contribution to chemical shift from amino NH intramolecular hydrogen bonding is much smaller (5 ppm). Thus the magnitude of the observed shifts parallel the acidity of the hydrogen bond donor. Clearly, ^{17}O NMR methodology is sensitive to intramolecular hydrogen bonding to carbonyl groups and provides a promising approach for future studies.

III. INTERMOLECULAR HYDROGEN BONDING (INCLUDING SOLVENT INTERACTIONS)

The pioneering ^{17}O NMR investigations of Christ and Diehl included a study of the dilution of acetone with water.[8] These workers demonstrated a large upfield shift (52 ppm) of the ^{17}O NMR acetone carbonyl signal at infinite dilution in water. Reuben confirmed and expanded upon the earlier work and suggested that multiple equilibria between water and acetone were required to explain the observed chemical shift acetone concentration curve (see Figure 2).[9] Reuben also reported the ^{17}O NMR chemical shift of water in a variety of protic and aprotic solvents. The effect of intermolecular hydrogen bonding by solvents on a variety of ^{17}O NMR data of carbonyl functional groups has been summarized.[30]

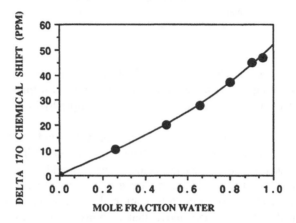

FIGURE 2. Plot of change in [17]O chemical shift for acetone vs. mole fraction water.[9]

Detailed studies of the influence of intermolecular hydrogen bonding with solvent on the [17]O chemical shift of formamide,[31] N-methyl formamide,[31] N,N-dimethylformamide,[31] 1-methyl-2-pyrrolidinone,[32] and 1-ethyl-2-pyrrolidinone[32] have been performed. Intermolecular hydrogen bonding by self-association for N-methylformamide has been demonstrated by the downfield shift of the amide carbonyl [17]O resonance on dilution in acetone.[31] Intermolecular interaction between water and N-methyl formamide has been demonstrated by an upfield shift of the amide carbonyl [17]O resonance upon dilution with water.[31] These studies show that on hydrogen bonding to solvent, the amide carbonyl signal is shifted upfield. A 52 ppm upfield shift was noted for the carbonyl signal of the non-hydrogen-bonded monomer of N-methylformamide (chemical shift determined at infinite dilution in diethylether) and the completely hydrogen-bonded carbonyl in water solution.[31]

The study of solvent effects on ^{17}O NMR chemical shift of 1-ethyl-2-pyrrolidinone (Table 10) showed that the stronger the proton-donating ability of the solvent, the greater the upfield shift of the amide carbonyl; of the solvents studied the largest shift noted was with 2,2,2-trifluoroethanol as the solvent.[32] A method of analysis for dissecting the ^{17}O NMR chemical shift of amides at infinite dilution in water into terms involving contributions from hydrogen bond formation at both carbonyl oxygen lone electron pairs and involving contributions from the N-H protons participating in hydrogen bonds has been described.[11] Solvent composition dependence of both ^{17}O NMR chemical shift and linewidth has been employed to detect intramolecular hydrogen vs. intermolecular hydrogen bond effects in peptides.[33]

An isopropylidine-uridine compound shows a preference for intermolecular hydrogen bonding with water at O-4 in comparison to O-2 in a study carried out in acetonitrile water

TABLE 10
Effect of Solvent on ¹⁷O NMR Chemical Shift Data for 1-Ethyl-2-Pyrrolidinone[32]

Solvent	δ/(C=O)ppm
Neat	300
Dimethylsulfoxide	298
Chloroform	289
Ethanol	286
Methanol	285
2,2,2-Trichloroethanol	285
Water	281
2,2,2-Trifluoroethanol	274

solutions.[34] Similarly, selective hydrogen bonding to nucleic acid bases has been demonstrated by upfield shifts of the carbonyl ¹⁷O NMR signals in DMSO solutions of the bases on addition of water.[35] In general, larger shifts were noted for O-4 than for O-2 resonances, and this result was tentatively interpreted in terms of multiple hydrogen bond formation at the O-4 carbonyl.

Intermolecular hydrogen bonding by solvent has been detected for *N,N*-dimethylmethanesulphinamide and *N,N*-dimethylmethanesulphonamide by ¹⁷O NMR spectroscopy.[36] Various protic solvents, alcohols of increasing acidity, cause shielding of the SO signal. The shifts of ¹⁷O resonance for the sulphinamide are greater than those for the sulphonamide consistent with the relative hydrogen bond formation abilities of the two functional groups.

It is clear from the studies cited above that carbonyl ¹⁷O NMR signals for various functional groups are sensitive to the medium. The solvent effects encountered include disruption of intermolecular interactions of solute by inert solvents and, in the case of protic solvents, formation of new intermolecular hydrogen bonds. The dilution of self-associated functional groups such as amides with inert solvents results in downfield shifts of the ¹⁷O resonance, whereas the signals for carbonyl groups are shifted upfield in protic solvents. Solvent effect studies by ¹⁷O NMR spectroscopy hold considerable promise for study of solute-solvent interactions, including detection of intramolecular hydrogen bonding by the lack of appreciable solvent effects. Extension of these types of studies to include studies of ions and metal-ion interactions also is likely to be a fruitful field of investigation.[37,38]

ACKNOWLEDGMENTS

Acknowledgment is made to the Donors of the Petroleum Research Fund, administered by the American Chemical Society, for partial support of this research, to NSF (CHE-8506665), to the NSF Instrumentation Program (CHEM-8409599), and to the Georgia State University Research Fund. A.L.B. was a fellow of the Camille and Henry Dreyfus Foundation, 1981 to 1986.

REFERENCES

1. (a) **Stewart, R.**, *The Proton: Applications to Organic Chemistry*, Academic Press, New York, 1985; (b) **Joesten, M. D. and Schaad, L. J.**, *Hydrogen Bonding*, Marcel Dekker, Inc., New York, 1974; (c) **Schuster, P., Zundel, G., and Sandorfy, C., Eds.**, *The Hydrogen Bond*, Vols. 1-3, North-Holland, New York, 1976.

2. **Boykin, D. W. and Baumstark, A. L.**, ^{17}O NMR spectroscopy: assessment of steric perturbations of structure in organic compounds, *Tetrahedron*, 45, 3613, 1989.

3. (a) **Kintzinger, J.-P.**, in Oxygen-17 NMR, *NMR of Newly Accessible Nuclei*, Vol. 2, Laszlo, P., Ed., Academic Press, New York, 1983, 79; (b) **Klemperer, W. G.**, in *The Multinuclear Approach to NMR Spectroscopy*, Proc. NATO Adv. Study Inst. on the Multinuclear Approach to the NMR Spectroscopy, Lambert, J. B. and Riddell, F. G., Eds., Riedel, Dordrecht, Holland, 1983, 245.

4. **Oakley, M. G. and Boykin, D. W.**, Relationship of torsion angle to ^{17}O chemical shift for aryl ketones, *J. Chem. Soc. Chem. Commun.*, p. 439, 1986.

5. **Boykin, D. W., Balakrishnan, P., and Baumstark, A. L.**, ^{17}O NMR studies of torsion angle relationships in aryl-alkyl ketones and aromatic aldehydes, *Magn. Reson. Chem.*, 25, 248, 1987.

6. **Baumstark, A. L., Balakrishnan, P., Dotrong, M., McCloskey, C. J., Oakley, M. J., and Boykin, D. W.**, ^{17}O NMR spectroscopy: torsion angle relationships in aryl carboxylic esters, acids, and amides, *J. Am. Chem. Soc.*, 109, 1059, 1987.

7. **Baumstark, A. L., Dotrong, M., Stark, R. R., and Boykin, D. W.**, Deshielding effects in rigid planar molecules, *Tetrahedron Lett.*, p. 2143, 1988.

8. (a) **Christ, H. A., Diehl, P., Schneider, H. R., and Dahn, H.**, Chemische Verschiebungen in der kernmagnetischen Resonanz von ^{17}O in organischen Verbindungen, *Helv. Chim. Acta*, 865, 44, 1961; (b) **Christ, H. A. and Diehl, P.**, Lösungsmitteleinflüssen und Spinkopplungen in der kernmagnetischen Resonanz von ^{17}O, *Helv. Phys. Acta*, 32, 170, 1963.

9. **Reuben, J.**, Hydrogen-bonding effects on oxygen-17 chemical shifts, *J. Am. Chem. Soc.*, 91, 5725, 1968.

10. **Boykin, D. W., Baumstark, A. L., and Balakrishnan, P.**, ^{17}O NMR spectroscopy of 4-substituted pyridine-N-oxides: substituent and solvent effects, *Magn. Reson. Chem.*, 23, 276, 1985.

11. (a) **Burgar, M. I., St. Amour, T. E., and Fiat, D.**, ^{17}O and ^{14}N NMR studies of amide systems, *J. Phys. Chem.*, 85, 502, 1981; (b) **Valentine, B., Steinschneider, A., Dhawan, D., Burgar, M. I., St. Amour, T. E., and Fiat, D.**, Oxygen-17 NMR of peptides, *Int. J. Pept. Protein Res.*, 25, 56, 1985.

12. **Balakrishnan, P., Baumstark, A. L., and Boykin, D. W.**, ^{17}O NMR spectroscopy: effect of substituents on chemical shifts for *p*-substituted benzoic acids, methyl benzoates, cinnamic acids and methyl cinnamates, *Org. Magn. Reson.*, 22, 753, 1984.

13. **Gorodetsky, M., Luz, Z., and Mazur, Y.**, Oxygen-17 nuclear magnetic resonance studies of equilibria between enol forms of β-diketones, *J. Am. Chem. Soc.*, 89, 1183, 1967.

14. **St. Amour, T. E., Burgar, M. I., Valentine, B., and Fiat, D.**, ^{17}O NMR studies of substituent and hydrogen bonding effects in acetophenones and benzaldehydes, *J. Am. Chem. Soc.*, 103, 1128, 1981.

15. (a) **Chandrasekaran, S., Wilson, W. D., and Boykin, D. W.**, ^{17}O NMR studies on polycyclic quinones and related cyclic ketones: models for anthracycline interacalators, *Org. Magn. Reson.*, 22, 757, 1984; (b) **Jaccard, G. and Lauterwein, J.**, Intramolecular hydrogen bonds of the C=O . . . H-O type as studied by ^{17}O-NMR, *Helv. Chim. Acta*, 69, 149, 1986.

16. **Boykin, D. W., Balakrishnan, P., and Baumstark, A. L.**, Natural Abundance ^{17}O NMR spectroscopy of heterocyclic N-oxides and di N-oxides. Structural effects, *J. Heterocycl. Chem.*, 22, 981, 1985.

17. **Jaccard, G., Carrupt, P.-A., and Lauterwein, J.**, Study of intramolecular hydrogen bonding in *o*-anisic acid and *o*-anisicamide by ^{17}O NMR and *ab initio* MO calculations, *Magn. Reson. Chem.*, 26, 239, 1988.

18. **Baumstark, A. L., Graham, S. S., and Boykin, D. W.**, ^{17}O NMR spectroscopy: intramolecular hydrogen bonding in 2'-amino-, 2'-acetamidoe-, and 2'-trifluoroacetamidoacetophenone, *J. Chem. Soc. Chem. Commun.*, p. 767, 1989.

19. **Baumstark, A. L., Vasquez, P. C., and Balakrishnan, P.**, 17-O-enriched α-azohydroperoxides: 17-O NMR spectroscopy, 17-O labeling reagents, *Tetrahedron Lett.*, p. 2051, 1985.

20. **Winter, W., Zeller, K. P., and Berger, S.**, Zur Struktur des Enols von Benzoylaceton, *Z. Naturforsch.*, 34b, 1606, 1979.

21. (a) **Lauterwein, J., Gerothanassis, I. P., and Hunston, R. N.**, A study of the *cis/trans* isomerism of *N*-acetyl-L-proline in aqueous solution by ^{17}O NMR spectroscopy, *J. Chem. Soc. Chem. Commun.*, p. 367, 1984; (b) **Hunston, R. N., Georthanassis, I. P., and Lauterwein, J.**, A study of L-proline, sarcosine, and the *cis/trans* isomers of *N*-acetyl-L-proline and *N*-acetylsarcosine in aqueous and organic solution by ^{17}O NMR, *J. Am. Chem. Soc.*, 107, 2654, 1985.

22. **Lambert, J. B. and Wharry, S. M.**, Nuclear magnetic resonance examination of organic dianions, *J. Am. Chem. Soc.*, 104, 5857, 1982.

23. **Lapachev, V. V., Mainagashev, I. Y., Svetlana, S. A., Fedotov, M., Krivopalov, V. P., and Mamaev, V. P.,** ^{17}O NMR studies of enol and phenol compounds with intramolecular hydrogen bonds, *J. Chem. Soc. Chem. Commun.,* p. 494, 1985.

24. **Mainagashev, I. Y., Lapachev, V. V., Fedotov, M. A., and Mamaev, V. P.,** Tautomerism of derivatives of azines. 16. Tautomerism of acylmethylpyrazines and quinoxalines, *Khim. Geterotsikl. Soedin.,* 12, 1663, 1987.

25. **Boykin, D. W., Deadwyler, G. H., and Baumstark, A. L.,** ^{17}O NMR studies on substituted N-arylacetamides and aryl acetates: torsion angle and electronic effects, *Magn. Reson. Chem.,* 26, 19, 1988.

26. **Baumstark, A. L., Graham, S., and Boykin, D. W.,** ^{17}O NMR spectroscopy: geometric effects on intramolecular hydrogen bonding in rigid carbonyl systems, *Tetrahedron Lett.,* p. 957, 1990.

27. **Boykin, D. W. and Baumstark, A. L.,** ^{17}O NMR spectroscopy: analysis of intramolecular hydrogen-bonding effects on chemical shifts of *ortho*-hydroxy substituted acetophenones, unpublished results.

28. **Baumstark, A. L. and Boykin, D. W.,** ^{17}O NMR spectroscopy: intramolecular hydrogen bonding of 1,8-dihydroxyanthraquinones, unpublished results.

29. **Prakash, A. Z.,** The crystal and molecular structure of chrysazin, *Kristallogr. Kristallgeom. Kristallphys. Kristallchem.,* 122, 5272, 1965.

30. **Fiat, D., Burgar, M. I., Dhawan, D., St. Amour, T., Steinschneider, A., and Valentine, B.,** ^{17}O NMR studies of labeled peptides and model systems, *Dev. Endocrinol.,* 13, 239, 1981.

31. **Burgar, M. I., St. Amour, T., and Fiat, D.,** ^{17}O NMR study of the hydration and association of N-methylformamide molecules, *Period. Biol.,* 82, 283, 1980.

32. **Ruostesuo, P., Hakkinen, A.-M., and Peltola, K.,** Carbon-13, nitrogen-15 and oxygen-17 NMR chemical shifts of 1-ethyl-2-pyrrolidinone and 1-methyl-2-pyrrolidinone in some solvents, *Spectrochim. Acta,* 41A, 739, 1985.

33. **Gilboa, H., Steinschneider, A., Valentine, B., Dhawan, D., and Fiat, D.,** Hydrogen bonds in the tripeptide pro-leu-gly-NH$_2$ ^{17}O and ^1H studies, *Biochim. Biophys. Acta,* 800, 251, 1984.

34. (a) **Schwartz, H. M., MacCoss, M., and Danyluk, S. S.,** Oxygen-17 NMR of nucleosides, *Tetrahedron Lett.,* p. 3837, 1980; (b) **Schwartz, H. M., MacCoss, M., and Danyluk, S. S.,** Oxygen-17 NMR of nucleosides. II. Hydration and self-association of uridine derivates, *J. Am. Chem. Soc.,* 105, 5901, 1983.

35. **Burgar, N. D. and Fiat, D.,** ^{17}O and ^{14}N spectroscopy of ^{17}O-labeled nucleic acid bases, *Org. Magn. Reson.,* 20, 184, 1982.

36. (a) **Häkkinen, A. M., Ruostesuo, P., and Kurkisuo, S.,** ^{17}O, ^{15}N and ^{13}C NMR chemical shifts of N,N-dimethylmethane sulphinamide in various solvents, *Magn. Reson. Chem.,* 23, 311, 1985; (b) **Häkkinen, A. M. and Ruostesuo, P.,** Carbon-13, nitrogen-15, oxygen-17 and sulphur-33 NMR chemical shifts of some sulphur amides and related compounds, *Magn. Reson. Chem.,* 23, 424, 1985.

37. **Baltzer, L. and Becker, E. D.,** Solvation of organic oxyanions studied by oxygen-17 NMR, *J. Am. Chem. Soc.,* 105, 5730, 1983.

38. **Fujkura, R. and Rode, B. M.,** ^{17}O, ^{13}C and ^1H nuclear magnetic resonance investigation on ion-solvent interactions in N,N-dimethylformamide, *Inorg. Chim. Acta,* 60, 99, 1982.

Chapter 6

¹⁷O-NMR AS A MECHANISTIC PROBE TO INVESTIGATE CHEMICAL AND BIOORGANIC PROBLEMS

Ronald W. Woodard

TABLE OF CONTENTS

I. INTRODUCTION

The oxygen atom is probably the most chemically and biologically important atom on earth. The earth's crust contains $\approx$50% by weight of oxygen. Oxygen forms compounds with all the elements except helium, neon, and possibly argon. It combines directly with all the other elements except the halogens, a few noble metals, and the noble gases, either at room temperature or at elevated temperatures.[1] The oxygen in these oxygen-containing compounds exists in various oxidation states as well as in various functional positions even with the same atom (i.e., with carbon, oxygen in the same oxidation state can functionally be an alcohol, ether, ester, epoxide, acid, anhydride, peracid, etc.). Despite the ubiquitousness and importance of the oxygen atom, relatively few chemical or biological studies based on the physical and chemical properties of the oxygen atom have appeared. The lack of studies is most likely due to one or more of several technical problems such as the magnetic inertness of ^{16}O $(I=0)$ and ^{18}O $(I=0)$ and the large electric quadrupole moment of ^{17}O $(I=\,^5/_2$, $Q = -2.6 \times 10^{-26}$ cm^2) as well as the low natural abundance of the two heavy isotopes of oxygen $^{18}O = 0.204\%$ and 0.037% for ^{17}O, respectively, as compared to $^{16}O = 99.759\%$.[2,3] Since the ^{18}O isotope is $\approx$ 70 times more abundant than ^{17}O, it has been utilized to a greater extent to study chemical mechanisms and bioorganic problems.[4] The detection methods, until recently, therefore relied solely on the use of mass spectroscopy to determine the presence of the heavy oxygen isotope, and, generally speaking, the exact location of the oxygen was rather difficult to assign with confidence in all but the more straight-forward cases. More recently, the ^{18}O isotope has been used because of the ^{18}O isotope shift-effect on the nuclear magnetic resonance (NMR) signal of various NMR-active atoms.[5] The application of ^{17}O NMR spectroscopy to directly study chemical and biochemical reactions has only recently[6,7] attracted attention, probably due to the difficulty associated with the measurement of the resonance that has been previously discussed in Chapter 1, namely the extremely low natural abundance and the large electric quadrupole moment. Efforts to obtain mechanistic information under such adverse technical conditions no doubt can be attributed to the importance of the oxygen isotope. It should be noted that the advantage of ^{17}O-labeling vs. ^{18}O-labeling is that ^{17}O-NMR is a nondestructive method (for that matter so is the ^{18}O isotope shift-effect methodology), however, ^{17}O-NMR allows one to obtain rather easily both the site of substitution and relative amount present at that site via integration of the ^{17}O signal.

In the work to be reviewed in this chapter, these technical problems are generally compensated for by two major factors: one, the ease with which isotopically enriched compounds may be synthesized, and two, the large chemical shifts observed for the ^{17}O resonances, which aid in the resolution of quadrupole-broadened resonances. In addition, the broad ^{17}O-NMR linewidths may be reduced by varying three simple experimental parameters, namely, the temperature, the sample concentration, and solvent viscosity. The theoretical basis[8] for these improvements is that the quadrupole relaxation time T_1 in the extreme motional narrowing condition is defined as follows

$$\text{linewidth} \approx 1/T_1 \approx 1/T_2 \approx 3/125\ [1 + \eta^2/_3]\ [e^2Q\ q/\hbar]^2\ \tau_c \qquad (1)$$

where q = electric field gradient, η = asymmetry factor of the electric field gradient (can vary from 0 to 1), $e^2Qq/\hbar$ = quadrupole coupling constant [eq = the field gradient due to electrons and eQ = the field gradient due to nucleus] and τ_C is the rotational correlation time for molecular reorientation. This equation holds as long as isotropic molecular tumbling is rapid on the NMR time scale (i.e., $\tau_c \ll \omega_O^{-1}$, ω_O = nuclear larmor frequency). Thus, for small molecules a reduction of τ_C should result from either increase in temperature (T) or decrease in solution viscosity (η) according to the Stokes-Einstein-Debye equation,

$$\tau_c = 4 \pi \eta a^3/3 \kappa T \qquad (2)$$

where a = molecular radius, κ = Boltzmann constant.

Therefore, when experimentally feasible, researchers have observed the ^{17}O-NMR spectra at an elevated temperature and low sample concentration in a solvent of low viscosity. The use of low sample concentration is, of course, not always possible due to sensitivity problems, unless ^{17}O enrichment is feasible and extensive signal accumulation using Fourier transform NMR is available. It is important to remember that even though the broad ^{17}O linewidths are undesirable in terms of quantitation and instrumentation, their cause (the fact that the quadrupolar relaxation process is the dominant relaxation mechanism) is often actually advantageous in many studies.

This chapter is divided into two major sections: (1) the study of chemical mechanisms utilizing the direct observation of ^{17}O via ^{17}O-NMR, and (2) the study of bioorganic problems via direct observation of ^{17}O via ^{17}O-NMR. An overview of each contribution, in terms of the methodology and rationale of the application of each paper to the advancement of the use of ^{17}O-NMR in the area of study, along with some insight into the mechanistic problem under study, has been given. Readers wishing to gain further insight into the studies are invited to consult the original literature.

II. CHEMICAL MECHANISMS

A. DECOMPOSITION REACTIONS

Iwamura et al.,[9] have reported the use of ^{17}O NMR to study the thermal decomposition of ^{17}O-labeled t-butyl O-methylthio- and O-phenylthioperoxybenzoate. Previous speculation as well as CIDNP and CNDO/2 MO calculations on the mechanism had suggested that the participation of the neighboring sulfur atom was responsible for the 10^4 enhanced rate factor of the homolytic O-O bond cleavage in t-butyl O-methylthioperoxybenzoate (1), however the possibility of a zwitterionic structure (3) had not been ruled out.

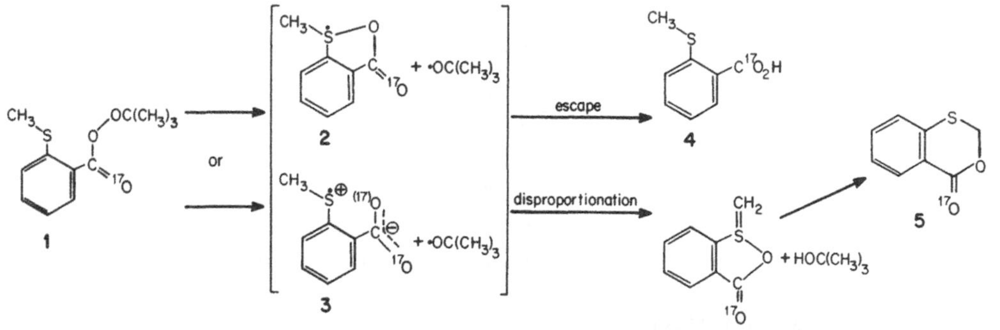

SCHEME 1. Thermal decomposition of ^{17}O-labeled t-butyl-O-phenylthioperoxybenzoates.

The peroxybenzoate (1) was synthesized from ^{17}O-labeled O-methylthiobenzoic acid, in which the carbonyl group had been activated by reaction of 1,1'-carbonyldiimidazole followed by reaction with t-butylhydroperoxide at low temperature. The natural abundance ^{17}O-NMR spectra of 5 displayed a resonance at δ = 365 ppm (from external ^2H_2O) assigned to the carbonyl oxygen and a highfield resonance at δ = 159 ppm assigned to the ether oxygen in the ratio 50 : 50. In contrast, oxathianone, 5, formed from the thermal decomposition (76.0°C) of ^{17}O-enriched 1, displayed resonances at δ = 365 ppm and 159 ppm but in the ratio 66 : 34, respectively. The ratio of ^{17}O carbonyl to ^{17}O ether, however, was a function of decomposition temperature with the ratio being 69 : 31 at 62°C and 74 : 26 at 40°C, respectively. If the reaction had proceeded exclusively through radical **2**, one would have predicted **5** to be labeled only at the carbonyl function; however, based on the integration ratios at the various temperatures, the authors suggest that a competitive channel is also available for scrambling of the ¹⁷O label but the alternative pathway has a higher activation energy of ≈ 2.3 kcal/mol. The zwitterionic radical **3** is probably an intermediate or transient species in the scrambling of the ¹⁷O label. The authors also thermally decomposed (52.0°C) thioperoxybenzoate [C=¹⁷O]**6** to give **7** as one of the major products.

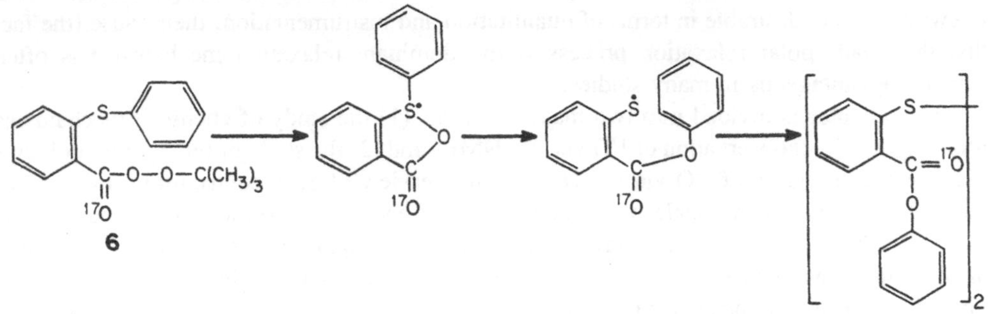

SCHEME 2. Thermal decomposition of ¹⁷O-labeled *t*-butyl-*O*-methylthioperoxybenzoates.

The ¹⁷O-resonance in **7** was too broad for analysis probably due to line broadening caused by the quadrupolar relaxation resulting from the increased size of the molecule. Based on this assumption the authors cleaved the sulfhydral bond in **7** to yield two equal sulfide halves which displayed ¹⁷O-resonances at δ = 363 and 191 ppm in the ratio of 71:29. This again suggests the intermediacy of a sulfuranyl radical structure such as **8** for the *O*-thiobenzoyloxy radical as shown below.

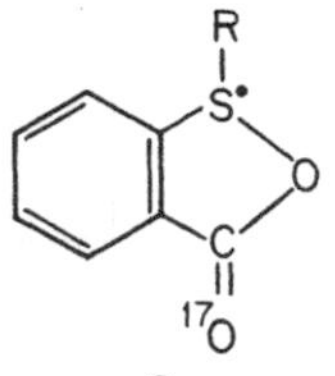

FIGURE 1.   The structure of the sulfuranyl radical.

B. OXIDATION REACTIONS

Yermakov et al.[10] have investigated the mechanism of the oxidation of ethylene (**9**) with lithium nitrate in acetic acid in the presence of Pd(OAc)$_2$ to ethylene glycol monoacetate **10** (EGMA) by use of ¹⁷O-NMR. Although little was known about this oxidation (see Equation 3), it was thought that the mechanism was different from the known mechanism of oxidation by Pd(II) salts to carbonyl compounds.

$$3\ C_2H_4 + 2\ LiNO_3 + 5\ HOAc \xrightarrow{\ Pd(OAc)_2\ } 3\ CH_2(OH)CH_2OAc$$
$$\mathbf{(9)} \hspace{8cm} \mathbf{(10)}$$

$$+2\ LiOAc + 2\ NO + H_2O \hspace{4cm} (3)$$

The authors carried out the oxidation of ethylene (**9**) at 50°C, under atmospheric pressure of ethylene, under three different labeling conditions: (**1**) natural abundance ^{17}O-reagents, (**2**) $LiN^{17}O_3$ at $\approx$ 20 times natural abundance levels, and (**3**) $CH_3C^{17}O_2H$ enriched fivefold over natural abundance. The ^{17}O-NMR spectrum of each product was obtained at 40.7 MHz. The natural abundance ^{17}O-NMR spectra of EGMA (**10**) displayed three nonequivalent resonances at δ = 364 ($C=^{17}O$), 162 ($CH_3{}^{17}O$) and -4 ($CH_3{}^{17}OH$) ppm with respect to external ^2H_2O with intensities in the ratio 1:1:1 respectively. The ^{17}O NMR spectrum obtained from ^{17}O-enriched $LiNO_3$ showed the oxygen atom of the carbonyl group (δ = 364 ppm) is derived exclusively from the NO^{-3} while the ^{17}O-NMR spectrum obtained from the ^{17}O-enriched CH_3CO_2H demonstrated that the other oxygen atoms (ether and alcohol functions) are incorporated from the acetate moiety. The authors hence described the formation of (**10**) by the following equation:

$$3\ C_2H_4 + 2\ LiN^{17}O_3 + 5\ CH_3CO_2H \rightarrow 3\ CH_2(OH)CH_2O(CH_3)C = {}^{17}O$$
$$\text{(9)} \qquad\qquad\qquad\qquad\qquad\qquad\qquad\qquad\qquad \text{(10)}$$

$$+ 2\ LiOCOCH_3 + H_2O + 2\ N^{17}O$$

$$\text{(4)}$$

It was further suggested by the authors that the ethylene interacts with the nitrite complex of palladium (actually a nitroso complex of palladium II) formed during the initial stages of the reaction and not directly with the nitrite as depicted in the scheme below:

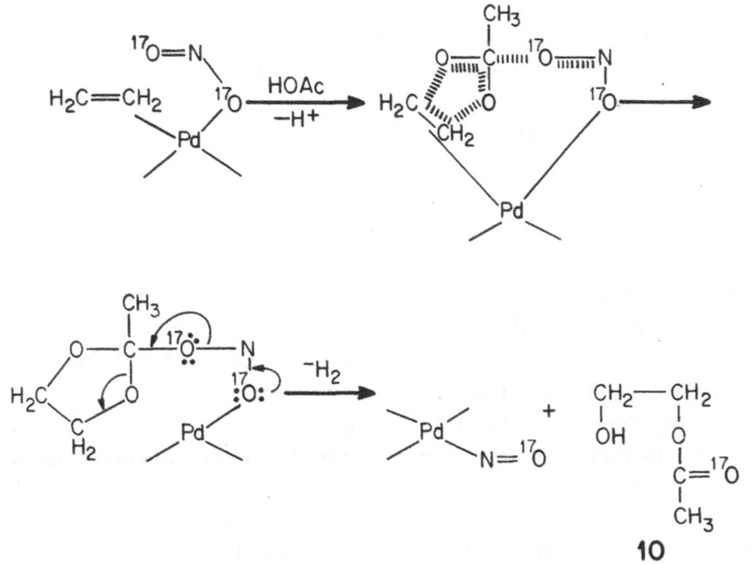

SCHEME 3. The proposed mechanism for the formation of EGMA (**10**).

The results suggest that the formation of EGMA is accompanied by oxygen atom transfer from the oxidant to the carbonyl of the product.

The magnetic nonequivalence of geminal groups close to a center of molecular asymmetry has been observed in ^1H, ^{13}C, and ^{19}F NMR spectra and utilized in a number of stereochemical mechanistic studies.[11] Since the sulfur atom of a sulfone exists in a tetrahedral geometry, the two oxygen atoms are enantiotopic and thus isochronous (cannot be distinguished via NMR) in a molecule containing no other centers of asymmetry; however, when they are diastereotopic (in molecules containing another center of chirality) they become anisochronous and thus should be distinguishable via NMR. Iwamura et al.[12] have synthesized the

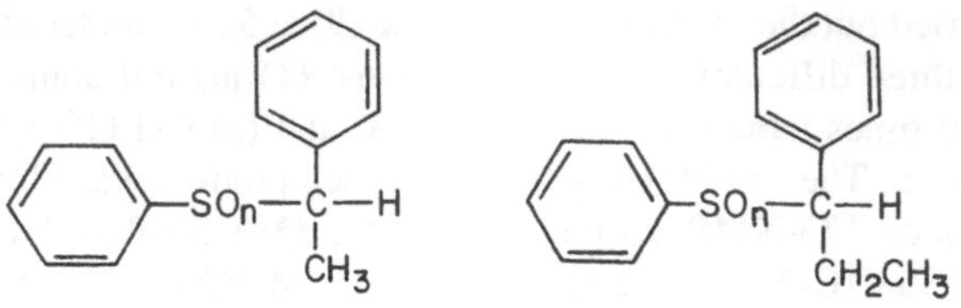

11 n = 1
12 n = 2

13 n = 1
14 n = 2

FIGURE 2.   The structures phenyl 1-phenylmethyl and phenyl 1-phenylethyl sulfoxides and sulfones.

diasteromeric mixtures of both phenyl 1-phenylethyl[^{17}O]sulfoxide (**11**) and phenyl 1-phenylpropyl[^{17}O]sulfoxide (**12**) (from H_2^{17}O, 10 atom %), separated them into their respective enantiomer pairs [(*RR/SS*)- and (*RS/RS*)-1-phenylethyl [^{17}O] sulfoxide and phenyl 1-phenylpropyl [^{17}O] sulfoxide, respectively], measured their ^{17}O NMR spectra, and further oxidized them both to their corresponding sulfones (*RR/SS*)- and (*RS/RS*)-phenyl 1-phenylethyl [^{16}O,^{17}O] sulfone (**13**) and (*RR/SS*)- and (*RS/RS*)-phenyl 1-phenylpropyl [^{16}O,^{17}O]sulfone (**14**). The diastereomeric oxygen atoms of the sulfones displayed distinct chemical shift nonequivalence which remained practically unchanged from the corresponding diastereomeric sulfoxides. From this observation the authors believe that there is little conformation population difference between the sulfoxides and sulfones. The authors note that fairly large upfield shifts of the ^{17}O resonances of the sulfones occur in protic solvents (probably due to hydrogen bonding or some other solvent interaction), but that the Δ ppm values are solvent ($CDCl_3$, C_6H_6, $CHCl_3$, CH_3OH, CF_3CO_2H) -independent indicating that conformation population to be independent of solvent. Attempts at evaluating intrinsic diastereomerism at lower temperatures failed due to line-broadening (efficient quadrupolar relaxation) of the ^{17}O signals; however, it should be noted that these experiments do demonstrate that the oxidation of sulfoxides (in this case by *m*-chloroperbenzoic acid) to sulfones proceeds stereospecifically, a phenomenon later substantiated by Lowe et al.[13] in the ruthenium (IV) oxide oxidation of cyclic sulfite diesters to cyclic sulfate diesters.

By use of the effect of lanthanide shift reagents on the ^{17}O NMR signals of the diastereomeric cyclic sulfite diesters (**16**) shown in Scheme 4, Lowe et al.[13] were able to demonstrate not only that the oxidation to the corresponding sulfones occurred stereospecifically, but with retention of configuration at sulfur. Their experimental evidence for these conclusions was as follows: the cyclization of *meso*-1,2-diphenyl-1,2-dihydroxylethane (**15**) with [^{16}O] thionyl chloride gave a 88 : 12 mixture of 2-oxo-4,5-diphenyl-1,3,2-dioxathiolanes (**16a** and **16b**, respectively) with the *trans*-cyclic analog (**16a**) predominating, and further oxidation to the cyclic sulfate diesters with ruthenium (IV) [^{17}O] oxide gave a mixture of diastereomers. The ^{17}O NMR spectra of this diastereomeric mixture displayed two signals, δ = 153.1 ppm (88%) and 167.5 ppm (12%). These results confirm that the oxidation of the sulfur atom has occurred stereospecifically. Since the authors[13] have shown (unpublished work) that the shift reagent, 1,1,1,2,2,3,3-heptafluoro-7,7-dimethyloctane-4,6-dionatoeuropium(III), binds tighter to the least hindered oxygen, i.e., the oxygen trans to the two phenyl groups (major isomer **16a**), they were able, by adding this shift reagent to the oxidation mixture, to show that the major ^{17}O resonance, 88% by integration, was shifted by only 0.5 ppm while the minor *cis*-isomer (**17b**) -enriched ^{17}O signal (12% by integration) was shifted 4.0 ppm. This demonstrated that the oxidation had occurred with retention of configuration at the sulfur atom.

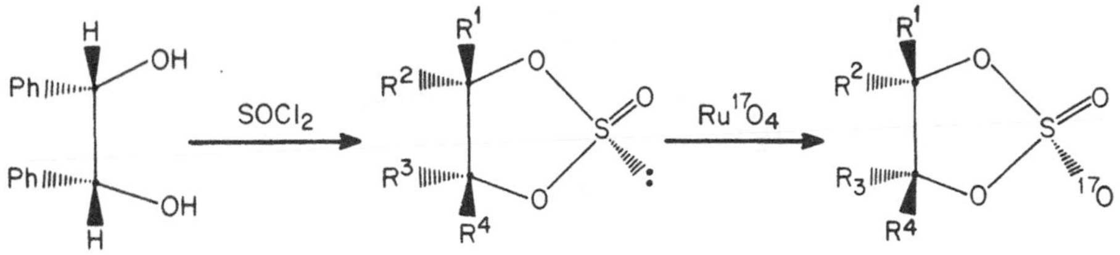

SCHEME 4. The synthesis of ^{17}O-labeled cyclic sulfate diesters.

C. REARRANGEMENT REACTIONS

King et al.[14] in their study of oxygen migration from carbon to sulfur have taken advantage of the fact that most ^{18}O-labeled compounds from commercial sources contain a considerable enrichment of ^{17}O, which gives one the possibility of not only gaining information from both the α and β ^{18}O isotope effects on ^{13}C-NMR spectra but also directly by detection of the ^{17}O isomers via ^{17}O-NMR. In order to detect the presence and/or absence of a carbon → sulfur oxygen migration, King et al.[14] used both ^{18}O-induced ^{13}C shifts and direct observation of the ^{17}O during the chlorination of mercapto alcohols.

$$HOCH_2CH_2SH + Cl_2/H_2O \longrightarrow HOCH_2CH_2SO_2Cl + ClCH_2CH_2SO_2Cl \qquad (5)$$
$$\textbf{(18)} \qquad\qquad\qquad \textbf{(19)} \qquad\qquad\qquad \textbf{(20)}$$

When the above reaction was carried out in $^2H_2{}^{18}O[^{18}O$-water not normalized, i.e., 98 atom % ^{18}O, 95 atom % ^2H and 0.5 atom % ^{17}O (≈ 12 × natural abundance)], the 2-hydroxyethanesulfonyl chloride **(19)** was formed in > 95% yield with only a trace of the acid chloride **(20)** and the reaction proceeded without intramolecular oxygen transfer, i.e., both the oxygens of the SO_2 were labeled equally from the solvent. However, when the aqueous chlorination of 3-mercapto-1-propanol was carried out under identical reaction conditions the heavy oxygen isotopes were found in the positions indicated in Scheme 5.

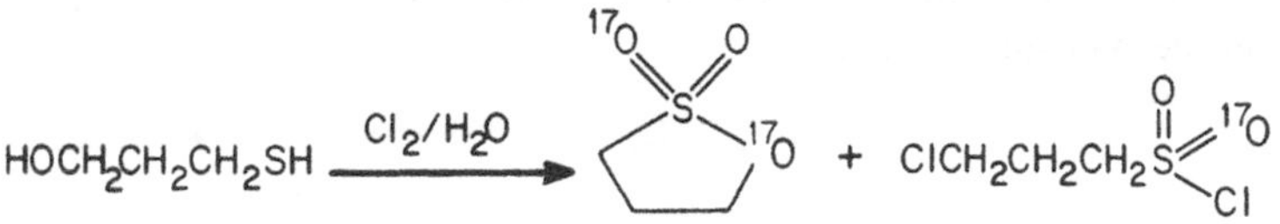

SCHEME 5. The aqueous chlorination of 2-mercapto-1-propanol.

These results were based on both ^{18}O isotope shifts and ^{17}O-NMR. The natural abundance ^{17}O spectra of **(22)** had signals at δ = 175 and 146 ppm assignable from their 2:1 ratio to the sulfonyl oxygen and the endocyclic oxygen, respectively, while the propyl derivative **(23)** had only one single peak at δ = 237 ppm. The ^{17}O-NMR spectra of the reaction mixture (ring: open chain formed in the 2:1) had peaks at δ = 146, 175, and 237 ppm in the ratio 2 : 2 : 1. Based on the results shown in Scheme 5 together with the ^{17}O-NMR results from the aqueous chlorination of sultine **(24)**, (Scheme 6), the researchers interpreted the aqueous chlorination of 3-mercapto 1-propanol to proceed via a carbon → sulfur oxygen migration.

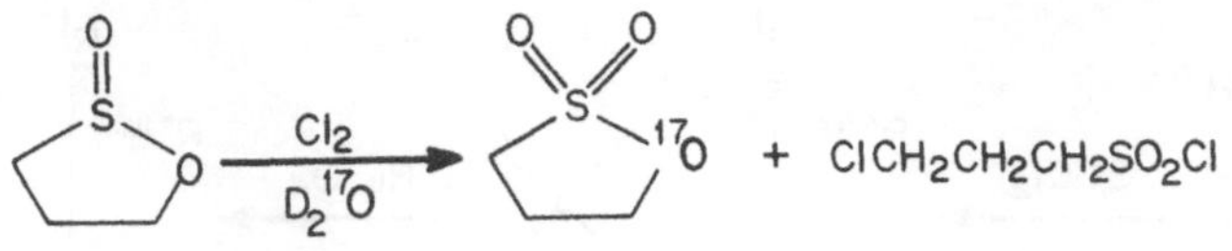

SCHEME 6. The aqueous chlorination of the sultine (24).

The difference in the apparent mechanism of chlorination between the propyl and ethyl derivatives was explained by the fact that, in the case of the propyl derivative, the requisite sultine is a stable five-membered ring and that for the corresponding mechanism for the same type of migration in the aqueous chlorination of the ethyl derivative, one would require the formation of a strained four-membered ring. The authors demonstrate that it is possible to obtain excellent and complementary data from both ^{18}O isotope shifts of ^{13}C-NMR and ^{17}O-NMR by using commercially available ^{18}O-water without the compound destruction normally associated with the mass spectral analysis of ^{18}O-labeled compounds.

In the first structural application of ^{17}O-NMR to carbohydrate chemistry, Lukacs et al.[15] have helped elucidate a controversial reaction mechanism, namely the mechanism of the conversion of 1-*O*-acetyl-2,3-di-*O*-benzyl-4,6-bis-*O*-(methylsulfonyl)-α-D-galactopyranose (25) to 1,4-anhydro-6-azido-2,3-di-*O*-benzyl-6-deoxy-β-D-galactopyranose (27) by treatment with sodium azide. The two possible mechanisms are shown in Scheme 7. As can be seen, mechanism (a) involves a ring contraction in which carbon atom 4 becomes bonded to the ring oxygen atom *O*-5 and pathway (b) in which carbon atom 4 becomes bonded to the anomeric oxygen atom *O*-1. This problem, therefore, could be solved by either or both ^{18}O isotope shifts on the ^{13}C-NMR and directly via ^{17}O-NMR. Since the ^{17}O chemical shifts of the *O*-5 ether oxygen of a number of hexopyranoses or hexopyranosides in the chair conformation and the ^{17}O chemical shift of the bridging ether oxygen in 7-oxanorbornane (δ = 85.9 ppm broad singlet) was known, the authors synthesized 1-*O*-acetyl-2,3-di-*O*-benzyl-4,6-bis-*O*-(methylsulfonyl)-α-D-galactopyranose (25) labeled with ^{17}O in the anomeric oxygen atom and subjected it to azide treatment. The ^{17}O-NMR of the resulting 1,4-anhydro sugar (27) gave one broad singlet at δ = 85.9 ppm. Since the ^{17}O-NMR of the hexopyranoses or hexopyranosides were available only for the chair conformation, the authors verified their result using the ^{18}O isotope shifts, from ^{18}O-labeled starting material, on the ^{13}C resonances of carbon atom C-1 and C-4. Both techniques confirm that pathway (b), the C-1 oxy anion mechanism, is correct.

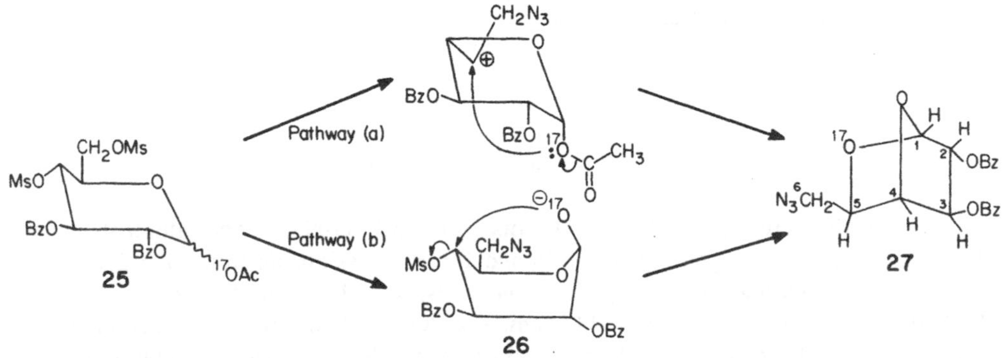

SCHEME 7. Possible mechanisms for the conversion of 25 to 27.

D. SOLVOLYSIS REACTIONS

Chang and le Noble[16] have prepared [^{17}O]-exo-2-norbornyl brosylate **(28)** and sulfonyl-^{17}O-exo-2-norbornyl brosylate **(29)** in order to directly detect and measure internal ion-pair return during solvolysis, *in situ* and without workup, and determine both the solvolysis and oxygen scrambling rate constants.

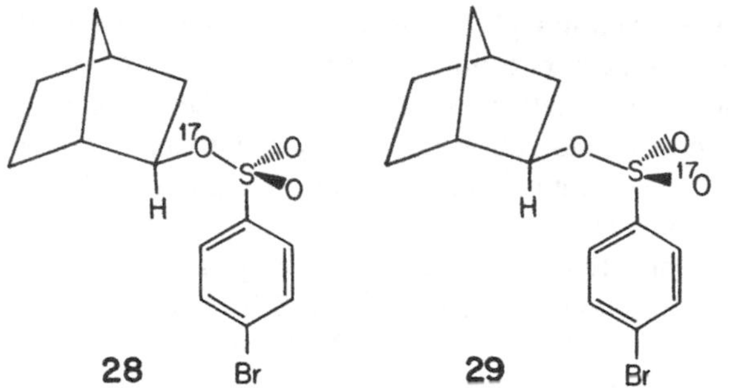

FIGURE 3. The structure of ^{17}O-labeled *exo*-2-norbornyl brosylates.

The ethanolysis of **(29)** at 25°C was directly observed by ^{17}O-NMR at various time points (good quality spectra were obtained at each time point in only a few minutes due to the enrichment of the brosylate). A decrease in the intensity of the sulfonyl ^{17}O resonance at δ = 160 ppm was found to be simultaneous with the appearance and growth of a ^{17}O-resonance at δ = 172. The resonance at δ = 172 ppm was due to the absorbance of both the ether oxygen and the sulfonic acid oxygens. The authors nicely resolved these signals by the addition of the shift reagent praseodymium (III) nitrate which shifted only the anionic oxygen atom signals to δ = 210 ppm and thus allowed both the solvolysis and scrambling of the ^{17}O oxygen to be observed separately. The ethanolysis of **(28)** was easier for the authors to follow and quantify by ^{17}O NMR since the conversion of the ether oxygen to the sulfone ^{17}O oxygen resulted in the change from a broad signal (ether oxygen) to a sharper signal (sulfone oxygen) during the scrambling portion of the reaction (the brosylates were completely scrambled after two half-lives).

The authors of this paper, in addition to introducing the use of shift reagents in ^{17}O-NMR, also suggest the recycling of the ^{17}O-labeled sulfonic acid liberated for other studies with the caveat that only one-third of the label will be available at the bridging position (statistically).

Le Noble and Chang[17] have also applied the same techniques to study the ethanolysis of the endo isomers of 2-norbornyl mesylate in order to ascertain a true ratio of solvolysis of the two epimeric 2-norbornyl derivatives. The authors were forced to change the leaving group from brosylate to mesylate since the ^{17}O-resonances of both the ether and sulfone oxygen in the brosylate were inseparable (both within 0.1 ppm of 161.5 ppm). The natural

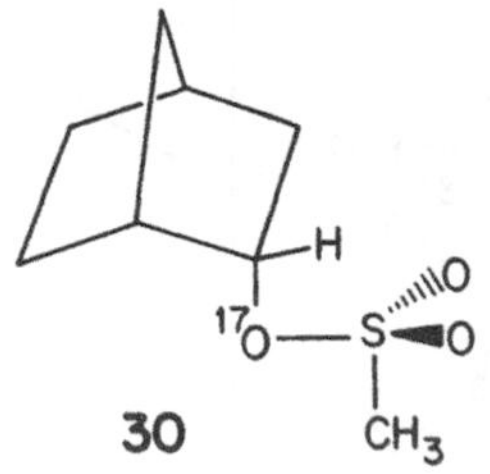

FIGURE 4. The structure of ^{17}O-labeled *endo*-2-norbornyl mesylate.

abundance-¹⁷O-NMR of the mesylate (30) displayed signals at $\delta = 158.3$ (ether oxygen) and 173.5 (sulfone oxygen) ppm which were assigned initially via linewidth and later by synthesis of the ¹⁷O-labeled ether. The oxygens of the free methane sulfonic acid which appeared at $\delta = 176.2$ ppm were again shifted out of the main observation region during ethanolysis by addition of the shift reagent praseodymium nitrate. During the course of the solvolysis only the ¹⁷O peak was observed due to the labeled ether. There was no sulfone oxygen signal observed even though the signal is ≈ 8 times sharper than the signal due to the ether absorption. The authors estimate that there was less than 0.5% scrambling in the endo case vs. complete (two half-lives) scrambling in the exo case previously discussed. The authors conclude that *endo*-sulfonates solvolyze without significant solvent assistance and the endo/exo solvolysis ratio is in the order 10^3 (assumes that internal return is fast, based on the NMR time scale).

In a different type of solvolysis experiment, Creary et al.[18] have attempted to determine the mechanism of the reaction shown in Scheme 8 in which the sulfur atom and oxygen atom, in the major product of the reaction (32), have formally interchanged positions as compared to starting material.

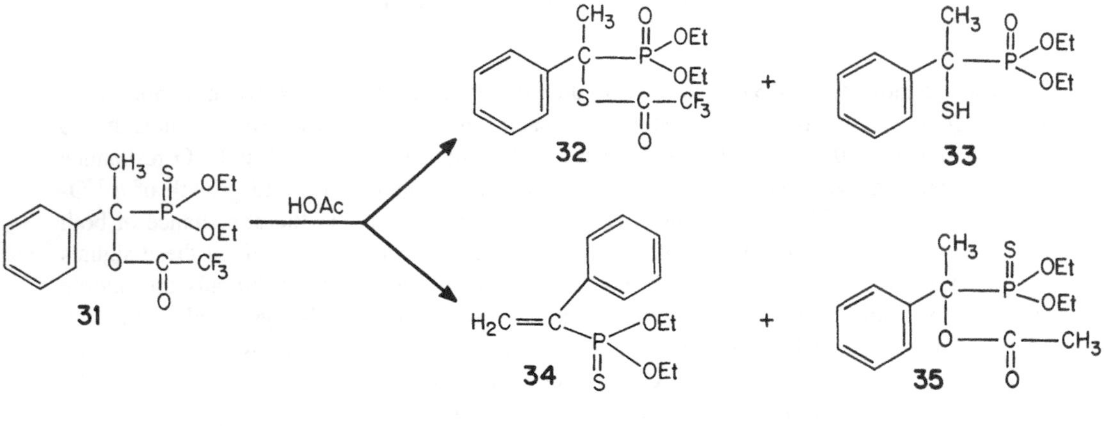

SCHEME 8. The products from the reaction of 31 with acetic acid.

The key cyclic intermediate in this rearrangement was thought to be (38) which could have been derived by either of the two different mechanisms shown in the scheme below.

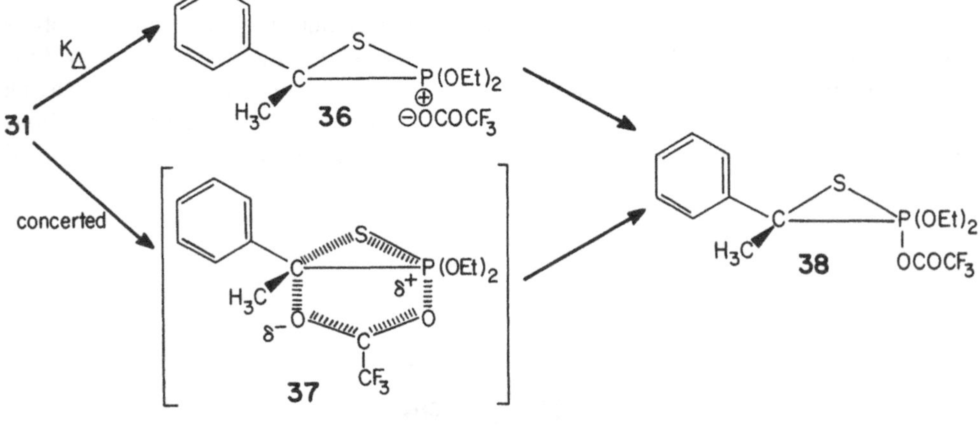

SCHEME 9. The possible processes involved in the conversion of 31 into the key intermediate 38.

There are at least two ways in which this cyclic intermediate may be converted into the observed product but neither involve breaking the newly formed P-O bond; therefore, rearrangement of an appropriately oxygen-labeled starting material (31) should allow one to distinguish between the tight ion pair and concerted pathways shown in Scheme 9. The paper reports that treatment of either [ester[17]O- or ^{18}O-labeled] α-thiophosphoryl trifluoroacetate with acetic acid leads to a product in which substantial quantities of labeled oxygen appear in the P=O group as well as an unequal amount in the C=O of the trifluoroacetyl group. These results allowed the authors to rule out the concerted pathway, since one would have expected no label in the P=O group if the reaction had gone through the concerted pathway. The unequal incorporation of label (20% from the ester oxygen of trifluoroacetate was incorporated into the carbonyl C=O group vs. 80% incorporation of label into the P=O group) was interpreted to mean that the two oxygens of the trifluoroacetate are functionally nonequivalent in the ion-pair intermediate. These authors also verified their results using [18]O isotope shifts on the ^{31}P-NMR.

The hydrolysis of 2-acetoxyethyl(pyridine)cobaloxime and the hydration of formylmethyl(pyridine)cobaloxime was studied by Golding et al.[19] using [17]O-enriched cobaloximes and ^{17}O NMR. The hydrolysis of 2-[[17]O]hydroxyethyl(pyridine)cobaloxime in unlabeled water showed a gradual decrease of the [17]O resonance at $\delta = 178$ ppm, due to the hydroxyl function, and the slow formation of a peak at $\delta = 277$ ppm, due to [17]O-labeled acetate. This mechanism is consistent with the loss of acetate from acetoxyethyl(pyridine)cobaloxime to yield a π-ethylene complex of cobaloxime(III) which either adds water to give 2-hydroxyethyl(pyridine)cobaloxime or loses ethylene to produce hydroxy(pyridine)cobaloxime, an example of the rather uncommon $B_{AL}1$ type hydrolysis.

In an effort to gain some insight into the adenosylcobalamine-dependent diol dehydratase-catalyzed conversion of ethane-1,2-diol to ethanal, the same authors have synthesized [[17]O]formylmethyl(pyridine)cobaloxime($\delta = 544$ ppm) by hydrolysis of 2,2,-diethoxyethyl (pyridine)cobaloxime with one equivalent of $H_2^{17}O$. The authors then compared the rates of loss of [17]O from [[17]O]pentanal ($\delta = 583$ ppm) in an aqueous solution via hydration-dehydration to the loss of [17]O from this [[17]O]formylmethyl(pyridine)cobaloxime, presumably through [hydroxy-[17]O]-2,2-dihydroxyethyl(pyridine)cobaloxime (i.e., hydration-dehydration) by ^{17}O NMR. The rate of loss of [17]O from [[17]O]pentanal was considerably faster.

III. BIOORGANIC PROBLEMS

A. BIOSYNTHETIC STUDIES

The use of ^{17}O NMR as a mechanistic probe in the area of biosynthesis is surprisingly limited, when one considers that many of the generally practiced labeling and multiple-labeling methodologies, as well as some of the more sophisticated 1D and 2D-NMR techniques, have evolved from the laboratories of researchers involved with the investigation of the biosynthetic pathways and the mechanism of the individual enzyme reactions catalyzing these transformations.

Two papers from the same group, Baldwin, Abraham, et al.[20,21] have described the use of ^{17}O-NMR to investigate the transformation of the tripeptide,δ-(L-α-aminoadipyl)-L-cysteinyl-D-valine (LLD-AVC)(39), into isopenicillin (40) by a cell-free extract of *Cephalosporium acremonium*. The biosynthesis of isopenicillin (shown in Scheme 10) involves the coupling of L-cysteine, L-valine, and L-α-aminoadipic acid to give the tripeptide, LLD-AVC, which then cyclizes to the product antibiotic 40.

Controversy arises in how the cyclization of 39 → 40 occurs. Several groups have suggested mechanisms, but most of the mechanisms involve a hydration/dehydration step of the carbonyl portion of the amide formed between aminoadipic acid, acid carboxyl donating, and cysteine, amino donating.

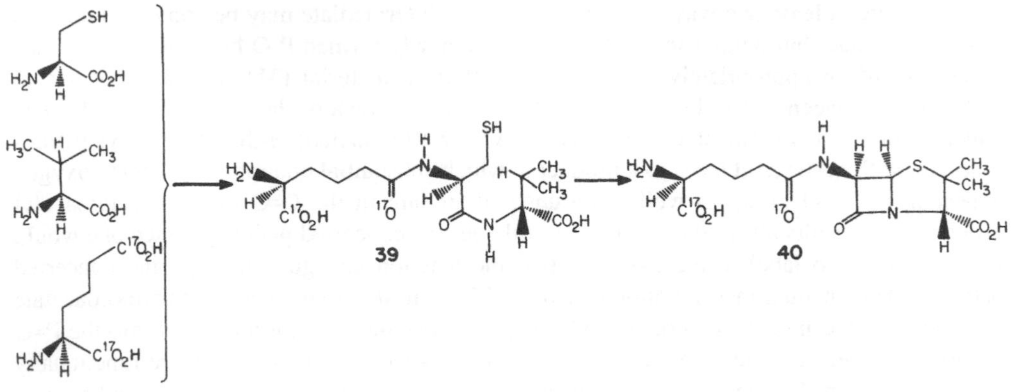

SCHEME 10. The biosynthesis of **40** from the amino acids, L- cysteine, L-valine and L-α-aminoadipic acid.

These authors of the present study have attempted to at least eliminate any mechanism or pathway that involves addition or elimination of water as a step in the transformation. They have synthesized $^{17}O/^{18}O$-labeled δ-(L-α-aminoadipyl)-L-cysteinyl-D-valine, enriched in the α- and δ-positions of the L-α-aminoadipic acid, and incubated it in a cell-free extract of *C. acremonium* in H_2O. In a complementary experiment they incubated the LLD-AVC with a cell-free extract of *C. acremonium* in labeled H_2O ($^{17}O/^{18}O$ enriched; 15 mol% ^{17}O). The ^{17}O-NMR spectrum of the labeled LLD-AVC showed only one peak δ = 280 ppm, $W_{1/2}$ = 760 Hz; pH = 1 which indicates a total enrichment of about 1.3 of ^{17}O per mole of aminoadipic acid, although both the α and δ carboxyl groups are labeled. The oxygen-labeled aminoadipic acid (one peak δ = 280 ppm) when dehydrated into the corresponding lactam, however, showed two peaks δ = 377 ppm (lactam amide carboxyl oxygen) and 253 ppm (acid carboxyl oxygen) in the ratio 40 : 60. Therefore, **39** was enriched in both the α- and δ-carboxyl groups in the ratio 40 : 60. The ^{17}O-NMR spectra of isopenicillin N isolated from the oxygen-17-labeled LLD-AVC showed no loss of ^{17}O from either position (this was also verified by C.I. mass spectroscopy), and the ^{17}O-NMR spectrum of the isopenicillin N isolated from the oxygen-17-labeled water experiment showed no detectable incorporation of ^{17}O. These results eliminate any mechanism that involves a dehydration/ hydration step and any intermolecular, covalently linked intermediate such as an enzyme-bound thioester, ester, etc., involving any of the oxygen sites of LLD-AVC. (Any linkage between the enzyme and peptide that could be formed and broken without oxygen exchange is, of course, still valid, i.e., mixed anhydrides.) Although these authors did not suggest a mechanism for the transformation under study, they were able to eliminate several previously viable mechanisms.

The biosynthesis of citrinin (**41**) in *Aspergillus terreus* was investigated using a combination of ^{13}C-NMR through ^2H-^{13}C coupling and the isotope shift induced by ^2H and ^{18}O with $[2-^{13}C,2-^2H_2],[1-^{13}C,^{18}O_2]$-acetate as well as with ^{17}O-NMR and the multiple labeled acetate, $[1-^{13}C,^{17}O]$-acetate by Seto et al.[22] The desired information (albeit indirect) concerning the incorporation of acetate was obtained from the ^{18}O-induced shift of the ^{13}C resonance. The same information is not available from ^{13}C-^{17}O coupling since the ^{13}C-^{17}O coupling constant is usually too small to be detected, as had been predicted from both theoretical and other experimental studies.[23] The ^{17}O-NMR spectrum of the biosynthetic material derived from the $[1-^{13}C,^{17}O]$-acetate feeding, however, displayed three signals at δ = 148, 179, and 279 ppm, which were tentatively assigned to the oxygen atoms attached to the C-2, C-8, and C-6 atoms, respectively. The ^{17}O-NMR spectra thus allowed one to trace unequivocally the metabolic fate of the oxygen atoms of the labeled acetate. It is also of some interest to note that the unusual chemical shifts of the two oxygen atoms attached to C-8 (low for a C-OH [179]) and C-6 (high for a C=O oxygen [279]) provided further information to the researchers concerning the actual solution structure of citrinin. Since the ^{17}O-chemical shift is very sensitive to a change of electron density at the oxygen atom, the workers were able to suggest that the structure may be visualized as a tautomeric equilibrium between the two structures **41** and **42** shown below in Figure 5.

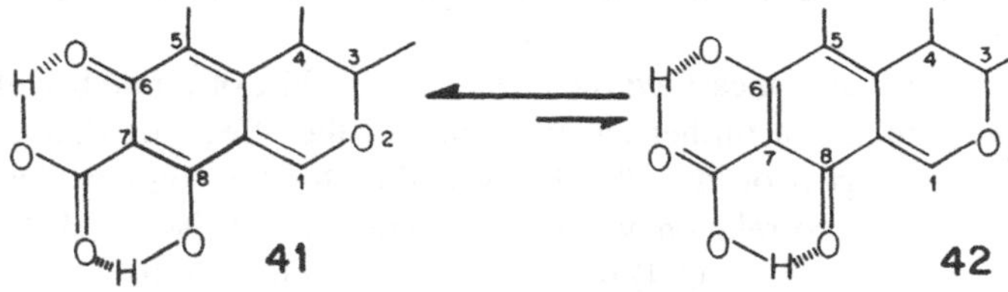

FIGURE 5. The possible tautorimeric forms of citrinin.

The results from these two studies clearly demonstrate the utility of ^{17}O-labeling and ^{17}O-NMR as tools for gaining information on biosynthetic problems not otherwise available.

B. BIOCHEMICAL STUDIES
1. Heterocycles

The combined techniques of ^{17}O and ^{14}N-NMR have been used to study the hydrogen bonding interactions between the ^{17}O-labeled nucleic acid bases, [2,4-$^{17}O_2$]thymine (**43**),[2,4-$^{17}O_2$]uracil (**44**), and [2-^{17}O]cytosine (**45**), and between the same ^{17}O-labeled bases and solvent. Hydrogen bonding between the nucleic acids plays a major role in determining the three-dimensional structure of the helical nucleic acids. In particular, the oxygen atom of the heterocyclic portion of these nucleic acids plays a major role in the hydrogen bonding process. Fiat and co-workers[24] have extended their studies from the area of investigating hydrogen bonding in amides, peptides, urea, etc. (to be discussed later in the section on amino acids) to the area of hydrogen bonding in nucleic acids. This extention is quite logical given that the -NH-CO-moiety is common to both groups of compounds, with the major difference being the degree of conjugation between the carbonyl group and adjacent functionality. Fiat et al.[24] were able to demonstrate that the ^{17}O chemical shift is quite sensitive as a structural probe in determining this degree of conjugation between the carbonyl group and adjacent functionality as well as being an indicator of site and degree of hydrogen bonding.

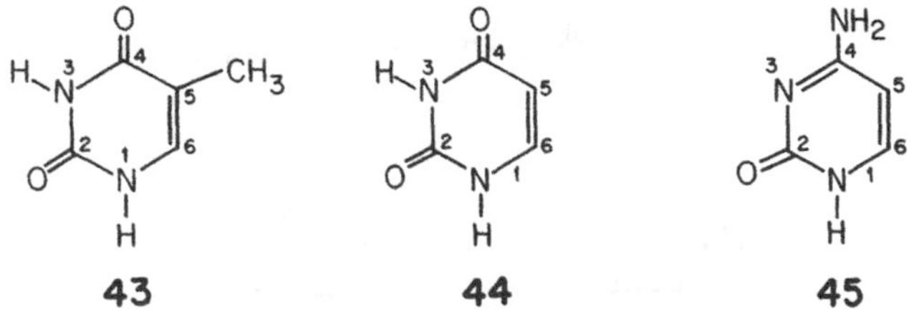

FIGURE 6. The structures of thymine (**43**), uracil (**44**), and cytosine (**45**).

Boykin et al.[25] have determined the ^{17}O chemical shifts of several quinones, hydroxyquinones, and related cyclic ketones in preparation for possible study of intercalation of these or more complex cyclic ketones into DNA. The authors believe that the large chemical shift range of ^{17}O and the sensitivity of ^{17}O to hydrogen bonding makes ^{17}O-NMR an attractive tool in the study of the interaction of various small oxygen-containing molecules and DNA which, of course, is made of various oxygen-containing and hydrogen-bond-forming heterocyclic bases. However, broadening of the signal of a small molecule, upon binding to a macromolecule, remains a difficulty to be resolved.[26] This approach will be of particular interest if it can be extended to anti-tumor compounds such as daunorubicin and adriamycin with DNA where intercalation is thought to be one of the principal modes of action. In a recent manuscript, Boykin et al.[26] have reported the use of ^{17}O-NMR to study the effect of intercalator structure, at 70°C, on the binding strength and base pair specificity in calf thymus DNA interactions.

Boykin et al.[27] have also measured the ^{17}O chemical shift of a number of 5-substituted uracils (44) in order to gain a further understanding of the effects of structural change (in this case electronic changes) on the ^{17}O chemical shift for this system. The ^{17}O data for O(2) of uracil gave a good correlation with both Hammett (r = 0.982) and DSP (r = 0.992) treatments, however, data for the O(4) did not correlate well with the Hammett correlation (r = 0.960). Information from these types of studies could lead to a possible understanding of the mechanism of how certain 5-substituted uracils might act as chemotherapeutic agents in various base-pairing type situations.

2. Nucleosides

Danyluk and coworkers,[28] as well as Gerlt et al.,[29] have reported the use of ^{17}O-NMR spectroscopy to obtain detailed information concerning water-nucleoside hydrogen-bonding equilibria. Danyluk first reported preliminary findings suggesting the feasibility of these types of studies in 1980.[30] Using labeled 2',3'-O-isopropylideneuridine (46) (IPU), ^{17}O-enriched in either the O(2) or O(4) position, or the corresponding uridine (47), the authors monitored the ^{17}O chemical shift in various water-acetonitrile mixtures as a function of temperature. From the analysis of these measurements the authors were able to obtain equilibrium constants for the hydration process. There is also a discussion of the various factors contributing to the ^{17}O chemical shift for carbonyl groups with the role of the dominant factor being assigned to the magnetic screening tensor, β_p. This term is dependent on the inverse energy of the low-lying electronic transition ($E^{-1}_{n-\pi}$), the inverse of the mean cube atomic p orbital radius ($\langle r_p^{-3} \rangle$), and other atomic parameters. Considerable oversimplification in previous works states that the ^{17}O chemical shift is determined by the π-bond of the carbon-oxygen double bond.[29] Therefore, increasing the π character of the carbonyl group shifts the ^{17}O resonance downfield and hydrogen bonding to carbonyl groups, which decreases the degree of π-bonding, resulting in an upfield shift. There is significantly more discussion on this topic in the manuscript.

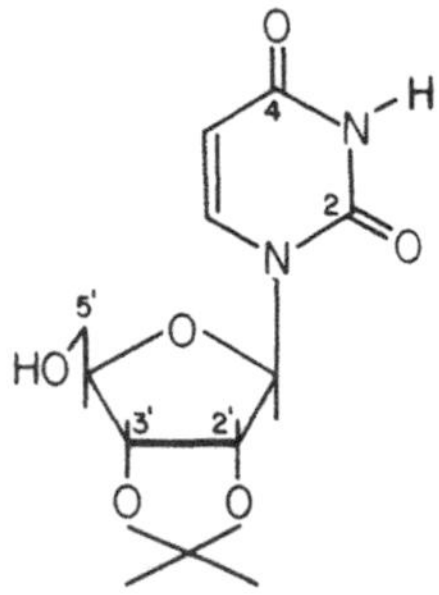

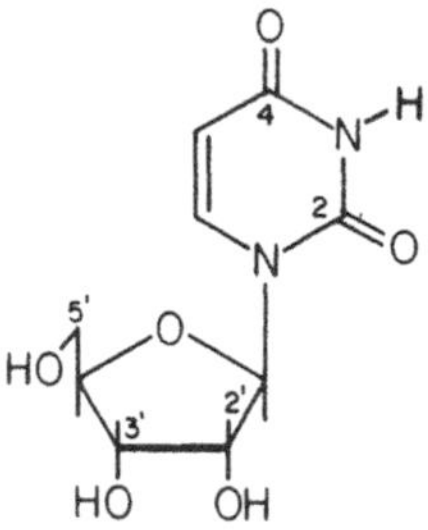

FIGURE 7. The structure of 2',3'-O-isopropyli-deneuridine (46) and uridine (47).

In their earlier paper Danyluk et al.[30] measured both the ^{17}O chemical shifts and linewidths of several ^{17}O-enriched carbonyls [O(2) and O(4)] of three pyrimidine derivatives at various temperatures. They noted marked linewidth decreases at higher temperature and in solvents with lower viscosity. Their results demonstrated significantly more hydrogen-bonding to ^2H_2O and a higher π-bond order for the O(4) in the uridine analogues studied.

3. Nucleotides

An excellent review by Tsai[31] on the use of the ^{17}O-nuclei in the investigation of bioorganic problems — in particular the use of $^{31}P(^{18}O)$, $^{31}P(^{17}O)$, and ^{17}O-NMR methods to study enzyme mechanisms involving phosphorus — has appeared, and the reader is encouraged to peruse the article for additional examples. In the discussion that follows, occasional mention is made of the indirect observation of the ^{17}O atom, but only in those instances where the methodology would provide either further or substantiating information available from the ^{17}O-NMR experiment. These alternate methods are generally available to the researcher since most instruments capable of recording ^{17}O-NMR spectra can also record ^{13}C- and ^{31}P-NMR spectra, and the compounds needed for these spectra are already in hand.

In 1979, Tsai[32] demonstrated that an ^{17}O nucleus directly bonded to a ^{31}P nucleus causes a quantitative decrease in the ^{31}P-NMR signal (it should be noted, as pointed out by Tsai,[31] that this methodology was only demonstrated by the biochemist, not discovered by the biochemist), and reported the successful use of this methodology to elucidate the stereochemical course of acetate activation at the phosphorus center catalyzed by acetyl-coenzyme A synthetase (see Scheme 11).

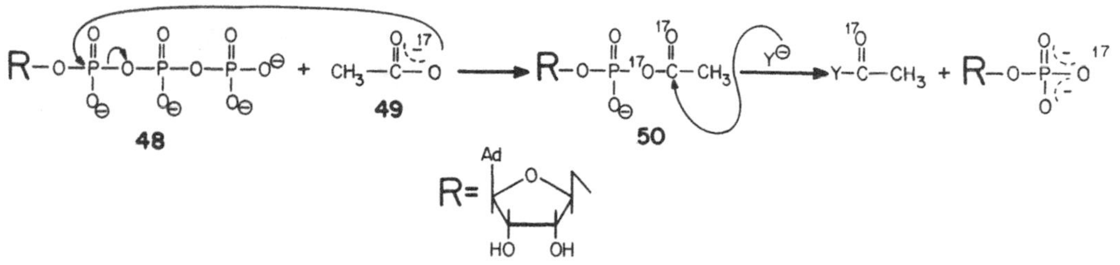

SCHEME 11. The mechanism of the acetate activation by acetyl coenzyme A synthetase.

Although no ^{17}O-NMR data were measured, this paper marks the beginning of the use of ^{17}O in the area of phosphorus biochemistry. The original idea of Tsai[31] in the study was to use ^{17}O-NMR to distinguish between the bridging and nonbridging ^{17}O. However, the synthetic $(S_p)ATP\alpha S$ (51) (Figure 8) labeled with ^{17}O at both the α-nonbridging and α,β-bridging positions gave only one signal, and model studies indicated that the signal of the nonbridging ^{17}O was too broad to be observed. This lack of signal forced Tsai to abandon the method since, even though the chemical shifts of the two ^{17}Os are different, he could not assign the absence of a peak to the fact that he had a bridging ^{17}O. The lack of success in utilizing ^{17}O-NMR as a direct probe, however, lead Tsai and his group to develop this methodogy [$^{31}P(^{17}O)$] and apply it and the direct measurement of the ^{17}O-NMR, as described in several of the following papers, to some interesting and important biological problems.

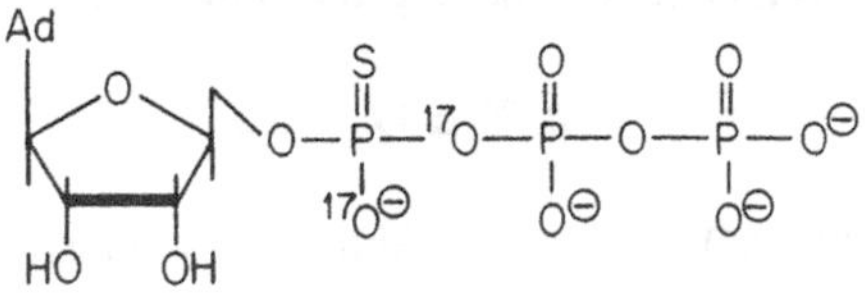

51

FIGURE 8. The structure of [$^{17}O_2$] adenosine 5'-(thiophosphate).

In a study analogous to Tsai's study, Lowe et al.[33] reported observing the same effect of the ^{17}O atom on the chemical shift and linewidth of the ^{31}P resonance in nucleotides comparable to those previously reported, plus several phosphate esters.

Tsai et al.,[34] in an extension of their ^{31}P(^{17}O) NMR methodology reported above, correlated the linewidths of the ^{17}O and ^{31}P-NMR signals with the line-broadening effect of ^{17}O on the ^{31}P signal to determine whether the ^{31}P(^{17}O)NMR method would be a useful tool to study the stereochemical and mechanistic aspects of other phosphoryl transfer reactions. For this study the authors synthesized trimethyl [^{17}O$_4$] phosphate (52), [^{17}O$_4$] phosphate (53), [α-^{17}O$_2$]adenosine 5'-(thiophosphate) (54), [α-^{17}O,α,β-^{17}O]adenosine 5'-(thiotriphosphate) (55), [γ-^{17}O$_3$]adenosine 5'-triphosphate (56), [αβ,βγ-^{17}O$_2$,β-^{17}O$_2$]adenosine 5'-triphosphate (57), magnesium [γ-^{17}O$_3$]adenosine 5'-triphosphate (58), and magnesium [αβ,βγ-^{17}O$_2$,β-^{17}O$_2$ adenosine 5'-triphosphate (59) (shown in Figure 9) and measured both their ^{17}O and

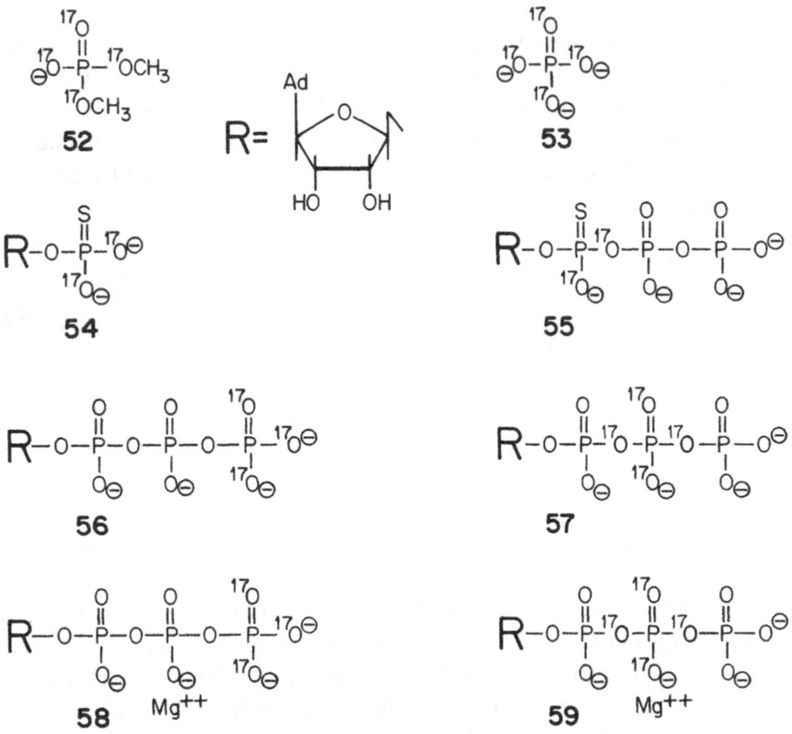

FIGURE 9.   The structure of various ^{17}O-labeled phosphates.

^{31}P NMR spectra. The ^{17}O line widths of the compounds range from small (W$_{1/2} \approx$ 50 Hz), as in compounds 52 and 53 which are low molecular weight and relatively symmetric, intermediate (W$_{1/2} \approx$ 400 to 450 Hz) as in 54, 56 and the nonbridging ^{17}O of 55 and 57, and very large (W$_{1/2} \approx$ 1000 to 1500 Hz) as in 58, 59 and the bridging ^{17}O of 55 and 57. The authors based their arguments on the approximate relationship given below:

$$\Delta \, X \, \Delta \, Q \approx a \, J^2 \qquad (6)$$

where $\Delta X = \,^{31}$P linewidth, $\Delta Q = \,^{17}$O linewidth, α = a constant, and J = ^{31}P-^{17}O coupling constant. It was assumed that when ^{31}P is the dipolar nucleus and ^{17}O is the quadrupolar nucleus, the method should generally hold true when ΔX is larger than the limiting value of 20 Hz. It was also found that Mg^{2+} caused the ^{17}O signal of 56 and 57 to broaden (ΔQ increases), thereby causing a sharpening of the ^{31}P signal (ΔX decreases). This observation led to other important ^{17}O-NMR studies which will be discussed later. The main emphasis of these studies, however, was still not to look at the oxygen atom which is directly involved in the inter- or intramolecular interactions but rather to look at the atoms attached to the oxygen.

In the course of characterizing the diastereomers of the cyclic [[17]O, ^{18}O]dAMP, Gerlt et al.[29] discovered that they could distinguish between the diastereomers (**60** and **61**, shown in Figure 10) on the basis of their ^{17}O NMR signals. Furthermore, they were able to show that in the ^{31}P decoupled ^{17}O-NMR spectra of a racemic sample, the [31]P-[17]O coupling constants and the linewidths of the separate diastereomers are significantly different, as can be seen in Table 1. Thus, ^{17}O NMR spectroscopy is the first spectroscopic technique that has been found to directly detect configurational differences in oxygen chiral phosphate esters.

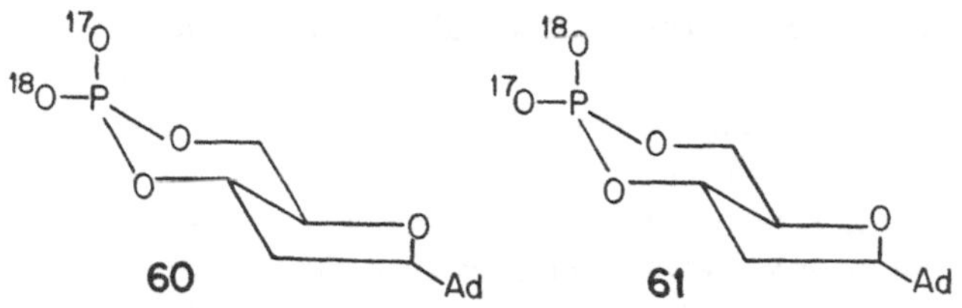

FIGURE 10. The structure of the two diastereomers of cyclic [[17]O, ^{18}O]deoxyadenosine 5'-phosphate.

TABLE 1
^{17}O-NMR Data for the Diastereomers of Cyclic [[17]O,[18]O]dAMP

Configuration	Chemical shift[a]	J_{po}(in Hz)	Linewidth (in Hz)
R_P(axial [17]O)	92.8	130	50
S_P(equatorial [17]O)	91.2	102	82

[a] In ppm from H_2[17]O, positive value indicates lower shielding.

In addition to the main findings reported in this paper, the authors contributed two things to further improve the resolution of ^{17}O-NMR spectra, namely, decoupling the ^{31}P nuclei with a specially designed probe to remove the $\approx$ 110 Hz broadening due to [31]P-[17]O coupling, and heating the sample to remove quadrupolar line broadening (probe temperature 95°C). As a result, the ΔO of [17]O-enriched nucleotides can be reduced to <150 Hz, and the chemical shifts and [31]P-[17]O spin-spin coupling constants can be accurately determined; however, 95°C is probably not a reasonable temperature for most biological studies. This method is a nice complement to the method of Tsai[32] which uses the line-broadening effects of [17]O on ^{31}P resonances to indirectly monitor changes in the [17]O linewidths.

Based on their initial experimental finding discussed above,[29] the Gerlt group[35] was the first to attempt to use high resolution ^{17}O-NMR, instead of the indirect probe ^{31}P-NMR, to investigate questions concerning the surroundings of the phosphoryl oxygen in phosphate esters and nucleotides, since it is, indeed, the oxygen atom and not the phosphorus atom involved in the various intra- and intermolecular interactions. Although Tsai et al.[32] had earlier suggested the possible use of ^{17}O-NMR for the investigation of such problems, they had access only to a low-field instrument (1.9 T) which did not have the capability to decouple phosphorus, and as a result they developed the methodology ([31]P([17]O)) described previously. The Gerlt laboratories were able to obtain excellent ^{17}O-NMR spectra by utilizing their specially designed probe, which was capable of heteronuclear ^{31}P decoupling to remove the large [31]P-[17]O coupling constant, using a magnetic field of 6.3 T and heating the aqueous solution of the isotopically enriched sample. They were able to show that the magnitude of the charge on the phosphoryl oxygen is important in determining the chemical shift and that the chemical shift values of the weakly acidic phosphoryl oxygen are significantly dependent on the pH; therefore, one should be able to determine the extent of charge neutralization of phosphoryl oxygens via the observation of the chemical shift of the ^{17}O resonance.

Based on their earlier studies in which they show that Mg^{2+} causes the ^{17}O signal of $[\gamma\text{-}^{17}O_3]$ATP to broaden, Tsai and Huang[36] have utilized ^{17}O NMR to investigate the controversial question of whether the α-phosphate of ATP is involved in the chelation process. Their findings indicate that Mg^{2+} interacts with both the α- and β-phosphates of ADP and all the α, β, and γ-phosphates of ATP, albeit to different extents. They conclude that the extent of a coordination in MgATP is small as compared to the β and γ coordination. These results were further substantiated from the ^{17}O-NMR of substitution-inert Co^{+3} complexes of ADP and ATP with known structures. It was also noted that the Co^{+3} coordination caused an upfield shift of 180-200 ppm on the ^{17}O-NMR signal. A detailed discussion on the mechanism of the line-broadening effect in ^{17}O-NMR is also presented. The main contribution of this paper to ^{17}O-NMR as it relates to mechanistic studies is that the method allows the direct observation of the metal binding to oxygen atoms in biologically important molecules.

A comparison of the three NMR techniques available to date — ^{31}P(^{17}O), ^{31}P(^{18}O), and direct observation of ^{17}O — which involve oxygen isotopes in the study of biochemical and physical problems involving biochemically interesting phosphates, is presented by Tsai et al.[37] This paper reports the feasibility of using ^{17}O as a label of both oxygen and phosphorus in macromolecular systems such as arginine kinase and dipalmitoylphosphatidylcholine, but as indirect probes via the quadrupolar broadening of the ^{31}P signal.

The chemical shifts of the ^{17}O resonances associated with the bridging and nonbridging oxygens of pyrophosphate were unambiguously assigned by Kenyon et al.[38] via regiospecific isotopic labeling, which then allowed the total assignment of the ^{17}O-resonances of all of the oxygens of ADP and ATP. Additionally, the ^{17}O chemical shift values of regio- or stereospecifically labeled inorganic phosphate, pyrophosphate (with and without bridging oxygen), ADP, ATP, AMPS, ATPγS **(62)**, phosphonates such as AMP-PCP **(63)**, PCP **(64)** and methyl phosphonate **(65)** were measured as a function of pH. They demonstrated that ^{17}O NMR spectroscopy can be used to quantitate the site and degree of charge neutralization in phosphate ester anions and related species. Their data were consistent with the hypothesis that the chemical shift of the ^{17}O resonance is determined by the magnitude of the charge

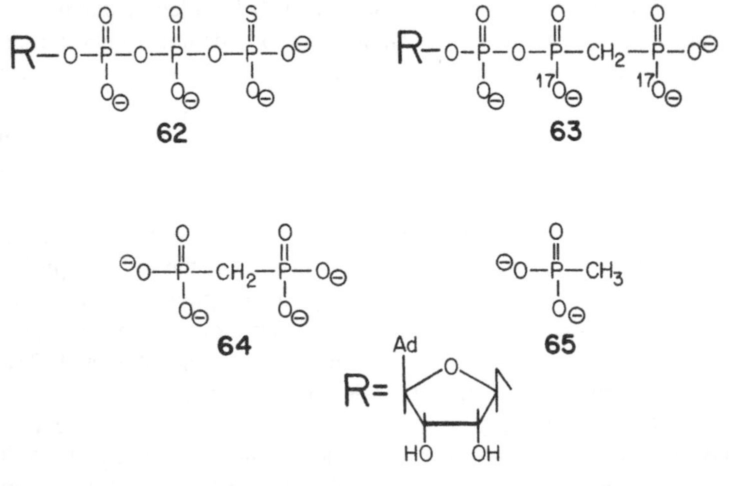

FIGURE 11.   The structure of several phosphonates, thiophosphates and their derivatives.

located on a phosphoryl oxygen. The protonation of a basic phosphoryl oxygen produced an upfield shift of approximately 50 ppm per charge neutralized. Although linewidths are broad, charge neutralization can be detected and quantified. The knowledge of the sites of coordination of nucleotides to protons, metal ions, and cationic sites in the enzyme active sites should allow one insight into the understanding of enzymic reactions (e.g., proton binding to compounds with uncertain tautomeric structure, or divalent metal ions binding to ADP or ATP, such as that explored by Tsai et al.[36] in the work described above).

A report[39] dealing with the optimization of techniques to observe the natural abundance ^{17}O-NMR spectra of inorganic and organic phosphates and polyphosphates was presented by Gerothanassis and Sheppard. In addition to presenting the [17]O chemical shifts of some biologically important phosphates, several of the corresponding J_{PO} coupling constants are also reported.

In a study on the proton binding sites in imidodiphosphate, Kenyon et al.[40] prepared ^{15}N and [17]O-enriched samples of imidodiphosphate (PNP) **(66)**, its tetraethyl ester and 5′-adenylyl imidodiphosphate (AMP-PNP) **(67)** (shown in Figure 12) and measured both their

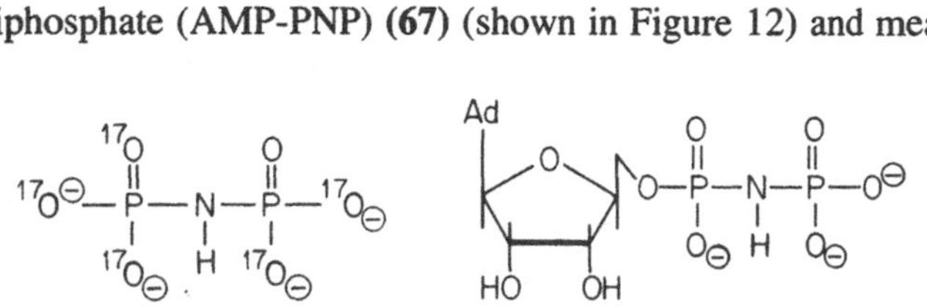

66 **67**

FIGURE 12. The structure of imidodiphosphate (**66**) and adenylyl imidodiphosphate (**67**).

^{15}N and ^{17}O-NMR spectra under various conditions. The ^{15}N-NMR spectra of both PNP and AMP-PNP showed a 70 Hz [1]H-^{15}N coupling constant for the fully ionized samples, which demonstrated an imido-type tautomeric structure in both cases. The ^{15}N-NMR spectra of AMP-PNP in the presence of a stoichiometric amount of Mg^{2+} also reveals the same coupling constants. The ^{17}O-NMR chemical shifts of the resonances of the phosphoryl oxygen were also assigned, and the effect of varying the pH on the chemical shifts was measured. The effect was very similar to the effect observed for ATP and its thiophosphate and phosphonate structural analogues in an earlier study by Kenyon et al.[38] The implication is that protonation of the tetraanion of PNP occurs exclusively on the oxygens and, furthermore, that protonation of the tetraanion of AMP-PNP occurs exclusively at the γ-phosphoryl oxygens. The ^{17}O-NMR of the monoanion of the tetraethyl ester of PNP shows that protonation occurs predominantly on the nitrogen however, the ^{15}N chemical shift change seems to be insensitive to change in ionization. The authors, therefore, concluded that ^{15}N-NMR cannot be used as a reliable method for the determination of the sites of protonation in the case of imidodiphosphates.

In a study to determine the utility of ^{17}O-NMR as a probe of nucleic acid dynamics, Prestegard et al.[41] were able to calculate the transverse relaxation times (T_2) of [17]O-labeled polyadenylic acids from their linewidth data at various temperatures, in the cases where they obtained resonances that appeared to be Lorentzian. They were able to conclude that further studies utilizing ^{17}O-NMR to study the nucleic acid dynamics should prove to be highly successful.

Kenyon et al.[42] have synthesized isotopically enriched pyrophosphate in the bridging oxygen positions to compare the various NMR methods available in order to determine the applicability of each to the positional isotope exchange (PIX) method in the study of enzymes that utilize pyrophosphate.

In order to investigate the stereochemical course of the reaction catalyzed by bovine pancreatic deoxyribonuclease I (DNase I), Mehdi and Gerlt[43] synthesized and determined the configuration of the diasteromers of the three acyclic phosphate diesters that are chiral at the phosphorus atom due to the substitution of the phosphoryl oxygen with ¹⁷O and ¹⁸O. The three diesters shown in Figure 13 (note that only one diestereomer of two of the diesters is drawn) — thymidine 3'-(4-nitrophenyl[¹⁷O,¹⁸O]phosphate) (**68** and **69**), thymidine 5'-(4-nitrophenyl-[¹⁷O,¹⁸O]phosphate) (**70**), and thymidine 3'-(4-nitrophenyl[¹⁷O,¹⁸O]phosphate)-5'-(4-nitrophenyl[¹⁷O,¹⁸O]phosphate) (**71**)-were chosen because their unlabeled counterparts have been shown to be substrates for a number of phosphodiesterases and nucleases. The configuration of the two diastereomers **68** and **69** was ascertained directly by measuring the ¹⁷O-NMR of each of the products formed (3',5'-cyclic nucleosides) after treating each isomer individually with potassium *tert*-butoxide, since these same authors have previously demonstrated[29] (discussed earlier in this section) that the axial and equatorial exocyclic phosphoryl oxygen of 3',5'-cyclic nucleosides have different chemical shifts and one-bond ³¹P-¹⁷O coupling constants. The configurations were also verified by ³¹P techniques. In this study, the hydrolysis of **71** as catalyzed by bovine pancreatic deoxyribonuclease I (DNase I) was shown to have proceeded with inversion of configuration at phosphorus.

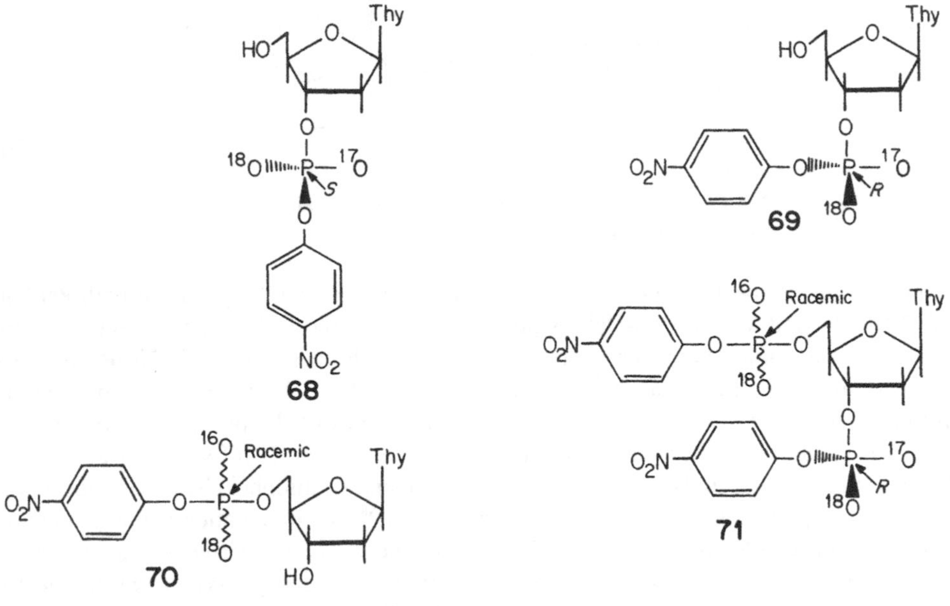

FIGURE 13.   The structure of various thymidine 4-nitrophenyl [¹⁷O,¹⁸O]phosphates.

In a 1985 paper Tsai et al.[44] have reported the further use of ¹⁷O-NMR to study the internal rotational freedom of an oxygen-containing substrate in the active site of an enzyme. In the case studied, the adenylate kinase-adenosine 5'-triphosphate binary complex, the measurement of the ¹⁷O-NMR spectra of [β-¹⁷O₂]ATP in the presence of increasing amounts of adenylate kinase indicated that the triphosphate moiety of ATP most likely has appreciable internal rotational freedom. Some warning concerning the interpretation of these types of data is also presented.

4. Amino Acids and Peptides

Amino acids and small peptides were some of the first biologically important compounds to be studied by ¹⁷O-NMR, no doubt due to the ease with which they may be isotopically labeled. Various authors have used ¹⁷O-NMR to study the conformation, solute solvent interacts, pH effects, and determination of the sites of protonation as well as metal binding. Some representative papers are reviewed here.

In an early investigation designed to demonstrate the pH-dependence of the ^{17}O chemical shift and linewidths on simple amino acids, Fiat et al.[45] isotopically labeled both glycine (18.5%) and alanine (18.2%) in the carboxyl group with ^{17}O and measured their ^{17}O chemical shifts and linewidths as a function of pH. A plot of the pH vs. the chemical shift showed an inflection point which corresponded to the pKa value of the carboxyl group. This pH-dependent $\approx$ 20 ppm downfield shift was explained on the basis of the difference between the change in the ratio of cationic to zwitterionic forms of the amino acid in solution. A plot of ^{17}O linewidths of the two amino acids vs. pH shows a minimum around pH = 7 and two inflection points at high and low pH around their respective pKa values. These results were rationalized by the fact that at high and low pH values, the ^{17}O-^1H coupling is averaged out due to fast proton exchange, whereas at neutral pH values the rate of exchange is slow enough to have an effect on the ^{17}O-^1H coupling values. Further explanations of these phenomena were given which invoked potential differences in the hydrogen bonding capabilities of the different charged species in solution with the solvent. The authors also note significant differences in both the chemical shifts and linewidths between the alanine and glycine, which is rather surprising when one considers that they are structurally so similar (-CH$_3$ vs. -H). Some mention of the π-bond nature of the amide bond was also given.

Fiat et al.,[46] in a model study designed to develop techniques to investigate the hydrogen-bonding properties of the carboxyl group of the amide portion of peptides using ^{17}O-NMR, measured the ^{17}O chemical shifts and linewidths of N-methylformamide in various solvents, which differed in their ability to hydrogen bond, as a function of molar ratio. The results clearly demonstrated a variation in the ^{17}O nuclear screening constant as a function of hydrogen bond-making and hydrogen bond-breaking. This paper ''demonstrates'' the advantages of ^{17}O-NMR as a technique to study hydrogen bonding. Since the ^{17}O chemical shift response to the hydrogen bond is highly specific for single and double bonds, the Δ chemical shift value between the hydrogen-bonded and nonbonded oxygen is large, and the ^{17}O relaxation rates also may provide complementary information on the presence of an intermolecular hydrogen-bonded complex. The study of both the ^{14}N and ^{17}O chemical shifts and linewidths of other amides as a function of solvent, concentration, and temperature has also been reported by these same workers.[47,48]

Several other groups[49-51] have also synthesized ^{17}O-enriched amino acids and measured the ^{17}O chemical shifts and linewidths as a function of solvent concentration and temperature. The common conclusion reached by each of these researchers is that ^{17}O-NMR is an underutilized tool for the investigation of the hydrogen-bonding phenomenon in amino acids and small peptides.

^{17}O-NMR has been used by Fiat and co-workers[52] to help solve the controversial question of the site(s) of protonation of urea, the simplest diamide, in various solvents such as water, HFSO$_3$, and HFSO$_3$/SbF$_5$.

In one of the first papers reporting on the hydrogen-bonding characteristics of each of the carboxyl groups in a peptide, Fiat et al.[53] synthesized ^{17}O-enriched Pro-Leu-Gly-NH$_2$ and measured the ^{17}O-NMR spectra as a function of mixed solvent, concentration, pH, and temperature. The chemical shifts of each of the oxygen sites were sensitive to the ratio of solvents (in this case water : acetonitrile), however, only the prolyl C=^{17}O was sensitive to changes in pH. Further results from the chemical shift and linewidth data collected suggest that in acetonitrile the prolyl C=^{17}O is involved in an intramolecular hydrogen bond, whereas in water there appears to be no intramolecular hydrogen bond. Studies by the same group[54] have involved the investigation of the effect of various carboxyl terminus derivatives of the above peptide tripeptides Pro-Leu-Gly-NH$_2$, namely Pro-Leu-Gly-OCH$_3$, Pro-Leu-Gly-NHCH$_3$, and Pro-Leu-Gly-N(CH$_3$)$_2$, on the properties of the ^{17}O-NMR spectra, as well as a study[55] on the ^{17}O-NMR properties of a series of cyclic and linear protected and nonprotected peptides.

A very useful compilation of the ^{17}O-NMR chemical shifts of the twenty amino acids at various pHs has been published by Lauterwein et al.[56] They also included the ^{17}O-NMR data for a few derivatives of the essential amino acids as well as the data for some naturally occurring unusual amino acids.

^{17}O-NMR has been used by Lauterwein et al.[57] to study the *cis/trans* isomerism of *N*-acetyl-L-proline and *N*-acetylsarcosine in both aqueous and organic solvents. Two resonances were observed for the carboxyl group of *N*-acetyl-L-proline and assigned to the *cis* and *trans* isomers based on the difference in their intensities at low pH. This chemical shift difference of the two isomers was independent of pH. The authors were able to use all the chemical shift values in different solvents to suggest various solution conformations, such as γ-turns, etc.

5. Proteins

Luchinat et al.[58] have studied the low-activity phenylglyoxal (PHG) modified copper zinc superoxide dismutase (SOD) from bovine via the active site water molecules utilizing both 1H and ^{17}O-NMR. Their results suggested that the water in the modified enzyme is still semicoordinated in the axil position but at a greater distance than in the native SOD.

6. Miscellaneous Whole Cell Experiments

In an early study on the transport phenomenon of water into and out of whole tissue, Shporer and Civan[59] utilized ^{17}O-NMR to investigate the flux of $H_2^{17}O$ in whole striated muscle from frog. By determining the spin-lattice relaxation time (T_1 or the longitudinal relaxation rate) using pulsed NMR techniques, the authors were able to show a nonexponential relaxation in fresh muscle and an exponential relaxation in deteriorated muscle. Therefore, in the fresh muscle a heterogeneous population of water molecules exists and in older tissue, a homogeneous population. Several arguments were given to explain their experimental results.

Shporer and Civan[60] have used ^{17}O-NMR to study the exchange process of $H_2^{17}O$ across simpler cell membranes as well as examine the different relaxation processes affecting the ^{17}O resonance both inside and outside the red blood cell. These studies were possible due to the short relaxation rate of the ^{17}O, which is of the same order of magnitude as the mean life time of water within the erythrocyte. Using two techniques, these authors demonstrated that the longitudinal relaxation of ^{17}O (T_1) is four to five times faster inside human erythrocytes than in the outside medium — first, by separately comparing the relaxation rates in isolated samples of cell pellets and supernatant, and, second, by adding Mn^{2+} to solutions of the cell suspension, greatly reducing the transverse relaxation time (T_2) of the extracellular ^{17}O and thereby eliminating the contribution of the ^{17}O of extracellular $H_2^{17}O$. The differences in intra- and extracellular longitudinal relaxation of ^{17}O (T_1) were attributed to the presence of the intracellular hemoglobin molecule which contains an iron atom. The rate constant for the water exchange was calculated to be 60 and 107 s^{-1} at 25° and 37°C, respectively. In contrast, the slow rate of proton relaxation does not allow one to differentiate between intra- and extracellular water using ^1H-NMR; therefore, the information obtained in this study was available only by the use of ^{17}O-NMR.

Shporer and Haran[61] have applied some of the same techniques as those described to determine the equilibrium water permeability through phospholipid vesicle membranes. By taking advantage of the effect of Mn^{2+} on the transverse relaxation times of ^{17}O and the fact that the membrane is impermeable to the transition metal, they were able to distinguish between intra- and extracellular water. Although the same information is extractable from comparable ^1H-NMR experiments, a 9 to 10 times higher molar amount of manganese is necessary since the effect of Mn^{2+} is $\approx$ 10 times greater on the T_2 of ^{17}O than on the T_2 of ^1H. The mean lifetime of water inside the vesicle was calculated to be $\approx$ 1 ms at 22°C.

Based on the ^{17}O T_1 values of the intra- and extracellular water (available since the concentration of metal used to affect T_2 was too low to affect T_1), the authors conclude that the intravesicular water maintains the same molecular mobility as bulk water. Again, the use of ^{17}O-NMR allowed one to obtain information not directly available from other measurements.

Gutowsky et al.[62] have used the same approach to study the exchange of water ($H_2^{17}O$) in and out of dark-adapted chloroplasts which contain intracellular Mn(II) and Mn(III). They, too, measured different T_1 values indicating both a slow exchange between the intra- and extracellular water and a relatively different chemical environment for the two different water species. In addition to the experiments described above, these authors treated these cells with a detergent which disrupts the membrane and allows a more free exchange of intra- and extracellular materials. The net result on the ^{17}O relaxation was, as predicted, a single value for T_1.

IV. CONCLUSIONS

Based on the information presented from the papers reviewed in this chapter, both in the chemical mechanism section and the biological problem section, the utility of ^{17}O-NMR as a probe in investigating the mechanism of chemical reactions and in solving various biological problems cannot be overlooked. With the continued advancement in instrumental hardware, such as new probe designs that have the capability for multinuclear decoupling, enhanced signal-to-noise ratio, and greater sensitivity, the future of ^{17}O-NMR as a probe to solve important questions will continue to bloom.

REFERENCES

1. **Cotton, F. A. and Wilkinson, G. W.,** *Advanced Inorganic Chemistry, A Comprehensive Text,* 4th ed., John Wiley & Sons, New York, 1967, 483.
2. **Rodger, C. and Sheppard, N.,** Group VI-oxygen, sulphur, selenium and tellurium, in *NMR and the Periodic Table,* Harris, R. K. and Mann, B. E., Eds., Academic Press, New York, 1978, chap. 12A.
3. **Klemperer, W. G.,** ^{17}O-NMR spectroscopy as a structural probe, *Angew. Chem. Int. Ed. Engl.,* 17, 246, 1978.
4. **Samuel, D. and Steckel, F.,** Research with the isotopes of oxygen (^{15}O, ^{17}O, and ^{18}O) during the period 1963—1966, *Int. J. Appl. Radiat. Isot.,* 16, 97, 1965.
5. **Risley, J. M. and Van Etten, R. L.,** ^{18}O-Isotope effect in ^{13}C nuclear magnetic resonance spectroscopy. 2. The effect of structure, *J. Am. Chem. Soc.,* 102, 4609, 1980.
6. **Fiat, D., St. Amour, T. E., Burgar, M. I., Steinschneider, A., Valentine, B., and Dhawan, D.,** ^{17}O Nuclear magnetic resonance and its biological application, *Bull. Magn. Reson.,* 2, 18, 1981.
7. **Fiat, D.,** Biophysical Application of ^{17}O NMR, *Bull. of Magn. Reson.,* 6, 30, 1984.
8. **St. Amour, T. and Fiat, D.,** ^{17}O Magnetic resonance. *Bull. of Magn. Reson.,* 1, 118, 1980.
9. **Nakanishi, W., Jo, T., Miura, K., Ikeda, Y., Sugawara, T., Kawada, Y., and Iwamura, H.,** Thermal decomposition of ^{17}O-labelled *t*-butyl *O*-methylthio- and *O*-phenyl-thioperoxy-benzoates studied by ^{17}O NMR. The sulfuranyl radical structure of the *O*-thiobenzoyloxy radicals, *Chem. Lett.,* p. 387, 1981.
10. **Kuznetsova, N. I., Likholobov, V. A., Fedotov, M. A., and Yermakov, Y. I.,** The mechanism of the formation of ethylene glycol monoacetate from ethylene in the system $MeCO_2H$ + $LiNO_3$ + $Pd(OAc)_2$ *J. Chem. Soc. Chem. Commun.,* p. 973, 1982.
11. **Jennings, W. B.,** Chemical shift nonequivalence in prochiral groups, *Chem. Rev.,* 75, 307, 1975.
12. **Kobayashi, K., Sugawara, T., and Iwamura, H.,** Diastereomeric differentiation of sulphone oxygen by ^{17}O N. M. R. spectroscopy, *J. Chem. Soc. Chem. Commun.,* p. 63, 1982.
13. **Lowe, G. and Salamone, S. J.,** Application of a lanthanide shift reagent in ^{17}O NMR spectroscopy to determine the stereochemical course of oxidation of cyclic sulphite diesters to cyclic sulphate diesters with ruthenium tetroxide, *J. Chem. Soc. Chem. Commun.,* p 1392, 1983.

14. **King, J. F., Skonieczny, S., Khemani, K. C., and Stothers, J. B.,** Combined ¹⁷O NMR spectra and ¹⁸O isotope effects in ¹³C NMR spectra for oxygen labelling studies. Carbon → sulfur oxygen migration in the aqueous chlorination of mercato alcohols, *J. Am. Chem. Soc.,* 105, 6514, 1983.

15. **Dessinges, A., Castillon, S., Olesker, A., Thang, T. T., and Lukacs, G.,** Oxygen-17 NMR and oxygen-18-induced isotopic shifts in carbon-13 NMR for the elucidation of a controversial reaction mechanism in carbohydrate chemistry, *J. Am. Chem. Soc.,* 106, 451, 1984.

16. **Chang, S. and le Noble, W. J.,** Study of ion-pair return in 2-norbornyl brosylate by means of ¹⁷O NMR, *J. Am. Chem. Soc.,* 105, 3708, 1983.

17. **Chang, S. and le Noble, W. J.,** Search for oxygen scrambling during ethanolysis of [ether-¹⁷O]-*endo*-2-norbornyl mesylate. The true exo/endo rate ratio and the nature of the rate-controlling step, *J. Am. Chem. Soc.,* 106, 810, 1984.

18. **Creary, X. and Inocencio, P. A.,** ¹⁷O and ¹⁸O Labeling studies by NMR. Mechanism of rearrangement of an α-thiophosphoryl trifluoroacetate to an α-phosphoryl thiotrifluoro-acetate, *J. Am. Chem. Soc.,* 108, 5979, 1986.

19. **Curzon, E. H., Golding, B. T., and Wong, A. K.,** Direct studies of reactions of ¹⁷O-labeled cobaloximes by ¹⁷O NMR spectroscopy: hydrolysis of 2-acetoxyethyl(pyridine)-cobaloxime and hydration of formyl-methyl(pyridine)cobaloxime, *J. Chem. Soc. Chem. Commun.,* 479, 1981.

20. **Adlington, R. M., Aplin, R. T., Baldwin, J. E., Field, L. D., John, E.-M. M., Abraham, E. R., and White, R. L.,** Conversion of ¹⁷O/¹⁸O-labeled δ-(L-α-aminoadipyl)-L-cysteinyl-D-valine into ¹⁷O/¹⁸O-labelled isopenicillin N in a cell-free extract of *C. acremonium, J. Chem. Soc. Chem. Commun.,* p. 137, 1982.

21. **Adlington, R. M., Aplin, R. T., Baldwin, J. E., Chakravarti, B., Field, L. D., John, E.-M. M., Abraham, E. P., and White, R. L.,** Conversion of ¹⁷O/¹⁸O-labelled δ-(L-α-amino-adipyl)-L-cysteinyl-D-valine into ¹⁷O/¹⁸O-labelled isopenicillin N in a cell-free extract of *C. acremonium.* A study by ¹⁷O-NMR spectroscopy and mass spectroscopy, *Tetrahedron,* 39, 1061, 1983.

22. **Sankawa, U., Ebizuka, Y., Noguchi, H., Isikawa, Y., Kitaghawa, S., Yamamoto, T., Kobayashi, T., Iitak, Y., and Seto, H.,** Biosynthesis of citrinin in *Aspergillus terreus,* incorporation studies with [2-¹³C, 2-²H₂], [1-¹³C, ¹⁸O₂] and [1-¹³C, ¹⁷O₂]-acetate, *Tetrahedron,* 39, 3583, 1983.

23. **Broze, M. and Lui, Z.,** Oxygen-17 spin-spin coupling with manganese-55 and carbon-13, *J. Phys. Chem.,* 73, 1600, 1969.

24. **Burgar, M. I., Dhawan, D., and Fiat, D.,** ¹⁷O and ¹⁴N Spectroscopy of ¹⁷O-labeled nucleic acid bases, *Org. Magn. Reson.,* 20, 184, 1982.

25. **Chandresekaran, S., Wilson, W. D., and Boykin, D. W.,** ¹⁷O NMR studies on polycyclic quinones, hydroxyquinones and related cyclic ketones: models for anthracyclic intercalators, *Org. Magn. Reson.,* 22, 757, 1984.

26. **Wilson, W. D., Wang, Y.-H., Kusuma, S., Chandresekaran, S., and Boykin, D. W.,** The effect of intercalator structure on binding strength and base pair specificity in DNA interactions, *Biophys. Chem.,* 24, 101, 1986.

27. **Chandresekaran, S., Wilson, W. D., and Boykin, D. W.,** ¹⁷O NMR studies on 5-substituted uracils, *J. Org. Chem.,* 50, 829, 1985.

28. **Schwartz, H. M., MacCross, M., and Danyluk, S. S.,** Oxygen-17 NMR of nucleosides. II. Hydration and self-association of uridine derivatives, *J. Am. Chem. Soc.,* 105, 5901, 1983.

29. **Coderre, J. A., Mehdi, S., Demou, P. C., Weber, R., Traficante, D. D., and Gerlt, J. A.,** Oxygen chiral phosphodiesters. III. Use of ¹⁷O NMR spectroscopy to demonstrate configurational differences in the diastereomers of cyclic 2'-deoxyadenosine 3',5'-[¹⁷O,¹⁸O]monophosphate, *J. Am. Chem. Soc.,* 103, 1870, 1981.

30. **Schwartz, H. M., MacCross, M., and Danyluk, S. S.,** ¹⁷O NMR spectroscopy of nucleoside derivatives; bonding characteristics of pyrimidine carbonyls, *Tetrahedron Lett.,* 21, 3837, 1980.

31. **Tsai, M.-D.,** Use of ³¹P(¹⁸O), ³¹P(¹⁷O), and ¹⁷O NMR methods to study enzyme mechanisms involving phosphorus, in *Methods of Enzymology,* Purich, D. L., Ed., Academic Press, New York, 1982, 235.

32. **Tsai, M.-D.,** Use of phosphorus-31 nuclear magnetic resonance to distinguish bridge and nonbridge oxygen of oxygen-17-enriched nucleoside triphosphate. Stereochemistry of acetate activation by acetyl coenzyme A synthetase, *Biochemistry,* 18, 1468, 1979.

33. **Lowe, G., Potter, B. V. L., Sproat, B. S., and Hull, W. E.,** The effect of ¹⁷O and the magnitude of the ¹⁸O-isotope shift in ³¹P nuclear magnetic resonance spectroscopy, *J. Chem. Soc. Chem. Commun.,* p. 733, 1979.

34. **Tsai, M.-D., Huang, S. L., Kozlowski, J. F., and Chang, C. C.,** Applicability of the phosphorus-31 (oxygen-17) nuclear magnetic resonance method in the study of enzyme mechanisms involving phosphorus, *Biochemistry,* 19, 3531, 1980.

35. **Gerlt, J. A., Demou, P. C., and Mehdi, S.,** ¹⁷O NMR spectral properties of simple phosphate esters and adenine nucleotides, *J. Am. Chem. Soc.,* 104, 2848, 1982.

36. **Huang, S. L. and Tsai, M.-D.,** Does the magnesium(II) ion interact with the α-phosphate of adenosine triphosphate? An investigation by oxygen-17 nuclear magnetic resonance, *Biochemistry,* 21, 951, 1982.

37. **Sammons, R. D., Frey, P. A., Bruzik, K., and Tsai, M.-D.**, Effects of ^{17}O and ^{18}O on 31P NMR: further investigation and applications, *J. Am. Chem. Soc.*, 105, 5455, 1983.

38. **Gerlt, J. A., Reynolds, M. A., Demou, P. C., and Kenyon, G. L.**, ^{17}O NMR spectral properties of pyrophosphate, simple phosphonates, and thiophosphate and phosphonate analogues of ATP, *J. Am. Chem. Soc.*, 105, 6469, 1983.

39. **Gerothanassis, I. P. and Sheppard, N.**, Natural abundance ^{17}O NMR spectra of some inorganic and biologically important phosphates, *J. Magn. Reson.*, 46, 423, 1982.

40. **Reynolds, M. A., Gerlt, J. A., Demou, P. C., Oppenheimer, N. J., and Kenyon, G. L.**, ^{15}N and ^{17}O NMR studies of the proton binding sites in imidodiphosphate, tetraethyl imidodiphosphate, and adenylyl imidodiphosphate, *J. Am. Chem. Soc.*, 105, 6475, 1983.

41. **Petersheim, M., Miner, V. W., Gerlt, J. A., and Prestegard, J. H.**, ^{17}O NMR as a probe of nucleic acid dynamics, *J. Am. Chem. Soc.*, 105, 6357, 1983.

42. **Reynolds, M. A., Oppenheimer, N. J., and Kenyon, G. L.**, Enzyme-catalyzed positional isotope exchange by phosphorus-31 nuclear magnetic resonance spectroscopy using either ^{18}O- or ^{17}O-β,γ-bridged-labeled adenosine 5''-triphosphate, *J. Am. Chem. Soc.*, 105, 6663, 1983.

43. **Mehdi, S. and Gerlt, J. A.**, Syntheses and configurational analyses of thymidine 4-nitrophenyl [$^{17}O,^{18}O$]phosphate and the stereochemical course of a reaction catalyzed by bovine pancreatic deoxyribonuclease I, *Biochemistry*, 23, 4844, 1984.

44. **Wisner, D. A., Steginsky, C. A., Shyy, Y.-J, and Tsai, M.-D.**, Mechanism of adenylate kinase. 1. Use of ^{17}O NMR to study the binding properties of substrates, *J. Am. Chem. Soc.*, 107, 2814, 1985.

45. **Valentine, B., St. Amour, T., Walter, R., and Fiat, D.**, pH Dependence of oxygen-17 chemical shifts and linewidths of L-alanine and glycine, *J. Magn. Reson.*, 38, 413, 1980.

46. **Burgar, M. I., St. Amour, T., and Fiat, D.**, ^{17}O NMR study of the hydration and the association of N-methylformamide molecules, *Period. Biol.*, 82, 283, 1980.

47. **Burgar, M. I. and Fiat, D.**, ^{17}O and ^{14}N NMR studies of amide systems, *J. Phys. Chem.*, 85, 502, 1981.

48. **Fiat, D., Burgar, M. I., Dhawan, D., St. Amour, T., Steinschneider, A., and Valentine, B.**, ^{17}O NMR studies of labeled peptides and model systems, *Dev. Endocrinol.*, 13, 239, 1981.

49. **Gerothanassis, I. P., Hunston, R., and Lauterwein, J.**, ^{17}O NMR of enriched acetic acid, glycine, glutamic acid and aspartic acid in aqueous solution. I. Chemical shift studies, *Helv. Chim. Acta*, 65, 1764, 1982.

50. **Gerothanassis, I. P., Hunston, R., and Lauterwein, J.**, ^{17}O NMR of enriched acetic acid, glycine, glutamic acid and aspartic acid in aqueous solution. II. Relaxation studies, *Helv. Chim. Acta*, 65, 1774, 1982.

51. **Hunston, R., Gerothanassis, I. P., and Lauterwein, J.**, ^{17}O Nuclear magnetic resonance studies of some enriched amino acids, *Org. Magn. Reson.*, 18, 120, 1982.

52. **Valentine, B., St. Amour, T. E., and Fiat, D.**, A ^{17}O NMR study of the protonation of urea, *Org. Magn. Reson.*, 22, 697, 1984.

53. **Gilboa, H., Steinschneider, A., Valentine, B., Dhawan, D., and Fiat, D.**, Hydrogen bonds in the tripeptide pro-leu-gly-NH$_2$, ^{17}O and ^{1}H Studies, *Biochim. Biophys. Acta*, 800, 6251, 1984.

54. **Steinschneider, A. and Fiat, D.**, Carbonyl-^{17}O N.M.R. of amino acid and peptide carboxamide and methyl ester derivaties, *Int. J. Pept. Protein Res.*, 23, 591, 1984.

55. **Valentin, B., Steinschneider, A., Dhawan, D., Burgar, M. I., St. Amour, T., and Fiat, D.**, Oxygen-17 NMR of peptides, *Int. J. Pept. Protein Res.*, 25, 56, 1985.

56. **Gerothanassis, I. P., Hunston, R. N., and Lauterwein, J.**, ^{17}O NMR chemical shifts of the twenty protein amino acids in aqueous solution, *Magn. Reson. Chem.*, 23, 659, 1985.

57. **Hunston, R., Gerothanassis, I. P., and Lauterwein, J.**, A study of L-proline, sarcosine, and the cis/trans Isomers of N-Acetyl-L-proline and N-acetylsarcosine in aqueous and organic solution by ^{17}O NMR, *J. Am. Chem. Soc.*, 107, 2654, 1985.

58. **Bertini, I., Lanini, G., and Luchinat, C.**, A water ^{1}H and ^{17}O NMR study on PHG-modified SOD, *Inorg. Chim.*, 93, 51, 1984.

59. **Civan, M. M. and Shporer, M.**, Pulsed NMR studies of ^{17}O from H$_2$$^{17}O$ in frog striated muscle, *Biochim. Biophys. Acta*, 343, 399, 1974.

60. **Shporer, M. and Civan, M. M.**, NMR study of ^{17}O from H$_2$$^{17}O$ in human erythrocytes, *Biochim. Biophys. Acta*, 385, 81, 1975.

61. **Haran, N. and Shporer, M.**, Study of water permeability through phospholipid vesicle membranes by ^{17}O NMR, *Biochim. Biophys. Acta*, 426, 638, 1976.

62. **Wydrzynski, T. J., Marks, S. B., Schmidt, P. G., Govindjee, and Gutowsky, H. S.**, Nuclear magnetic relaxation by the manganese in aqueous suspensions of chloroplasts, *Biochemistry*, 17, 2155, 1978.

Chapter 7

¹⁷O NMR SPECTROSCOPY OF SINGLE BONDED OXYGEN: ALCOHOLS, ETHERS, AND THEIR DERIVATIVES

S. Chandrasekaran

TABLE OF CONTENTS

I. INTRODUCTION

Since the pioneering work of Christ et al.[1] and the subsequent investigations[2-5] which soon followed the early studies, alcohols and ethers have been studied fairly extensively by [17]O nuclear magnetic resonance (NMR) spectroscopy.[6] A survey of the literature in this area reveals that of all the organic functionalities containing disubstituted oxygen, alcohols, ethers, and their derivatives have been the subject of more systematic investigations than any other compounds in this category. In addition to the routine characterization of the ^{17}O NMR parameters unique to the functional groups,[6-8] a large number of studies on alcohols and ethers has concentrated on the effects of changes in molecular structure, conformation, and environment on the ^{17}O NMR parameters, especially chemical shifts. This chapter describes the above-mentioned effects on [17]O chemical shifts of alcohols, ethers, and their derivatives. Whenever applicable, substituent-induced chemical shift effects (SCS), and their dependence on various stereochemical arrangements of oxygen atoms with respect to the substituent, conformational, and hydrogen-bonding effects are discussed. The coverage of the material is not completely exhaustive, but a major number of publications up to September 1989 are included. Interpretations of the observed effects on [17]O chemical shifts are made, as much as possible, in terms of the formulation of paramagnetic screening $\sigma_p{}^A$ by Karplus and Pople[9] shown in Equation 1:

$$\sigma_p^A = -\frac{e^2h^2}{2m^2c^2} (\Delta E^{-1}) \langle r^{-3}\rangle_{2_p} (Q_{AA} + \overset{\Sigma}{\underset{B=A}{}} Q_{AB}) \tag{1}$$

where ΔE is the mean or effective electronic excitation energy, $\langle r^{-3}\rangle_{2p}$ is the mean inverse cube radius for the oxygen 2p orbital, Q_{AA} relates to the local charge density on atom A in question (which is oxygen in this case), and Q_{AB} represents the bond order between atom A and adjacent atoms B. In addition, several empirical correlations of the [17]O chemical shifts with physicochemical properties, such as calculated electron charge densities, first ionization potentials, IR stretching frequencies, and Hammett constants are also described. Correlations of [17]O chemical shifts with ^{13}C chemical shifts of the corresponding carbon-containing compounds are also discussed.

II. SATURATED ALIPHATIC ALCOHOLS AND ETHERS—ACYCLIC DERIVATIVES

A comparison of the [17]O chemical shifts of aliphatic alcohols and ethers[10-12] found in Table 1 reveals that the substitution of the hydroxyl hydrogen by a methyl group results in a substantial upfield shift of the [17]O chemical shift of the oxygen involved. This upfield shifting α-effect is in contrast to what is observed in ^{13}C spectra, where α-methyl substitution results in downfield shifts. Eliel and co-workers[12] attributed this effect to the result of the electrostatic nature of the C^+-O^- dipole in alcohols and ethers as opposed to the homopolar C-C bond found in the corresponding hydrocarbons. However, the ^{13}C shifts of carbon atoms situated in a potentially polar C-C bond, viz., $C-C_\alpha-N_\beta^{+}$[13] or $C-C_\alpha-C_\beta^{+}$[14] show effects similar to the α-SCS (CH$_3$) effect found in ^{17}O NMR. In these cases, the carbon adjacent to C-α experiences the similar upfield shifting effect as the ether oxygen in C-O.

The shielding effect of α-methyl substitution in ^{17}O NMR can be readily understood from an examination of the Karplus-Pople equation for the paramagnetic term, Equation 1. The charge polarization in the C^+-O^- dipole arising from α-methyl substitution would enlarge the oxygen 2p orbitals, thereby decreasing the $\langle r^{-3}\rangle_{2p}$ term. The reduction of this term should lower $\sigma_p{}^A$ and shift the [17]O resonance upfield. This explanation assumes that ΔE, Q_{AA}, and Q_{AB} do not change appreciably in going from alcohols to ethers, which may be justified in the case of compounds closely related in structure. Apparently, the charge polarization effect acts in the opposite way for α-methyl effects in ^{13}C NMR as noted by Cheney and Grant.[15] These authors suggest that the resulting charge polarization of the C^A-H bond upon substitution of the hydrogen atom by methyl contracts the C^A_{2p} orbitals and increases the $\langle r^{-3} \rangle_{2p}$ term. This increase is accompanied by the well-documented deshielding α-methyl effect in ^{13}C NMR.

It is seen from Table 1 that the α-effect of a methyl group on the ^{17}O shift is dependent on the type of ether. α-Methyl substitution of a hydrogen atom in water (H-O-H to H-O-CH_3) results in an upfield shift of -37 ppm for methanol, whereas the α-SCS (CH_3) is -15.5 ppm for the methyl ether of methanol (dimethyl ether), -26 to -29 ppm for methyl ethers of secondary alcohols, and -54 ppm for methyl *tert*-butyl ether. This diverse upfield effect in ethers might be due to differences in the degree of 2p orbital enlargement of the ether oxygen atoms. The characteristic ^{17}O chemical shifts of the various types of alcohols and ethers can serve as useful indicators of identification among primary, secondary and tertiary structures of unknown alcohols.

TABLE 1
α-SCS(CH_3) Effects in ^{17}O NMR from a
Comparison of ^{17}O Shifts of Aliphatic Alcohols
(ROH) and Their Methyl Aliphatic Ethers
(ROCH$_3$)

R	^{17}O Shifts[a]		α-SCS(CH_3)
	δ_{ROH}[b]	δ_{ROCH_3}[c]	$(\delta_{ROCH_3} - \delta_{ROH})$
CH_3	-37.0	-52.5	-15.5
CH_3CH_2	5.9	-22.5	-28.4
$CH_3CH_2CH_2$	-0.5	-28.5	-28.0
$CH_3CH_2CH_2CH_2$	0.0	-28.5	-28.5
$CH_3CH_2CH_2CH_2CH_2$	-0.8	-27.5	-26.7
$(CH_3)_2CHCH_2$	-2.0	-30.0	-28.0
$(CH_3)_2CHCH_2CH_2$	-0.6	-29.5	-28.9
$(CH_3)_3CCH_2$	-6.8	-32.5	-25.7
$(CH_3)_3CCH_2CH_2$	1.9	-29.5	-27.6
$(CH_3)_2CH$	39.8	-2.0	-41.8
$CH_3CH_2(CH_3)CH$	34.0	-8.5	-42.5
$CH_3CH_2CH_2(CH_3)CH$	35.3	-16.5	-51.8
$(CH_3)_3C$	62.3	8.5	-53.8

The effect of β-methyl substitution β-SCS (CH_3) on ^{17}O chemical shifts of aliphatic alcohols (Crandall and Centeno)[10] and ethers (Delseth and Kintzinger)[11] is illustrated in Tables 2 and 3, respectively. These β effects produce downfield shifts, a trend which is observed in ^{13}C NMR also, although the effects on the oxygen nucleus are considerably larger in magnitude (11 to 43 ppm for ^{17}O vs. 3 to 9 ppm for ^{13}C). The β-methyl effect for alcohols seems to be larger by about 10 ppm than ethers. An interesting feature which emerges from Tables 2 and 3 is that successive β-methyl substitution results in a diminution of the β effect, an observation which is prevalent in ^{13}C NMR.[16,17]

Iwamura et al.[18] showed that the downfield shifts due to the β-methyl effect in a limited series of simple alcohols (Compounds **1** to **4**, Table 2) correspond to a decrease in their ionization potentials. These authors concluded that if the ionization potentials of these

TABLE 2
β-SCS(CH$_3$) Effects in ^{17}O NMR of Aliphatic Alcohols

No.	Compound	δ_{17O}[a]	β-SCS(CH$_3$)
1	CH$_3$OH	−37.0	—
2	CH$_3$CH$_2$OH	5.9	42.9
3	(CH$_3$)$_2$CHOH	39.8	33.9
4	(CH$_3$)$_3$COH	62.3	22.5
5	CH$_3$CH$_2$CH$_2$OH	−0.5	—
6	CH$_3$CH$_2$(CH$_3$)CHOH	34.0	34.5
7	CH$_3$CH$_2$(CH$_3$)$_2$COH	56.5	22.5
8	CH$_3$CH$_2$CH$_2$CH$_2$OH	0.0	—
9	CH$_3$CH$_2$CH$_2$(CH$_3$)CHOH	35.3	35.3
10	CH$_3$CH$_2$CH$_2$(CH$_3$)$_2$COH	61.5[b]	26.2

[a] Taken from Reference 10.
[b] S. Chandrasekaran, unpublished results.

TABLE 3
β-SCS(CH$_3$) Effects in ^{17}O NMR of Aliphatic Ethers

No.	Compound	δ_{17O}[a]	β-SCS(CH$_3$)
11	CH$_3$OCH$_3$	−52.5	—
12	CH$_3$CH$_2$OCH$_3$	−22.5	30
13	(CH$_3$)$_2$CHOCH$_3$	−2.0	20.5
14	(CH$_3$)$_3$COCH$_3$	8.5	11.0
15	CH$_3$CH$_2$CH$_2$OCH$_3$	−28.5	—
16	CH$_3$CH$_2$(CH$_3$)CHOCH$_3$	−8.5	20.0
17	CH$_3$CH$_2$(CH$_3$)$_2$COCH$_3$	9.1[b]	17.6
18	CH$_3$CH$_2$CH$_2$CH$_2$OCH$_3$	−28.5	—
19	CH$_3$CH$_2$CH$_2$(CH$_3$)CHOCH$_3$	−16.5	12.0
21	CH$_3$OC$_2$H$_5$	−22.5	—
22	CH$_3$CH$_2$OC$_2$H$_5$	6.5	29.0
23	(CH$_3$)$_2$CHOC$_2$H$_5$	28.0	21.5
24	(CH$_3$)$_3$COC$_2$H$_5$	40.5	12.5
25	CH$_3$CH$_2$CH$_2$OC$_2$H$_5$	1.7	—
26	CH$_3$CH$_2$(CH$_3$)CHOC$_2$H$_5$	24.5	22.8
28	CH$_3$CH$_2$CH$_2$CH$_2$OC$_2$H$_5$	−1.5	—
29	CH$_3$CH$_2$CH$_2$(CH$_3$)CHOC$_2$H$_5$	15.0	16.5

[a] Taken from Reference 11.
[b] S. Chandrasekaran, unpublished results.

compounds were taken as a measure of the average electronic excitation energy term ΔE in Equation 1, a decrease in ionization potentials due to β-methyl substitution should increase the ΔE^{-1} term, thereby producing the observed downfield effects. Theoretical calculations of $\sigma_p{}^A$ for compounds **1** to **4** by Takasuka,[19] assuming ΔE in Equation 1 to be the negative of the highest occupied orbital energy, also indicated the absolute value of the paramagnetic term, $\sigma_p{}^A$ to increase with β-methyl substitution. These calculations and the experimental findings of Iwamura et al.[18] suggest that, at least for simple alcohols, the β-methyl effects are mainly determined by the ΔE^{-1} term. Delseth and Kintzinger[11] postulated a similar explanation for β-methyl effects in aliphatic ethers. It is interesting to note that these β effects in ^{13}C NMR have also been shown, by Baird and Teo,[20] to be governed by variations in the ΔE^{-1} term. These authors derived the ^{13}C effective paramagnetic shielding terms $\sigma_p{}^A$

for a number of organic compounds from the difference between the calculated diamagnetic shieldings (by MINDO/1) and the experimental total shieldings. Using the values for $\langle r^{-3} \rangle_{2p}$ and Q terms obtained from MINDO/1 calculations and the values for $\sigma_p{}^A$ terms, ΔE was calculated from Equation 1. The wide variation found in these ΔE values was attributed to be the source of the observed ^{13}C chemical shifts.

Substituent effects for the replacement of γ hydrogens by methyl groups on the ^{17}O chemical shifts of aliphatic alcohols[10] and ethers[11] are summarized in Table 4. These γ-SCS (CH$_3$) effects result in upfield shifts of the ^{17}O resonances of alcohols and ethers. It is interesting to note that substitution by the first and third methyl groups induce larger upfield shifts than the second methyl. Similar observations of the γ effect are well known in ^{13}C NMR.[16] The origin of the γ-methyl effect in ^{17}O NMR may well be analogous to what has been proposed for corresponding effects in ^{13}C, ^{15}N and ^{31}P NMR.[16,59] Thus, interpretations as diverse as the sterically induced polarization of the O-H (in alcohols) and the O-C (in ethers) bonds leading to a net negative charge on the oxygen atoms; bond angle and torsional angle changes which cause significant variations in heavy-atom (^{13}C, ^{19}F, ^{31}P) chemical shifts; and attractive van der Waals interactions producing an expansion of orbitals in heavy-atom nuclei with a concomitant reduction in the $\langle r^{-3} \rangle_{2p}$ term in Equation 1 and a shielding effect, may well all apply.

TABLE 4
γ-SCS(CH$_3$) Effects in ^{17}O NMR of Aliphatic
Alcohols and Ethers

No.	Compound	$\delta_{^{17}O}$*	γ-SCS(CH$_3$)
2	CH$_3$CH$_2$OH	5.9	
5	CH$_3$CH$_2$CH$_2$OH	−0.5	−6.4
31	(CH$_3$)$_2$CHCH$_2$OH	−2.0	−1.5
32	(CH$_3$)$_3$CCH$_2$OH	−6.8	−4.8
12	CH$_3$CH$_2$OCH$_3$	−22.5	
15	CH$_3$CH$_2$CH$_2$OCH$_3$	−28.5	−6.5
33	(CH$_3$)$_2$CHCH$_2$OCH$_3$	−30.0	−1.5
34	(CH$_3$)$_3$CCH$_2$OCH$_3$	−32.5	−2.5

* Taken from References 10 and 11.

Substitution of δ hydrogens by methyl groups does not produce any significant changes in the ^{17}O chemical shifts of simple aliphatic alcohols[10] and ethers[11] as outlined in Table 5. This phenomenon is again very similar to what is observed in ^{13}C NMR. The effect of the introduction of polar electronegative substituents in the alkyl moiety of simple aliphatic alcohols and ethers on their ^{17}O chemical shifts has not been investigated in detail, except for a report by Arbuzov et al.,[21] who studied a series of simple chloroethers. The β-SCS (Cl) effects which these authors found for aliphatic ethers are summarized in Table 6. It is evident that the introduction of a single chlorine atom at the β-position to the oxygen atom produces a consistent downfield shift of ~71 ppm for this atom. Similar to what is observed for β-SCS (CH$_3$) effects, substitution by a second β-Cl atom results in a significant diminution of the downfield shift (only 44 to 48 ppm). The second chlorine atom produces less than half the downfield shift as the first one. The nature of the chlorine substituent seems to determine the magnitude of the effect, the primary chlorine atom producing a 71 ppm shift and the secondary chlorine atom a 66 ppm shift. The deshielding effect of the β-chlorine atom is probably dominated by inductive effects resulting from the higher electronegativity of chlorine compared to hydrogen. Due to inductive electron withdrawal by the chlorine atom, the 2p orbitals of oxygen contract resulting in an increase in the $\langle r^{-3} \rangle_{2p}$ term in Equation 1, thereby producing the observed downfield shift. As seen from the rather large

TABLE 5
δ-SCS(CH₃) Effects in ¹⁷O NMR of Aliphatic Alcohols and Ethers

No.	Compound	δ_{17O}[a]	δ-SCS (CH₃)
5	$CH_3CH_2CH_2OH$	−0.5	
8	$CH_3CH_2CH_2CH_2OH$	0.0	0.50
35	$(CH_3)_2CHCH_2CH_2OH$	−0.6	−0.60
36	$(CH_3)_3CCH_2CH_2OH$	1.9	1.90
15	$CH_3CH_2CH_2OCH_3$	−28.5	
18	$CH_3CH_2CH_2CH_2OCH_3$	−28.5	0.00
37	$(CH_3)_2CHCH_2CH_2OCH_3$	−29.5	−1.00
38	$(CH_3)_3CCH_2CH_2OCH_3$	−29.5	0.00
6	$CH_3CH_2CH(CH_3)OH$	34.0	
39	$CH_3CH_2CH_2CH(CH_3)OH$	35.3	1.30

[a] Taken from References 10 and 11.

TABLE 6
β-SCS(Cl) Effects in ¹⁷O NMR of Aliphatic Ethers

No.	Compound	δ_{17O}	β-SCS(Cl)
11	CH_3OCH_3	−52.5[a]	
41	$ClCH_2OCH_3$	18.9[b]	71.4
42	Cl_2CHOCH_3	65.8[b]	46.9
12	$CH_3OCH_2CH_3$	−22.5[a]	
43	$ClCH_2OCH_2CH_3$	48.1[b]	70.6
44	$Cl_2CHOCH_2CH_3$	95.5[b]	47.4
15	$CH_3OCH_2CH_2CH_3$	−28.5[a]	
45	$ClCH_2OCH_2CH_2CH_3$	43.0[b]	71.5
46	$Cl_2CHOCH_2CH_2CH_3$	90.3[b]	47.3
13	$CH_3OCH(CH_3)_2$	−2.0[a]	
47	$ClCH_2OCH(CH_3)_2$	69.0[b]	71.0
48	$Cl_2CHOCH(CH_3)_2$	113.3[b]	44.3
15	$CH_3OCH_2CH_2CH_3$	−28.5[a]	
49	$ClCH_2OCH_2CH_2CH_2CH_3$	43.1[b]	71.6
50	$Cl_2CHOCH_2CH_2CH_2CH_3$	91.4[b]	48.3
12	$CH_3CH_2OCH_3$	−22.5[a]	
51	$CH_3CHClOCH_3$	43.6[b]	66.1
22	$CH_3CH_2OCH_2CH_3$	6.5[a]	
52	$CH_3CHClOCH_2CH_3$	73.0[b]	66.5
25	$CH_3CH_2OCH_2CH_2CH_3$	1.7[a]	
53	$CH_3CHClOCH_2CH_2CH_3$	68.4[b]	66.7

[a] Taken from Reference 11.
[b] Taken from Reference 21.

effects of chlorine substitution on ¹⁷O chemical shifts of ethers, a study of the effects produced by other electronegative substituents at β, γ and δ positions relative to the oxygen atom of interest should prove to be a fruitful area of research.

III. STEREOCHEMICAL AND CONFORMATIONAL DEPENDENCE OF SUBSTITUENT EFFECTS — ALICYCLIC ALCOHOLS AND ETHERS

Eliel and co-workers[12,22] investigated the effect of conformation on the ¹⁷O chemical shifts of alcohols and ethers by examining a series of conformationally rigid cyclohexanols, hydroxynorbornanes, and the corresponding ethers shown in Scheme 1. The ^{17}O chemical shifts of the axial hydroxyl and ether groups are consistently upfield of their equatorial counterparts for all the compounds studied. Similar effects have been noted in ^{13}C NMR spectra of axial and equatorial methyl cyclohexanes. Eliel et al.,[12,22] in pointing out this similarity between ^{17}O and ^{13}C NMR, have noted that a scale factor of 2.2 is operative in going from ^{13}C to ^{17}O NMR, and that the effects in ^{17}O are larger by this magnitude. This finding may suggest that the origin of the conformational effects may be similar in both ^{13}C and ^{17}O NMR.

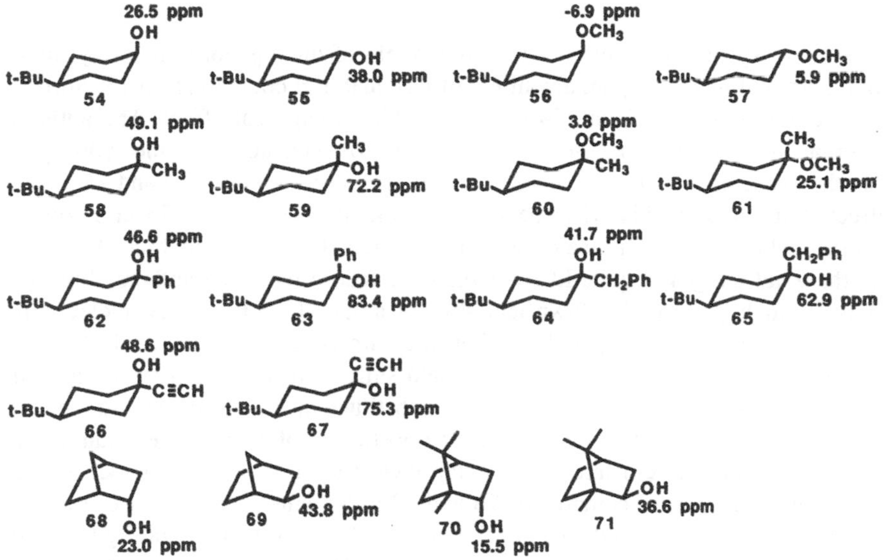

26.5 ppm — OH — t-Bu — **54**

OH 38.0 ppm — t-Bu — **55**

-6.9 ppm — OCH$_3$ — t-Bu — **56**

OCH$_3$ 5.9 ppm — t-Bu — **57**

49.1 ppm — OH — CH$_3$ — t-Bu — **58**

CH$_3$ — OH 72.2 ppm — t-Bu — **59**

3.8 ppm — OCH$_3$ — CH$_3$ — t-Bu — **60**

CH$_3$ — OCH$_3$ 25.1 ppm — t-Bu — **61**

46.6 ppm — OH — Ph — t-Bu — **62**

Ph — OH 83.4 ppm — t-Bu — **63**

41.7 ppm — OH — CH$_2$Ph — t-Bu — **64**

CH$_2$Ph — OH 62.9 ppm — t-Bu — **65**

48.6 ppm — OH — C≡CH — t-Bu — **66**

C≡CH — OH 75.3 ppm — t-Bu — **67**

68 OH 23.0 ppm

69 OH 43.8 ppm

70 OH 15.5 ppm

71 OH 36.6 ppm

SCHEME 1

The difference in the ^{17}O chemical shifts of the axial and equatorial oxygen atoms varies with the type of hydroxyl oxygen examined. The difference in ^{17}O shifts is 11 ppm for the secondary alcohols **54** and **55**, 21 to 27 ppm for the tertiary alcohol pairs **58/59**, **64/65**, and **66/67**, and 37 ppm for the phenylcyclohexanols **62** and **63**. Similar differences are noted for the ethers also. The axial-equatorial difference for the secondary ethers **56** and **57** is 13 ppm, whereas that for the tertiary ethers **60** and **61** is almost twice as much (21 ppm). The variation found for the upfield shifts of the axial oxygen atoms must be due to different β-effects of the −CH$_3$, −Ph, −CH$_2$Ph groups discussed later. The bornane and the norbornane derivatives **68** to **71** also show effects similar to what is observed for cyclohexanols **54** and **55**. The endo-OH resonates upfield of the exo-OH, the upfield shift being 20-21 ppm in both cases.

The phenomenon of the axial hydroxyl oxygen resonating upfield of the equatorial oxygen is reversed when second-row heteroatoms are present in a γ-gauche orientation with respect to the observed oxygen atom. Eliel et al[12] noted this effect in the hydroxy-1,3-dioxane (**72**) and the methoxy-1,3-dioxane (**74**) (Scheme 2).

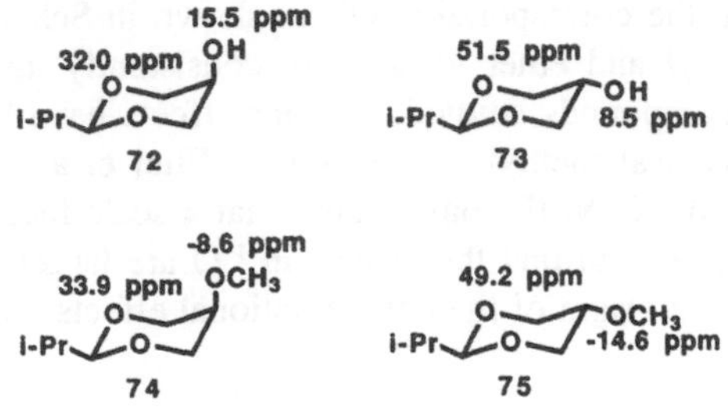

SCHEME 2

The axial hydroxyl in **72** resonates 7 ppm downfield of the equatorial hydroxyl in **73**, and the axial methoxyl in **74** is 6 ppm downfield of its equatorial counterpart in **75**. Surprisingly, the ring oxygen atoms in **72** and **74** do not exhibit a reciprocal effect. The authors ruled out intramolecular hydrogen bonding between the ring oxygen atoms and the hydroxyl oxygen in **72** as the cause of this unusual downfield shift because the ethers **74** and **75** also exhibit this effect. Furthermore, this effect persisted for the alcohols **72** and **73** in a solvent such as pyridine, which would eliminate the intramolecular hydrogen bonding. The downfield shift of the axial oxygen is quite likely explained, according to the authors,[12] by the same mechanism which governs the downfield shift of axial ¹³C, ¹⁹F and ¹H nuclei at the C-5 position in 1,3- dioxanes compared to when they are equatorially oriented. These effects have been attributed to the γ-gauche (downfield shift) and γ-antiperiplanar (upfield shift) substituent effects of second-row heteroatom substituents, which are well-documented in ¹³C NMR. An example[22] of a γ-gauche heterosubstituent of rows higher than second-row producing the normal upfield shift of the axial oxygen atom relative to the equatorial counterpart is seen in the pair of compounds **76** and **77** in Scheme 3. In this case, the third-row heteroatom sulfur, just like carbon in cyclohexanols **54**, causes the axial hydroxyl oxygen to resonate upfield of the equatorial isomer. Similar effects are known in ¹³C NMR.

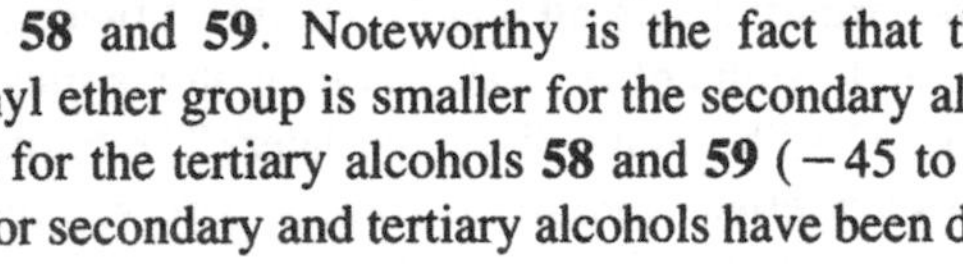

SCHEME 3

The α-SCS (CH₃) effects observed in simple acyclic alcohols are also prevalent in the cyclohexanols **54**, **55**, **58** and **59**. Noteworthy is the fact that the α-effect due to the substitution of the methyl ether group is smaller for the secondary alcohols **54** and **55** (-32 to -34 ppm) than that for the tertiary alcohols **58** and **59** (-45 to -47 ppm). Analogous variations in α-effects for secondary and tertiary alcohols have been described in the previous section.

Manoharan and Eliel[22] investigated the dependence of β-effects of substituents in ¹⁷O NMR on their stereochemical orientation (Table 7). The β-SCS (CH₃) effect in alcohols **58** and **59** (23 to 34 ppm) is larger than the corresponding effect in ethers **60** and **61** (11 to 19 ppm). It is quite striking that all the β-substituents listed in Table 7 produce larger effects on ¹⁷O shifts of alcohols at the axial orientation than when they are equatorially situated. Not surprisingly, parallel differences between axial and equatorial β-effects have also been noted in ¹³C NMR.[23] In ¹³C spectra of cyclohexane derivatives, the β-effect of axial methyl is 10 to 11 ppm and phenyl 12 to 13 ppm, whereas the β-effect of equatorial methyl is 6 to 8 ppm and that of phenyl 4.5 to 6.5 ppm.

TABLE 7

β-SCS Effects in ^{17}O NMR of Cyclohexanols and the
Corresponding Ethers.
Dependence of β-Effects on Substituent Stereochemical
Orientation

Compound No.	β-Substituent	δ_{17O}[a]	β-SCS	Δ[b]
54	—	26.5	—	
58	Eq. CH₃	49.1	22.6	11.6
62	Eq. Φ	46.6	20.1	25.3
64	Eq. CH₂Φ	41.7	15.2	9.7
66	Eq. C≡CH	48.6	22.1	15.2
56	—	−6.9		
60	Eq. CH₃	3.8	10.7	8.5
55	—	38.0	—	
59	Ax. CH₃	72.2	34.2	11.6
63	Ax. Φ	83.4	45.4	25.3
65	Ax. CH₂Φ	62.9	24.9	9.7
67	Ax. C≡CH	75.3	37.3	15.2
57	—	5.9		
61	Ax. CH₃	25.1	19.2	8.5

[a] Taken from Reference 22.
[b] Difference between axial and equatorial substituent effects.

Even though the δ-effect of a methyl group on the ^{17}O shifts of alcohols and ethers is negligible as seen in Table 5, this effect can be quite large when the methyl group and the oxygen atom are oriented in syn-axial position. Eliel and co-workers[12] were the first to draw attention to this phenomenon in compounds **76** and **77** (Scheme 4). This downfield shift of 6.9 ppm has been explained as arising from steric compression, similar to the deshielding of ^{13}C resonances of methyl groups, which have syn-axially oriented substituents.[24,25]

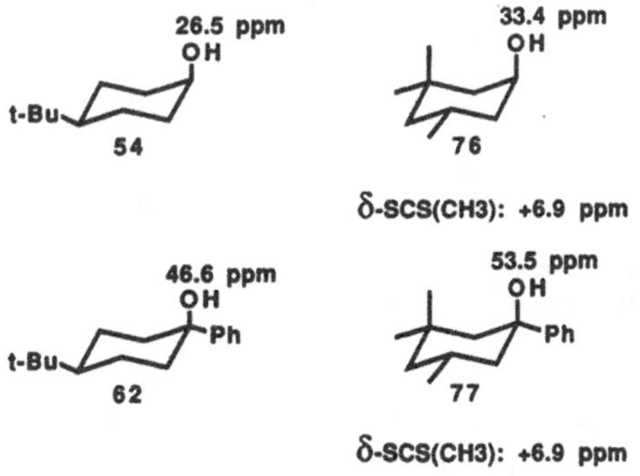

δ-SCS(CH3): +6.9 ppm

δ-SCS(CH3): +6.9 ppm

SCHEME 4

IV. ELECTRONIC EFFECTS IN ^{17}O NMR OF ALCOHOLS AND ETHERS

Reports of investigations of substituent effects in ^{17}O NMR of alcohols, ethers, and their derivatives propagated through unsaturated or aromatic frameworks have been few. Iwamura et al.[26] were the first to study electronic effects in ^{17}O NMR when they observed that in a series of *para-* and *meta-* substituted anisoles (**78** and **79**, Scheme 5), the ¹⁷O chemical shifts of the methoxy ether oxygen were quite sensitive (30 ppm range) to the electronic character of the substituents (Table 8).

X = electron withdrawing and attracting
substituents

SCHEME 5

TABLE 8
¹⁷O NMR Chemical Shifts of *p*- and *m*-Substituted
Anisoles

Compound No.	Substituent	$\delta_{^{17}O}$ (ppm)
80	*p*-NH$_2$	36[a]
81	*p*-OCH$_3$	38[a]
82	*p*-C(CH$_3$)$_3$	44[a]
83	*p*-CH$_3$	44[a]
84	*p*-F	45[a]
85	*p*-CH$_2$=CH	44[b]
86	H	48[a]
87	*p*-Cl	48[a]
88	*p*-Br	51[a]
89	*p*-CF$_3$	54[a]
90	*p*-SH	54[b]
91	*p*-CN	62[a]
92	*p*-NO$_2$	67[a]
93	*m*-NH$_2$	46[a]
94	*m*-Cl	51[a]
95	*m*-Br	52[a]
96	*m*-F	53[a]
97	*m*-NO$_2$	63[a]

[a]   Taken from Reference 26.
[b]   Taken from Reference 27.

Electron donation by the substituent should increase electron density through conjugative overlap at the ether oxygen, whereas electron withdrawal would decrease electron density. Thus, increased electron density produced shielding of the oxygen nucleus and vice-versa for decreased electron density. This effect is in accord with what can be expected from Equation 1 for paramagnetic screening in which the $\langle r^{-3}\rangle_{2p}$ term would decrease with increased electron density, thereby resulting in shielding of the oxygen nucleus. Good linear correlation was obtained between the methoxyl oxygen chemical shifts and the Hammett σ^- values with a correlation coefficient of 0.963. Application of the dual substituent parameter approach to the data gave better correlation (correlation coefficient = 0.98), and revealed that resonance effects govern the chemical shift, which is not surprising. The resonance term was almost twice as large in magnitude as the inductive term in the correlation. The ¹⁷O shifts of the ether oxygen in anisoles correlated well with the calculated (CNDO/2) π-electron densities at oxygen and the π-bond order between oxygen and the ipso aromatic carbon, supporting the view that it is the charge density at the oxygen atom that determines its chemical shift.

When the methyl group of the methoxy moiety in anisole is replaced by the vinyl group, which is capable of conjugative interactions with the oxygen p orbital, as in the case of *para* and *meta* substituted aryl vinyl ethers, the correlation of the ^{17}O chemical shifts of the ether oxygen with the σ_I and σ_R^0 constants is not as good (correlation coefficient = 0.93).[27] However, the chemical shifts for the *p*-substituted aryl vinyl ethers encompass a large range of 32 ppm, and showed the same response to substituent effects as those for anisoles (Table 9). The poor correlation is probably due to the interaction of the oxygen p-orbitals with the π system of the vinyl group, resulting in the transmission of the substituent effects to the terminal vinyl carbon. This is indeed supported by the excellent correlation found for the ^{13}C chemical shifts of the β-vinyl carbon with the substituent constants, σ_I and σ_R^0.[27]

TABLE 9
^{17}O NMR Chemical Shifts for *p*-
and *m*-Substituted Aryl Vinyl
Ethers, *p*-X-C$_6$H$_4$-O-CH=CH$_2$
and *m*-X-C$_6$H$_4$-O-CH=CH$_2$

Compound No.	Substituent X	δ_{17O}[a]
98	*p*-NH$_2$	107
99	*p*-N(CH$_3$)C$_3$H$_7$	114
100	*p*-CH$_3$	118
101	*p*-OCH$_3$	120
102	*p*-O-CH=CH$_2$	118
103	H	125
104	*p*-Cl	119
105	*p*-NO$_2$	139
106	*m*-CH$_3$	122
107	*m*-Cl	135
108	*m*-NO$_2$	138

[a] Taken from Reference 27.

Substituent effects in the reverse trend to that found for anisoles and aryl vinyl ethers were discovered for oxygen atoms situated one carbon removed from the aromatic ring. Boykin and co-workers[28a] found that for a series of *p*-substituted benzyl alcohols (Scheme 6), electron-attracting substituents produced upfield shifts of the oxygen located β to the aromatic ring while electron-donating substituents gave rise to downfield shifts (Table 10).

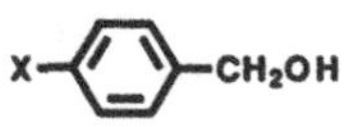

X—⟨ ⟩—CH$_2$OH

109

X = electron withdrawing and
attracting substituents

SCHEME 6

Correlations of chemical shifts with Hammett type substituent constants and a dual substituent parameter method of analysis of data revealed that both resonance and inductive effects are important in the determination of chemical shifts (correlation coefficients are 0.997 for alcohols and 0.999 for acetates). The authors,[28a] by successfully correlating the ^{17}O chemical

TABLE 10
^{17}O NMR Chemical Shifts for p-Substituted Benzyl Alcohols

Compound No.	Substituent	In acetone[a]	In toluene[a]
110	p-NO$_2$	−4.5	−1.0
111	p-CN	−4.2	−0.8
112	p-CF$_3$	−3.3	1.2
113	p-Cl	−0.5	4.3
114	p-H	0.7	6.0
115	p-F	0.7	6.0
116	p-CH$_3$	2.0	7.7
117	p-OCH$_3$	4.7	10.3
118	p-N(CH$_3$)$_2$	7.3	13.0

[a]   Taken from Reference 28a.

shifts versus the ^{19}F shifts of p-substituted benzyl fluorides[28b] (Figure 1), and ^{13}C shifts of the methyl groups of p-substituted isopropylbenzenes,[28c] conclude that the mechanism governing the ^{17}O shifts should be similar to the one determining the ^{19}F and ^{13}C shifts.

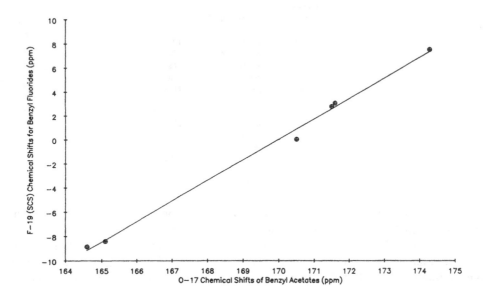

FIGURE 1.   Plot of the ^{17}O chemical shifts of the ester oxygen in p-substituted benzyl acetates[28a] vs. the ^{19}F substituent chemical shifts of the analogous benzyl fluorides.[28b]

Fliszar's explanation[29] of increased downfield shifts of ^{17}O resonances of aliphatic ethers with increased electron density of the oxygen atoms may well apply in the case of benzyl alcohols. Fliszar reasons that increased electron density leads to a more negative paramagnetic term (Equation 1), which may be a manifestation of decreased ΔE term in Equation 1. Iwamura et al.[18] and Kintzinger et al.[11] have identified the effective excitation energy ΔE as the dominant factor in determining the ^{17}O shifts of aliphatic alcohols and ethers. The ^{17}O shifts of benzyl alcohols, which could be considered as aliphatic alcohols, may be governed by the ΔE term in Equation 1. In the absence of available calculated oxygen 2p electron densities and ionization potentials for benzyl alcohols, it is difficult to assess the validity of this hypothesis. However, as Iwamura et al. pointed out,[26] the change in the effective excitation energy ΔE is the governing factor in the paramagnetic screening of aliphatic alcohols and ethers (with only the n orbitals being affected), but in the case of aromatic ethers, viz. anisoles, both the n and π^* orbitals would be involved, making the ΔE term negligible, and the $\langle r^{-3} \rangle_{2p}$ term predominant. This explanation is in accord with the contrasting electronic substituent effects observed in anisoles and benzyl alcohols. Furthermore, in anisoles, the ether oxygen, by conjugatively interacting with the aromatic ring, renders more double bond character to the ipso carbon-oxygen bond, thereby making the C-O bond similar to the carbonyl bond in benzaldehyde and acetophenone. This may be the reason why the anisole ether oxygen responds in the same way as a carbonyl oxygen to electronic substituent effects. Such conjugative interactions being absent in benzyl alcohol, the C-O bond in this compound retains the single bond character, and the ether oxygen displays the opposite behavior.

The failure of the electronic character of substituents in determining the ^{17}O chemical shift of an ether oxygen in an aromatic system was amply demonstrated by Clennan et al.[30] when they observed that the ^{17}O chemical shifts of the ether oxygen in 2-substituted and 2,5-disubstituted furans (Scheme 7) did not correlate with Hammett constants of the substituents (Table 11).

X,Y = electron donating and attracting
substituents

SCHEME 7

TABLE 11
**^{17}O NMR Chemical Shifts for 2-Substituted
and 2,5-Disubstituted Furans**

Compound No.	2-Substituent	5-Substituent	$\delta_{^{17}O}$[a]
121	CH₃	H	244
122	Br	H	249
123	t-Bu	H	236
124	CN	H	242
125	CO₂C₂H₅	H	238
126	CO₂CH₃	H	238
127	CHO	H	236
128	OCH₃	H	219
129	NO₂	H	230
130	CH₂OH	H	238
131	H	H	236
132	CH₃	CH₃	252
133	Br	Br	259
134	t-Bu	t-Bu	238
135	CN	CN	245
136	CO₂C₂H₅	CO₂C₂H₅	238
137	CO₂CH₃	CO₂CH₃	238
138	CHO	CHO	231
139	CH₃	CN	248
140	CH₃	CHO	241

[a] Taken from Reference 30.

Such a lack of correlation is known for ^{13}C chemical shifts of ortho carbons in mono substituted benzenes. However, a plot of the ^{17}O chemical shifts of the substituted furans vs. the Q values of the substituents defined by Schaefer and co-workers,[31] later determined experimentally by Smith and Proulx,[32] revealed a good correlation as seen in Figure 2. For a substituent X, Q is defined as P/Ir^3, where P is the C-X bond polarizability, I is the first ionization potential of X, and r is the C-X bond distance. It is interesting to note that the cyano and methyl substituents do not follow the Q relationship similar to what is observed in correlations of ^{13}C and ^{19}F chemical shifts with Q in aromatic systems. In explaining the dependence on the Q parameter of the ^{13}C chemical shifts of ortho carbons in monosubstituted benzenes, and the ^{19}F chemical shifts in ortho-substituted fluorobenzenes, it has been suggested[32] that the ionization potential of the substituent atom or group, which determines Q, contributes to the ΔE term in the expression for the paramagnetic screening (Equation 1). A similar explanation may apply to the ^{17}O chemical shifts of the furans. However, this rationalization differs from the usual interpretation of ^{17}O chemical shifts, invoking the ΔE term as being governed by the ionization potential of the oxygen atom, and not of the substituents.

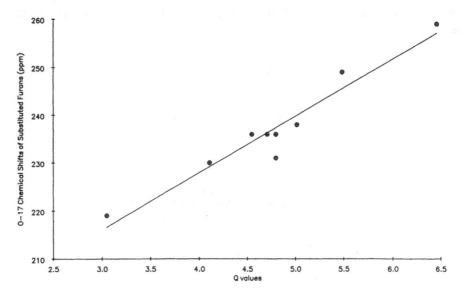

FIGURE 2.   Plot of the ^{17}O chemical shifts of 2- and 2,5-substituted furans[30] vs. the substituent Q values.[31,32]

V. CYCLIC ETHERS—EPOXIDES

Cyclization produces fairly small effects on the ^{17}O shifts of the aliphatic ether in going from acyclic to monocyclic ethers[33] (Table 12). However, cyclization resulting in the formation of bicyclic ethers deshields the ether oxygen by a large magnitude. For instance, diisopropyl ether, upon cyclization to 7-oxanorbornane, undergoes a downfield shift of 33 ppm[33] (Scheme 8). Kintzinger and co-workers[33] have attributed this effect in strained ethers and in the monocyclic ethers to changes resulting from cyclization in hybridization of the oxygen atom from sp^3, as measured by variations in the ΣA_{AB} term in Equation 1 for the paramagnetic contribution.

TABLE 12
Effect of Cyclization on the ^{17}O NMR Chemical Shifts of
Aliphatic Ethers

No.	Compound	$\delta_{^{17}O}{}^a$	$\Delta\delta_{Cyclization}{}^d$
11		-52.5^a	
141		-56.0^a	-3.5
		-49.0^b	$+3.5$
12		-22.5^a	
142		-20.5^a	$+2.0$
22		6.5^a	
143		14.0^a	$+7.5$
25		1.5^a	
144		5.0^a	$+3.5$
		8.8^c	$+7.3$
145		-3.5^a	
146		0.0^a	$+3.5$

a Taken from Reference 33.
b Taken from Reference 34.
c Taken from Reference 38.
d Shift for cyclic ether minus shift for acyclic ether.

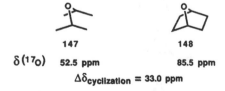

	147	**148**
$\delta(^{17}O)$	52.5 ppm	85.5 ppm

$$\Delta\delta_{cyclization} = 33.0 \text{ ppm}$$

SCHEME 8

Epoxides have been the most extensively investigated compounds of all the cyclic ethers. The oxygen in epoxides is the most highly shielded of all the ether oxygen nuclei, being similar in this respect to proton and carbon nuclei of cyclopropanes. Substitution of β hydrogen atoms by methyl groups induces a sizable downfield shift for the oxirane oxygen atom,[34] analogous to what is observed for acyclic ethers. However, in contrast to effects for the acyclic derivatives, successive methyl substitutions lead to an additive deshielding in oxiranes. These effects are exemplified in Table 13, from which it can be seen that the downfield shift per methyl group is approximately 30 ppm, and that this shift does level off to some extent for the polysubstituted oxirane. The usual shielding γ-methyl effect found in acyclic ethers is encountered in epoxides also, as shown in a comparison of the ¹⁷O shifts of propylene oxide (− 16.0 ppm) and 1,2-butylene oxide (− 18.0 ppm, Scheme 9). A β,γ-double bond or replacement of a β-methyl group by a phenyl substituent deshields the oxirane oxygen by 8.0 ppm, whereas the corresponding effects for acyclic ethers is much smaller.[34] Participation of the Walsh orbitals of the epoxide ring and the π-system of the

TABLE 13
β-SCS(CH₃) Effects in ¹⁷O NMR of Epoxides

No.	Compound	δ_{17O}	β-SCS(CH₃)
141		− 49.0[a] − 56.0[b]	
149		− 16.0[a]	33.0, 40.0
150		3.0[a]	
151		28.0[a]	25.0
149		− 16.0[a]	
152		10.0[a]	26.0
153		14.0[a]	30.0
154		17.0[a]	33.0
155		57.0[a]	40.0[c]

[a] Taken from Reference 34.
[b] Taken from Reference 33.
[c] Shift for the tetramethyl epoxide minus shift for the *trans* dimethyl epoxide.

substituents in conjugative interactions has been suggested to be responsible for these downfield shifts. An endo γ,δ-double bond in the bicyclic epoxide, cyclohexa-1,4-diene oxide (1.0 ppm) has a negligible effect on the ^{17}O shift of the oxirane ring relative to its saturated analog.[34]

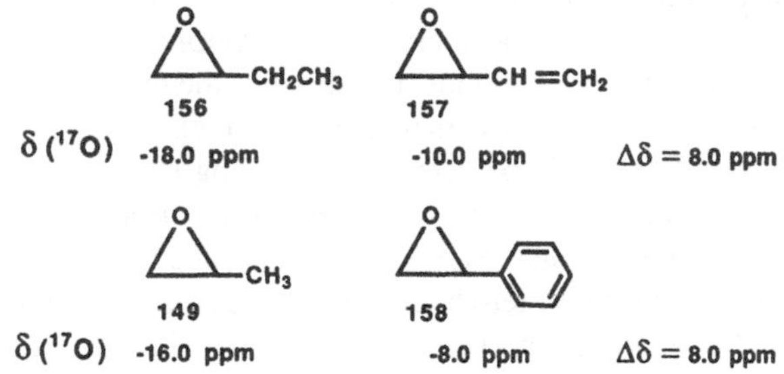

CH₂CH₃
CH =CH₂
156
157
δ (¹⁷O) -18.0 ppm
-10.0 ppm
Δδ = 8.0 ppm
CH₃
149
158
δ (¹⁷O) -16.0 ppm
-8.0 ppm
Δδ = 8.0 ppm

SCHEME 9

Monti et al.[35,36] investigated the ^{17}O spectra of a large number of mono-, di-, and polysubstituted oxiranes, containing methyl, ethyl, vinyl, isopropenyl, and phenyl substituents. Utilizing the direct additivity parameters of substituents derived from the monosubstituted epoxides, the authors calculated the ^{17}O shifts of the di- and polysubstituted compounds. The difference between the calculated and the observed ^{17}O shifts was termed crossed-additivity parameters resulting from pairwise contributions of substituents, and was explained as due to electronic interaction of the oxirane ring with the substituents, or to steric effects from the substituents.[36]

Iwamura et al[34] observed that the oxirane oxygen in norbornene exo-oxide was considerably shielded (−18 ppm, Scheme 10) relative to its counterpart in cyclohexene oxide, and attributed this upfield shift to a γ-effect from the methylene bridge carbon atom in the norbornane compound. However, this effect is reversed in norbornane systems containing an endo γ,δ-double bond or a fused aromatic ring at the γ,δ position. The epoxide oxygens in these cases are deshielded by large magnitudes (52 and 45 ppm, respectively, Scheme 10). Iwamura and co-workers[34] have explained this effect as due to a mixing of the π*

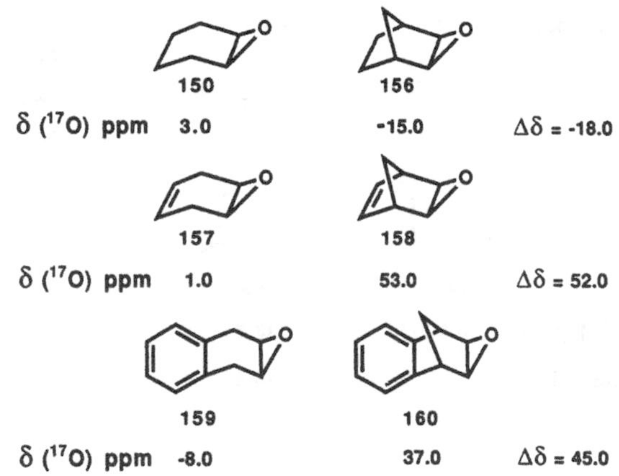

150
156
δ (¹⁷O) ppm 3.0
-15.0
Δδ = -18.0
157
158
δ (¹⁷O) ppm 1.0
53.0
Δδ = 52.0
159
160
δ (¹⁷O) ppm -8.0
37.0
Δδ = 45.0

SCHEME 10

orbital of the double bond or the aromatic ring with the antibonding Walsh orbitals of the oxirane ring leading to a decrease in the energy of the π^* orbital and the $n \rightarrow \pi^*$ transition (ΔE), thereby resulting in a deshielding effect. Similar deshielding effects have been noted in the ^{13}C spectra of the corresponding carbon compounds.[34]

Kas'yan et al.[37] noted that in a series of strained epoxides (Scheme 11), the ^{17}O shifts correlated marginally with the one bond C-H coupling constant of the epoxide fragment (correlation coefficient, 0.964). With an increase in the coupling constant, which is a measure of the s-character of the C-H bond and, thus, the strain of the epoxide ring, the ^{17}O resonance shifted upfield. The authors concluded that in a set of epoxides, in which other factors remain similar, the ^{17}O shifts could be determined by ring strain.

The deshielding β-methyl effect observed for epoxides[34] is also seen for the cyclic ethers tetrahydrofuran and tetrahydropyran β-SCS(CH$_3$) = 28 and 25 ppm, respectively,[38] (Scheme 12). Second methyl substitution in tetrahydrofuran produces deshielding which is close to being additive. γ- and δ-methyl effects in tetrahydrofuran and tetrahydropyran are negligible, which in this respect is similar to acyclic ethers[38] (Scheme 13).

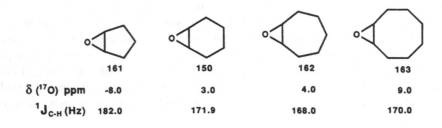

	161	150	162	163
δ (^{17}O) ppm	-8.0	3.0	4.0	9.0
$^1J_{C-H}$ (Hz)	182.0	171.9	168.0	170.0

SCHEME 11

	143	164	165
δ (^{17}O) ppm	16.2	43.9	66.7 (cis)
			74.5 (trans)
β - SCS (CH$_3$)		27.7	22.8 (cis)
			30.6 (trans)

	144	166
δ (^{17}O) ppm	8.8	33.6
β - SCS (CH$_3$)		24.8

SCHEME 12

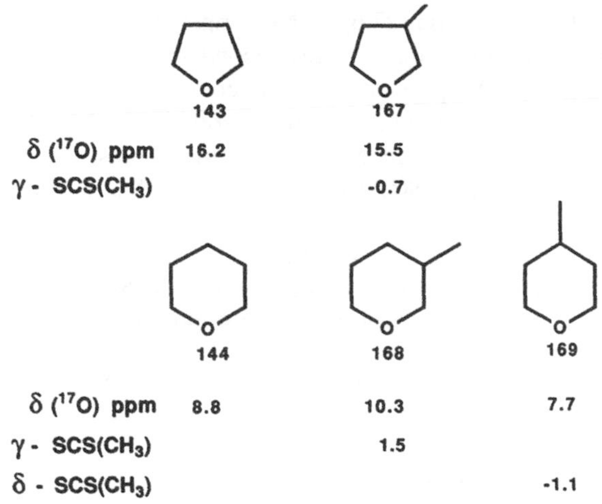

δ (^{17}O) ppm   16.2   15.5

γ - SCS(CH₃)   -0.7

143   **167**

144   **168**   **169**

δ (^{17}O) ppm   8.8   10.3   7.7

γ - SCS(CH₃)   1.5

δ - SCS(CH₃)   -1.1

SCHEME 13

VI. ACETALS, KETALS, ORTHOESTERS, AND THE CYCLIC DERIVATIVES 1,3-DIOXOLANES AND 1,3-DIOXANES

Replacement of a β-hydrogen atom in dialkyl ethers by an alkoxy group results in strong deshielding for the oxygen atoms, which are β to each other. This effect, after correcting for the γ-alkyl (-5 ppm) and δ-alkyl (1 ppm) contributions from the carbon atom in the alkoxy group, is strongly dependent on the nature of the carbon atom separating the two oxygen atoms. As can be seen from Table 14, the mean value of the β-SCS (alkoxy) effect due to the first substituent decreases in the order: secondary carbon, 61 ppm > tertiary

TABLE 14
β-SCS(Alkoxy) Effects as a Function of the Carbon Atom Separating the Alkoxy Oxygen Atoms

No.	Compound	δ_{17O}[a]	β-SCS(alkoxy)[b]
	Secondary carbon atom		
11	H₃COCH₃	−52.5	
170	H₃COCH₂OCH₃	3.0	60.5
12	H₃CH₂COCH₃	−22.5	
171	H₃CH₂COCH₂OCH₂CH₃	34.0	60.5
	Tertiary carbon atom		
12	H₃CH₂COCH₃	−22.5	
172	H₃CHC(OCH₃)₂	18.0	45.5
22	H₃CH₂COCH₂CH₃	6.5	
173	H₃CHC(OCH₂CH₃)₂	48.0	45.5
	Quaternary carbon atom		
13	(H₃C)₂CHOCH₃	−2.0	
174	(H₃C)₂C(OCH₃)₂	19.5	26.5
23	(H₃C)₂CHOCH₂CH₃	28.0	
175	(H₃C)₂C(OCH₂CH₃)₂	52.0	28.0

[a] Taken from Reference 39.

[b] Corrections due to γ-CH₃ (-5 ppm) and δ-CH₃ (1 ppm) effects are included.

carbon, 46 ppm > quaternary carbon, 27 ppm. Kintzinger et al.[39] suggested that the decrease in the β-SCS(alkoxy) effect, in going from the secondary carbon to the quaternary carbon case, is due to a decrease in the population of the conformers, in which at least two lone pairs of electrons, one from each oxygen, adopt a parallel orientation. Such conformations should become increasingly less dominant on successive substitutions by alkoxy groups leading to a diminution of the β-SCS(alkoxy) shifts. This effect is exemplified in Table 15,

TABLE 15
**β-SCS(Alkoxy) Effects as a Function of the Second
and Third Alkoxy Substituents**

No.	Compound	δ_{17_O}[a]	β-SCS(alkoxy)
	Second substituent		
170	$H_3COCH_2OCH_3$	3.0	
176	$H_3COCH(OCH_3)_2$	24.0	26.0
171	$H_3CH_2COCH_2OCH_2CH_3$	34.0	
177	$H_3CH_2COCH(OCH_2CH_3)_2$	53.0	23.0
172	$H_3CHC(OCH_3)_2$	18.0	
178	$H_3CC(OCH_3)_3$	30.5	17.5
173	$H_3CHC(OCH_2CH_3)_2$	48.0	
179	$H_3CC(OCH_2CH_3)_3$	58.5	14.5
	Third substituent		
180	$H_3COCH(OCH_3)_2$	24.0	
181	$H_3COC(OCH_3)_3$	22.0	3.0
182	$H_3CH_2COCH(OCH_2CH_3)_2$	53.0	
183	$H_3CH_2COC(OCH_2CH_3)_3$	56.5	7.5

[a] Taken from Reference 39.
[b] Corrections due to γ-CH_3(-5 ppm) and δ-CH_3 (1 ppm) effects are included.

with the second substituent producing a 25 ppm and a 16 ppm shift for the tertiary and quaternary derivatives, and the third substituent leading to a fairly small deshielding of 5 ppm.[39] The magnitude of the β-SCS(CH_3) effect in acetals, ketals and orthoesters is quite similar (~31 ppm, Table 16) to that encountered in ethers, suggesting that the ΔE term in Equation 1 might determine this large deshielding shift.[39]

TABLE 16
**β-SCS(CH_3) Effects in ^{17}O NMR of Acetals, Ketals
and Orthoesters**

No.	Compound	δ_{17_O}[a]	β-SCS(CH_3)
170	$CH_2(OCH_3)_2$	3.0	
171	$CH_2(OCH_2CH_3)_2$	34.0	31.0
180	$CH(OCH_3)_3$	24.0	
182	$CH(OCH_2CH_3)_3$	53.0	29.0
181	$C(OCH_3)_4$	22.0	
183	$C(OCH_2CH_3)_4$	56.5	34.5
178	$CH_3C(OCH_3)_3$	30.5	
179	$CH_3C(OCH_2CH_3)_3$	58.5	28.3
174	$(CH_3)_2C(OCH_3)_2$	19.5	
175	$(CH_3)_2C(OCH_2CH_3)_2$	52.0	32.5

[a] Taken from Reference 39.

Cyclization of acetals and ketals to 1,3-dioxolanes results in further deshielding of the oxygen atoms (Table 17) relative to the acyclic compounds.[39] This effect is in accord with the expectation that cyclization should restrict the oxygen lone pairs to a parallel orientation to a greater degree than for the noncyclic derivatives. However, cyclization to produce 1,3-dioxanes generates negligible effects suggesting conformational similarity between the acyclic and cyclic acetals involving analogous orientations of oxygen lone pairs of electrons.[38] Ring substitution by methyl groups at the 2- and 4- positions in 1,3-dioxolanes produces the usual deshielding β-effects (21.2 and 27.1 ppm, respectively).[38]

TABLE 17
Cyclization Effects in ^{17}O NMR of Dioxolanes and Dioxanes

No.	Compound	$\delta_{^{17}O}$	$\delta_{Corrected}$[a]	$\Delta\delta_{Cyclization}$
170		3.0[b]		
184		33.0[b]		30.0[d]
185		35.3[c]	5.3	2.3[e]
172		18.0[b]		
186		53.0[b]		35.0[d]
187		52.5[c]	22.5	4.5[e]
174		20.0[b]		
188		61.0[b]		41.0[d]
189		51.8[c]	21.8	1.8[e]

[a] $\delta_{^{17}O}$ minus 30 ppm due to β-alkyl effect.
[b] Taken from Reference 39.
[c] Taken from Reference 38.
[d] Shift for cyclic compound minus shift for acyclic compound.
[e] $\delta_{corrected}$ minus shift for acyclic compound.

Exocyclic equatorial and axial β-oxygen atoms in a six-membered ring deshield each other by the same magnitude (20 to 21 ppm,[22] Scheme 14). Similar deshielding effects are noted for the endocyclic oxygen atoms also from a comparison of oxane with 1,3-dioxane[22] (Scheme 15).

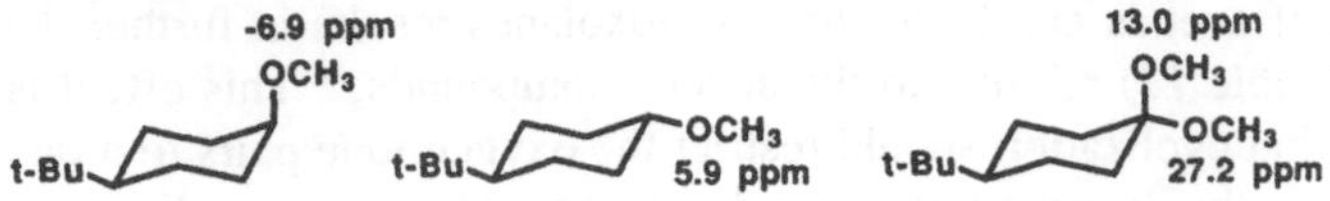

SCHEME 14

δ (^{17}O) ppm   8.8   35.3   Δδ = 26.5

SCHEME 15

Eliel et al.,[38] in an investigation of a large number of methyl-substituted 1,3-dioxanes, computed additive substituent parameters for ring methyl substitution. These values, listed in Table 18, indicate that the shifts induced by methyl groups are dependent on the stereochemical orientation of the substituent and its location for a given effect, thus pointing out their usefulness in the conformational analysis of 1,3-dioxanes. The β-methyl effect in 1,3-dioxanes is deshielding, and its magnitude is higher when the substituent is at the equatorial vs. the axial position. This effect is also larger for the methyl group at the 2-position relative to position 4. The γ-methyl effect is strikingly dependent on the substituent's orientation with the axial group causing a −11.4 ppm upfield shift, whereas the equatorial one produces a negligible effect (0.3 ppm). δ-Effects are expectably small for both possible orientations of the substituent. The authors,[38] in comparing these substitution parameters with those for ^{13}C shifts in analogous oxanes, note that the values for ^{17}O are proportional to the ones for ^{13}C with a factor of 2.5, indicating that the mechanism determining the substituent effects should be similar. Analogous ring methyl substituent effects on ^{17}O shifts are noted for the 1,3,2-dioxaphosphorinane system[40] in which the range for the $β_e$–CH_3, $β_a$–CH_3, $δ_e$–CH_3, and $δ_a$–CH_3 effects are 24 to 29, 13 to 16, −1 to −4, and −5 to −7 ppm, respectively. Methyl-substituted 2-oxo-1,3,2-dioxathiane derivatives exhibit the following substituent effects:[41] $β_e$–CH_3: 24; $β_a$–CH_3: 8 to 10; $γ_e$–CH_3: 3; $γ_a$–CH_3: −10; $δ_e$–CH_3: −1 to −3; $δ_a$–CH_3: −6.

In six-membered cyclic systems containing β-oxygen atoms, the downfield shift ex-

TABLE 18
Methyl Substitution
Parameters (SCS-CH₃) in ^{17}O
Spectra of 1,3-Dioxanes

Parameter	Position	SCS(CH₃)[a]
$β_e$	2	17.1
$β_e$	4	26.4
$β_a$	2	5.0
$β_a$	4	12.2
$γ_e$	5	0.3
$γ_a$	5	−11.4
$δ_e$	6	−3.4
$δ_a$	6	−2.6

[a]   Taken from Reference 38.

perienced by the ring oxygen is greater when the exocyclic β-oxygen is equatorial than when it is axial. Table 19 shows that in oxanes, the difference in the deshielding effects of the equatorial and axial methoxy oxygen atoms on the ring oxygen is 10 to 14 ppm.[38,42] This value is quite similar to what is seen for β-methyl effects in 1,3-dioxanes (Table 20), for which the axial-equatorial difference is 12 ppm[38,22]

TABLE 19
β-SCS(OCH₃) Effects in ¹⁷O NMR Spectra of Oxanes.
Dependence of Substituent Effects on Stereochemical
Orientation of Substituent

No.	Compound	$\delta_{17O}^{(1)}$	$\Delta\delta(1)^a$
169		7.7[b]	
191		39.0[c]	31.3
192		49.0[c]	41.3
166		33.6[b]	
193		60.0[c]	26.4
194		74.0[c]	40.4

[a] Shift for methoxy compound minus shift for unsubstituted compound.
[b] Taken from Reference 38.
[c] Taken from Reference 42.

The attenuation of the deshielding effect due to successive introduction of β-oxygen atoms in acyclic acetals is also encountered in the cyclic derivatives.[22,42] This effect is illustrated for the exocyclic oxygen in Table 21, and for the endocyclic oxygen in Table 22. Similar saturation of substituent effects due to β-methyl groups are observed for acyclic ethers and 1,3-dioxanes.[38]

An examination of Table 23, which contains the ¹⁷O chemical shifts of conformationally rigid 1,3-dioxanes,[22] and the structurally analogous six-membered cyclic phosphorus compounds,[40] allows the following generalizations about the various effects governing the ¹⁷O shifts in these systems to be made: (1) the axial exocyclic oxygen consistently resonates upfield of the equatorial one, being similar in this respect to hydroxyl and ether oxygen nuclei; (2) ring oxygen atoms are considerably deshielded relative to the exocyclic oxygen; this is probably due to the ring oxygen lone pairs being in a parallel orientation, thus leading to an enhanced mutual β-deshielding effect; and (3) the difference in the deshielding effect of the exocyclic oxygen on the ring oxygen nuclei is much smaller in phosphites and phosphates than in 1,3-dioxanes and thiophosphates.

TABLE 20

β-SCS(OCH₃) Effects in ¹⁷O NMR Spectra of 1,3-Dioxanes. Dependence of Substituent Effects on Stereochemical Orientation of Substituent

No.	Compound	$\delta_{^{17}O}{}^{(1)}$	$\delta_{^{17}O}{}^{(3)}$	$\Delta\delta(1)$[a]	$\Delta\delta(3)$[a]
195		61.7[b]	32.6		
196		72.9[c]	44.5	11.2	11.9
197		86.3[c]	57.0	24.6	24.4
198		62.3[c]	71.9		
199		71.5[c]	82.3	9.2	10.4
200		83.8[c]	92.1	21.5	20.2

[a] Shift for methoxy compound minus shift for unsubstitued compound.
[b] Taken from Reference 38.
[c] Taken from Reference 22.

TABLE 21

Attenuation of β-SCS(Oxygen) Effects Due to Successive Oxygen Substitutions on the ¹⁷O NMR Shifts Of Exocyclic Oxygen Atom in Oxanes and 1,3-Dioxanes

No.	Compound	δ_{OCH_3}	β-SCS(oxygen)
56		−6.9[a]	
193		23.0[b]	29.9
196		29.8[c]	6.8
57		5.9[a]	
194		33.0[b]	27.1
197		34.6[c]	1.6

[a] Taken from Reference 12.
[b] Taken from Reference 42.
[c] Taken from Reference 22.

TABLE 22
Attenuation of β-SCS(Oxygen) Effects Due to Successive
Oxygen Substitutions on the ^{17}O NMR Shifts of Ring Oxygen
Atom in Oxanes and 1,3-Dioxanes

No.	Compound	$\delta_{O(1)}$	β-SCS(oxygen)
166		33.6[a]	
193	OCH$_3$	60.0[b]	26.4
196	OCH$_3$	72.9[c]	12.9
166		33.6[a]	
194	OCH$_3$	74.0[b]	40.4
197	OCH$_3$	86.3[c]	12.3
169		7.7[a]	
191	OCH$_3$	39.0[b]	31.3
196	OCH$_3$	44.5[c]	5.5
169		7.7[a]	
192	OCH$_3$	49.0[b]	41.3
197	OCH$_3$	57.0[c]	8.0

[a] Taken from Reference 38.
[b] Taken from Reference 42.
[c] Taken from Reference 22.

The deshielding effect for hydroxyl and ether oxygen nuclei due to a syn- axially oriented δ-CH$_3$ group is also prevalent for the acetal oxygen in 1,3-dioxanes[22] and 1,3,2-dioxaphosphorinanes.[40] The magnitude of this effect (11 ppm), illustrated in Table 24, is larger relative to that in the cyclohexyl series, and has been attributed to a greater steric compression in the dioxa series of compounds, which contain the shorter C-O-X chain (X = C or P) in comparison to the C-C-C sequence in cyclohexanes.[40]

TABLE 24
δ-SCS(CH₃) Effects in ¹⁷O NMR Spectra of 1,3-Dioxanes and 1,3,2-Dioxaphosphorinanes

No.	Compound	δ_{OCH_3}	δ-SCS(CH₃)
196		29.8[a]	
207		40.3[a]	10.5
208		49.0[b]	11.0
209		60.0[b]	11.0
210		23.0[b]	
211		34.0[b]	11.0
212		48.0[b]	11.0
213		59.0[b]	11.0

[a] Taken from Reference 22.
[b] Taken from Reference 40.

TABLE 23
¹⁷O NMR Chemical Shifts of Conformationally Homogeneous 1,3-Dioxanes and 1,3,2-Dioxaphosphorinanes

No.	Compound	δ_{OCH_3}	$\delta_{e,1}$	$\delta_{e,3}$	Ref.
196		29.8	44.5	72.9	a
201		48.0	58.0	87.0	b
202		22.0	45.0	76.0	b
203		48.0	65.0	93.0	b
197		34.6	57.0	86.3	a
204		58.0	62.0	88.0	b
205		29.0	46.0	74.0	b
206		54.0	73.0	101.0	b

a Taken from Reference 22.
b Taken from Reference 40.

In a study of conformationally locked 2-alkoxytetrahydropyrans (Scheme 16), Iwamura and co-workers[42] found that in the axial isomers, the ring oxygen was more shielded and the alkoxy oxygen was less shielded relative to their counterparts in the equatorial isomers.

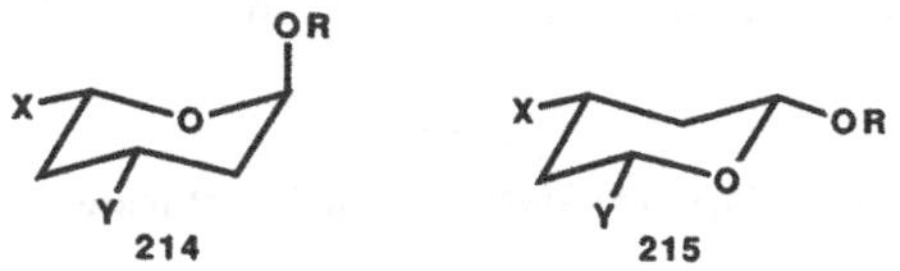

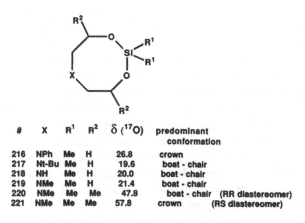

X = H, CH$_3$, HC(CH$_3$)$_2$

Y = H, CH$_3$

R = CH$_3$, C$_2$H$_5$

SCHEME 16

The authors interpreted this result to be a manifestation of the anomeric effect operating in the axial isomer, in which the backdonation of ring oxygen electron density to the σ* orbital of the exocyclic C-O bond should be more significant. The resulting increase in the mean excitation energy (ΔE) of the ring oxygen and a concomitant decrease in that of the exocyclic oxygen produce the observed effects. Anomeric effect has also been suggested to be the origin of the 7 to 10 ppm shielding observed for 1,3-dioxa-2-silacyclooctanes (Scheme 17) in the boat-chair conformation compared to those in the crown conformation.[43]

#	X	R^1	R^2	δ (^{17}O)	predominant conformation	
216	NPh	Me	H	26.8	crown	
217	Nt-Bu	Me	H	19.6	boat - chair	
218	NH	Me	H	20.0	boat - chair	
219	NMe	Me	H	21.4	boat - chair	
220	NMe	Me	Me	47.8	boat - chair	(RR diastereomer)
221	NMe	Me	Me	57.8	crown	(RS diastereomer)

SCHEME 17

VII. UNSATURATED ETHERS—EFFECT OF THE DOUBLE BOND ON [17]O CHEMICAL SHIFTS OF ETHER OXYGEN

The effect of the introduction of a double bond at the α,β-position to an ether oxygen on its [17]O chemical shift was investigated by Kintzinger et al.[39] and Kalabin et al.[27,44] These authors found that in a large number of unsaturated ethers (Table 25), α,β unsaturation produced a consistently large deshielding effect ($\Delta\delta$ in Table 25) on the ether oxygen [17]O

TABLE 25
^{17}O NMR Chemical Shifts of α,β-Unsaturated Ethers

No.	Compound	$\delta_{^{17}O}$	$\delta_{^{17}O}$[a] (saturated)	$\Delta\delta$[b]
222	$CH_2=CH-O-CH_3$	59.0[c]	−22.5[e]	81.5
223	$CH_2=CH-O-CH_2CH_3$	88.0[c]	6.5[e]	81.5
224	$CH_2=CH-O-CH_2CH_2CH_3$	84.0[c]	1.7[e]	82.3
225	$CH_2=CH-O-CH_2CH_2CH_2CH_3$	83.0[c]	−1.5[e]	84.5
226	$CH_2=CH-O-CH_2CH(CH_3)_2$	83.0[c]	−1.0[e]	84.0
227	$CH_2=CH-O-CH(CH_3)_2$	107.0[c]	28.0[e]	79.0
228	$CH_2=CH-O-CH(CH_3)CH_2CH_3$	104.0[c]	24.5[e]	79.5
229	$CH_2=CH-O-C(CH_3)_3$	117.0[c]	40.5[e]	76.5
230	$CH_2=CH-O-CH_2C_6H_5$	82.0[c]	9.0[e]	73.0
231	$CH_2=CH-O-C_6H_5$	123.0[c]	68.0[d]	55.0
232	$CH_2=C(CH_3)-O-CH_3$	62.5[d]	−2.0[e]	64.5
233	$CH_2=C(CH_3)-O-CH_2CH_3$	98.0[c]	28.0[e]	70.0
234	$CH_3-CH=CH-O-CH_2CH_3$	70.0[c]	1.7[e]	68.3
235	2,3-Dihydro-γ-Pyran	53.5[d]	8.8[f]	44.7

[a] [17]O chemical shifts for the corresponding saturated compound.
[b] Shift for the unsaturated compound minus shift for the saturated compound.
[c] Taken from Reference 27.
[d] Taken from Reference 39.
[e] Taken from Reference 11.
[f] Taken from Reference 38.

chemical shift in comparison with their saturated analogs. This deshielding has been explained as due to a depletion of π-electron charge density on the ether oxygen atom resulting from the polarization of the unsaturated system as follows: $C=C-O-C \leftrightarrow C^--C=O^+-C$. A decrease in the charge density of the oxygen atom increases the $\langle r^{-3}\rangle_{2p}$ term in Equation 1, and results in the observed deshielding. Consistent with this explanation is the correlation of the [17]O deshielding of the ether oxygen with the ^{13}C shielding of the terminal vinyl carbon in these unsaturated ethers.[27] By relating the difference in the calculated π-electron charges on the oxygen atoms in vinyl butyl ether (**225**) and ethyl butyl ether to the difference in their [17]O chemical shifts, Kalabin et al.[27] calculated the response of the [17]O shift in alkyl vinyl ethers to π-electron density changes to be 1200 ppm per unit of electronic charge. Kalabin et al.[27] also observed that for the vinyl ethers **223** to **229**, which differ in the alkyl moieties attached to the oxygen atom, their [17]O chemical shifts correlated well with the [17]O shifts of the saturated alkyl ethyl ether counterparts (Figure 3, correlation coefficient = 0.995). This correlation led the authors to conclude that in this limited series of unsaturated ethers, the contribution of the p-π-interaction to the [17]O chemical shifts is unaffected by the alkyl groups. Similar to the case of aryl vinyl ethers[27] (see section on electronic effects), large deshielding of the ether oxygen in vinyl ethers of pyridine (Scheme 18) and quinoline (Scheme 19) is observed.[45] This effect has been attributed to a balancing of the two competing processes of delocalization of the ether oxygen lone pairs of electrons into the π systems of the heteroaromatic ring and the vinyl group, respectively.

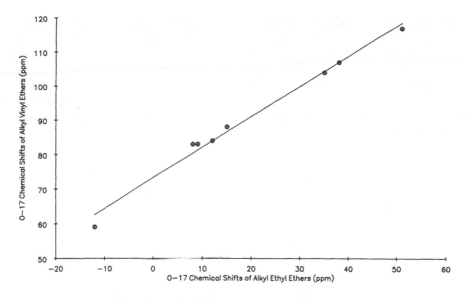

FIGURE 3.   Plot of the ^{17}O chemical shifts of alkyl vinyl ethers **223** to **229**[27] vs. the ^{17}O chemical shifts of the corresponding alkyl ethyl ethers.

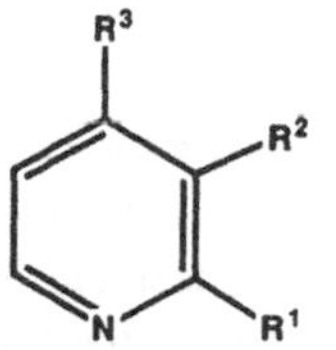

#	R^1	R^2	R^3	δ (^{17}O)
236	OCH=CH$_2$	H	H	142
237	H	OCH=CH$_2$	H	112
238	H	H	OCH=CH$_2$	129
239	H	CH$_2$OCH=CH$_2$	H	79
240	H	H	CH$_2$OCH=CH$_2$	73

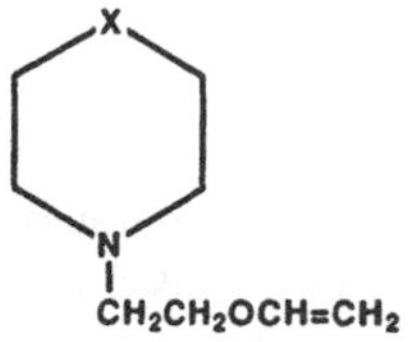

#	X	δ (^{17}O)
241	CH$_2$	79
242	O	80

SCHEME 18

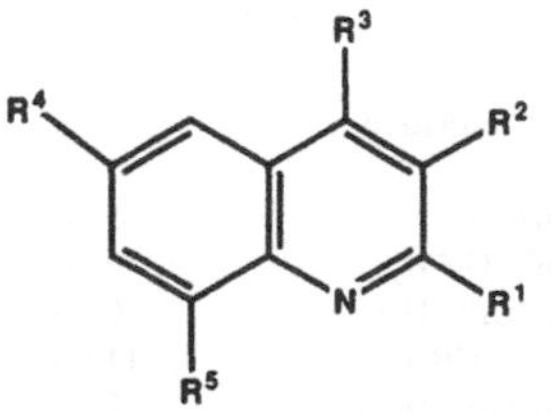

#	R^1	R^2	R^3	R^4	R^5	δ (^{17}O)
243	OCH=CH$_2$	H	H	H	H	146
244	H	OCH=CH$_2$	H	H	H	112
245	CH$_3$	H	OCH=CH$_2$	H	H	108
246	H	H	H	OCH=CH$_2$	H	127
247	H	H	H	H	OCH=CH$_2$	113

SCHEME 19

The effect of the introduction of successive double bonds at the α,β position to the ether oxygen on its ^{17}O chemical shift is not additive.[27,44] For instance, the deshielding effect of the first α,β-double bond on the ^{17}O shift of the oxygen in vinyl ethyl ether (**223**), in comparison with diethyl ether (**22**) is 82 ppm, but the introduction of the second double bond produces a smaller effect of only 41 ppm (**248** compared to **223**, Scheme 20). An extended system of conjugated double bonds, however, deshields the ether oxygen dramatically,[44] as evidenced by a 182 ppm difference in the ^{17}O chemical shifts of ethers **248** and **249**, (Scheme 20).

$CH_3CH_2OCH_2CH_3$	$CH_2=CHOCH_2CH_3$	
22	223	
δ (^{17}O) ppm 6.5	88.0	Δδ = 81.5

$CH_2=CHOCH_2CH_3$	$CH_2=CHOCH=CH_2$	
223	248	
δ (^{17}O) ppm 88.0	129.0	Δδ = 41.0

$CH_2=CHOCH=CH_2$	$CH_2=CHCH=CHOCH=CHCH=CH_2$	
248	249	
δ (^{17}O) ppm 129.0	241.0	Δδ = 112.0

SCHEME 20

The β-CH_3 substitution effects observed in the case of dialkyl ethers and aliphatic alcohols are operative in unsaturated ethers also[27] (Table 26). For alkyl vinyl ethers, single β-methyl substitution in the alkyl moiety results in a considerable deshielding (20 to 30 ppm) of the ether oxygen.[27] This effect is comparable to what is found for dialkyl ethers (12 to 30 ppm),[11] but smaller than β-CH_3 effects for aliphatic alcohols (35 to 43 ppm).[10] Deshielding of ether oxygen in alkyl vinyl ethers diminishes with successive β-CH_3 substitution, the second methyl group producing a 19.0 ppm change and the third one only a 10 ppm change.[27] The corresponding effects for dialkyl ethers are similar in magnitude (21 to 23 ppm and 11 to 13 ppm, respectively). However, for the aliphatic alcohols, the downfield shifts are larger: 23 to 24 ppm and 23 ppm, respectively. Even though Kalabin et al.[27] point out that π-

TABLE 26
β-SCS (CH_3) Effects in ^{17}O NMR of Unsaturated
Ethers

No.	Compound	$δ_{17O}$	β-SCS(CH_3)
222	$CH_2=CHOCH_3$	59.0[a]	
223	$CH_2=CHOCH_2CH_3$	88.0[a]	29.0
227	$CH_2=CHOCH(CH_3)_2$	107.0[a]	19.0
229	$CH_2=CHOC(CH_3)_3$	117.0[a]	10.0
224	$CH_2=CHOCH_2CH_2CH_3$	84.0[a]	
228	$CH_2=CHOCH(CH_3)CH_2CH_3$	104.0[a]	20.0
222	$CH_2=CHOCH_3$	59.0[a]	
232	$CH_2=C(CH_3)OCH_3$	62.5[b]	3.5
223	$CH_2=CHOCH_2CH_3$	88.0[a]	
233	$CH_2=C(CH_3)OCH_2CH_3$	98.0[a]	10.0

[a] Taken from Reference 27.
[b] Taken from Reference 39.

electron interactions, and, thus, the charge density on the oxygen atom relating to the $\langle r^{-3}\rangle_{2p}$ term in Equation 1, is important in determining the ^{17}O shifts of unsaturated ethers, they observed that the contribution of p,π-interactions to the net ^{17}O shielding of alkyl vinyl ethers **222** to **224** and **227** to **229** is quite independent of alkyl moieties. The explanation involving changes in the average excitation energies (ΔE^{-1} term in Equation 1), postulated for the β-CH$_3$ effects in dialkyl ethers,[11] may well apply in the case of β-CH$_3$ substitution in the alkyl moiety of vinyl alkyl ethers.

When the β-CH$_3$ substitution occurs in the vinyl group of vinyl alkyl ethers, for instance, **222→232; 223→233**, the deshielding effect (4 to 10 ppm)[27,39] is considerably reduced in comparison with the corresponding dialkyl ethers (21 to 22 ppm).[11] This effect is also smaller relative to changes in ^{17}O shifts for **222→223** and **223→227** (29 ppm and 19 ppm, respectively). Kalabin et al.[27] reasoned that the decrease in the deshielding effect in ethers **232** and **233** might be due to two opposing factors. The authors conclude that a moderate steric inhibition of conjugative interactions of the ether oxygen with the double bond, induced by the β-CH$_3$ substituent in the vinyl group, actually increases the charge density on the oxygen atom. The shielding produced as a consequence compensates for the deshielding due to β-CH$_3$ substitution, resulting in an overall smaller effect.

The usual deshielding of the ether oxygen due to β-methyl substitution (21 ppm) and a diminution of this effect (17 ppm) on further substitution are observed in the case of the bicyclic ethers **251** and **252**[33] (Scheme 21). It is interesting to note that this effect is intermediate between those seen for the acyclic and the monocyclic model compounds **147**, **253**, **254**[11] and **143**, **164**, **165**,[38] respectively.

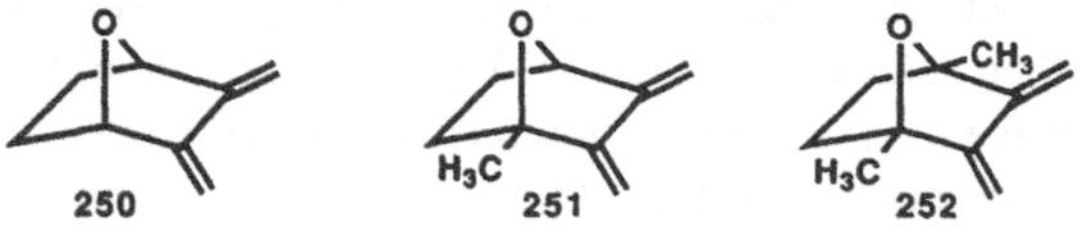

	250	251	252
δ (^{17}O) ppm	86.5	107.0	123.5
β - SCS(CH$_3$)		20.5	16.5

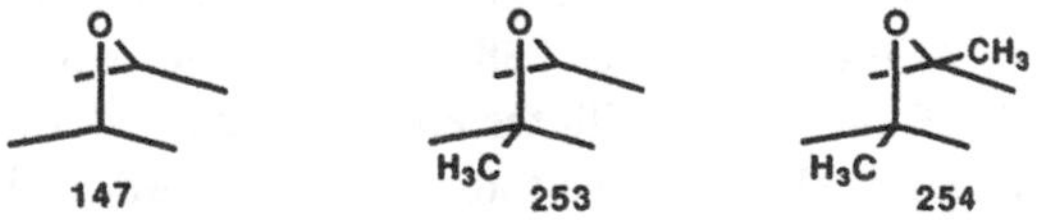

	147	253	254
δ (^{17}O) ppm	52.5	62.5	76.0
β - SCS(CH$_3$)		10.0	13.5

	143	164	165
δ (^{17}O) ppm	16.2	43.9	66.7 (cis), 74.5 (trans)
β - SCS(CH$_3$)		27.7	22.8 (cis), 30.6 (trans)

SCHEME 21

The γ and δ effects of methyl substitution in the alkyl moiety of vinyl alkyl ethers are unexceptionally small[27] (Table 27), being similar to what is observed for dialkyl ethers.[11] Contrastingly, γ-methyl substitution in the vinyl group of the ether **223**(**223→234**) produces an unusually large upfield shift of -18 ppm.[27] It is difficult to meaningfully interpret this result since this example appears to be the only one of this substitution effect reported in the literature.

TABLE 27
γ- and δ-SCS (CH$_3$) Effects in ^{17}O NMR of
Unsaturated Ethers

No.	Compound	$\delta_{^{17}O}$	SCS(CH$_3$)
		γ-CH$_3$ Effect	
223	CH$_2$=CHOCH$_2$CH$_3$	88[a]	
224	CH$_2$=CHOCH$_2$CH$_2$CH$_3$	84[a]	-4.0
226	CH$_2$=CHOCH$_2$CH(CH$_3$)$_2$	83[a]	-1.0
223	CH$_2$=CHOCH$_2$CH$_3$	88[a]	
234	CH$_3$CH=CHOCH$_2$CH$_3$	70[a]	-18.0
		δ-CH$_3$ Effect	
224	CH$_2$=CHOCH$_2$CH$_2$CH$_3$	84[a]	
225	CH$_2$=CHOCH$_2$CH$_2$CH$_2$CH$_3$	83[a]	-1.0

[a] Taken from Reference 27.

When the double bond is located at the β,γ position to an ether oxygen, there is negligible effect on its ^{17}O chemical shift, as seen for the vinyl ethers **255** and **256** (Table 28) in comparison with their saturated analogs.[39] Apparently, the double bond in these compounds is too remote from the oxygen atom to interact with it. However, β,γ-unsaturation produces marginal but opposite effects on the ^{17}O shifts of allyl phenyl ether **257** (-7 ppm)[27] and the oxirane **157** (8.5 ppm).[35]

For unsaturated ethers, in which transannular interactions are possible between the oxygen atom and the β,γ-double bond, large downfield effects on the ^{17}O shifts of the ether oxygen are observed. 7-Oxanorbornanes **258**, **259**, and **260**, which contain endocyclic β,γ-double bonds capable of interacting with the ether oxygen, exhibit large deshielding effects ranging from 24 to 82 ppm relative to the saturated ethers **148, 250**, and **258**, respectively.[33] However, when the double bond is exocyclic and cannot interact with the oxygen atom, as in **250** and **261**, the deshielding effect becomes insignificant.[33] These effects have been qualitatively rationalized[33] as arising from a change in the average excitation energy, ΔE term in Equation 1, which follows from a charge transfer between the double bond and the oxygen function. Such a charge transfer should be more efficient with an endocyclic double bond than an exocyclic one. It should be noted that effects similar to the ones discussed above have been noted for ^{13}C shifts of C-7 carbon in norbornanes.[46,47]

Unsaturation at the γ,δ position to the ether oxygen produces small effects on its ^{17}O chemical shift, if the double bond cannot interact with the oxygen atom, as in the acyclic ether **265**[39] and the cyclic ether **157**[34] (Table 29). However, when such through-space interaction is possible, as in the case of **158**,[34] significant deshielding of the oxygen function (67 ppm) is caused by the remote γ,δ-double bond. Iwamura et al.[34] have attributed this effect to a mixing of the π^* orbital of the double bond with the antibonding Walsh orbitals of the epoxide ring in **158**. The resulting decrease of the π^* orbital energy should lower

TABLE 28
^{17}O NMR Chemical Shifts of β,γ-Unsaturated Ethers

Compound no.	Unsaturated ethers	$\delta_{^{17}O}$	Compound no.	Saturated ethers	$\delta_{^{17}O}$	$\Delta\delta^a$
255	$CH_2=CHCH_2OCH_2CH_3$	1.5[b]	25	$CH_3CH_2OCH_2CH_3$	1.7[b]	−0.5
256	$CH_2=CHCH_2OCH_2CH=CH_2$	−5.5[b]	263	$CH_3CH_2CH_2OCH_2CH_2CH_3$	−3.5[b]	2.0
257	$CH_2=CHCH_2OC_6H_5$	68.0[c]	264	$CH_3CH_2CH_2OC_6H_5$	75.0[c]	−7.0
157	(epoxide)–$CH=CH_2$	−9.5[d]	156	(epoxide)–CH_2CH_3	−18.0[f]	8.5
258	[structure]	110.0[e]	148	[structure]	85.5[e]	24.0
259	[structure]	113.5[e]	250	[structure]	86.5[e]	27.0
260	[structure]	192.5[e]	258	[structure]	110.0[e]	82.5
250	[structure]	86.5[e]	148	[structure]	85.5[e]	1.0
261	[structure]	89.0[e]	250	[structure]	86.5[e]	2.5
262	[structure]	−7.0[e]	143	[structure]	14.0[e]	−21.0

a Shift for the unsaturated ether minus shift for the saturated ether.
b From Reference 39.
c From Reference 27.
d From Reference 35.
e From Reference 33.
f From Reference 34.

TABLE 29
17O NMR Chemical Shifts of γ,δ - Unsaturated Ethers

Compound no.	Unsaturated ethers	δ_{17O}	Compound no.	Saturated ethers	δ_{17O}	$\Delta\delta$[a]
265	$CH_2{=}C(CH_3)CH_2CH_2OCH_3$	−22.5[b]	37	$CH_3CH(CH_3)CH_2CH_2OCH_3$	−29.5[b]	7.0
157		1.0[c]	150		3.0[c]	−2.0
266		−1.0[d]	156		−14.0[d]	13.0
158		53.0[c]	156		−14.0[d]	67.0

[a] Shift for the unsaturated ether minus shift for the saturated ether.
[b] From Reference 11.
[c] From Reference 34.
[d] From Reference 33.

the n→π* transition energy (ΔE), and produce a downfield effect according to Equation 1. These authors have also noted a large deshielding effect (19 ppm) for the ^{13}C shift for the similarly situated carbon of the hydrocarbon analog of **158**, in comparison with the corresponding saturated compound.[34] Compared to **158**, which contains an endocyclic double bond, it is difficult to explain the downfield shift of 13 ppm experienced by the oxirane **266**, which has γ,δ-double bonds exocyclic to the norbornane ring.[33] Also puzzling is the fact that the γ,δ-double bond in **266** gives rise to a larger deshielding effect than the β,γ-double bond in **250** which is closer spatially to ether oxygen atom.

Deshielding of the ether oxygen due to a cyclopropyl moiety[27] is smaller than the effect produced by an isopropyl substituent[11] (Table 30), strongly suggesting the nonparticipation of the cyclopropyl ring in σ,p-conjugation with the ether oxygen atom. This result is in accord with the effects observed in ^{13}C spectra of cyclopropyl ethers.[48]

TABLE 30
A Comparison of the Effects of Cyclopropyl and Isopropyl Groups on the ^{17}O NMR Chemical Shifts of Ether Oxygen Atoms

No.	Compound	δ_{17O}	$\Delta\delta$[a]
25	$CH_3CH_2CH_2OCH_2CH_3$	1.7[b]	
23	$(H_3C)_2CH-OCH_2CH_3$	28.0[b]	26.3
267	▷–OCH₂CH₃	20.0[c]	18.3
268	$CH_3CH_2CH_2OCH(CH_3)_2$	24.0[b]	
269	$(H_3C)_2CH-OCH(CH_3)_2$	52.5[b]	28.5
270	▷–OCH(CH₃)₂	40.0[c]	16.0
224	$CH_3CH_2CH_2OCH{=}CH_2$	84.0[c]	
227	$(H_3C)_2CH-OCH{=}CH_2$	107.0[c]	23.0
271	▷–OCH=CH₂	87.0[c]	3.0

[a] Shift for the isopropyl compound or the cyclopropyl compound minus the shift for the corresponding n-propyl compound.
[b]. Taken from Reference 11.
[c] Taken from Reference 27.

VIII. SINGLE-BONDED OXYGEN IN ACIDS, ESTERS, LACTONES, AND ANHYDRIDES

The deshielding effect of a carbon-oxygen double bond on the ^{17}O chemical shift of a single-bonded oxygen directly attached to it is much larger than what is observed for a carbon-carbon double bond on a similarly situated oxygen atom.[49] A comparison of the two effects is made in Table 31 using the ^{17}O shifts of vinyl alkyl ethers[27] and alkyl esters[49]

TABLE 31

A Comparison of the ^{17}O NMR Chemical Shifts
of the Single-Bonded Oxygen in Esters and the
Ether Oxygen in Vinyl Alkyl and Dialkyl
Ethers

No.	Compound	δ_{17O}	$\Delta\delta^a$
12	$CH_3CH_2-O-CH_3$	-22.5^b	
222	$CH_2=CH-OCH_3$	59.0^c	81.5
272	$O=CH-OCH_3$	138.0^c	160.5
13	$CH_3CH(CH_3)-OCH_3$	-2.0^b	
232	$H_2C=C(CH_3)-OCH_3$	62.5^d	64.5
273	$O=C(CH_3)-OCH_3$	134.0^c	136.0
22	$CH_3CH_2OCH_2CH_3$	6.5^b	
223	$CH_2=CH-OCH_2CH_3$	88.0^c	81.5
274	$O=CH-OCH_2CH_3$	169.0^c	162.5
23	$CH_3CH(CH_3)-OCH_2CH_3$	28.0^b	
233	$CH_2=CH(CH_3)OCH_2CH_3$	98.0^c	70.0
275	$O=C(CH_3)OCH_2CH_3$	165.0^c	137.0

a   Shift for the vinyl alkyl ether or the ester minus shift for the dialkyl ether.
b   Taken from Reference 11.
c   Taken from Reference 27.
d   Taken from Reference 39.
e   Taken from Reference 49.

relative to those of the dialkyl ethers.[11] The downfield shifts produced by the C=O bond (103 to 161 ppm) are almost twice the effect due to the C=C bond. Electron density of the single-bonded oxygen should be delocalized to a higher degree into a C=O bond (esters) than a C=C bond (vinyl alkyl ethers). As a consequence, the larger depletion of electronic charge on the ester single-bonded oxygen, accompanied by an increase in the $\langle r^{-3} \rangle_{2p}$ term, leads to the greater deshielding effect relative to the oxygen atom in vinyl ethers. Several systems containing single-bonded oxygen atoms attached to a C=O bond, viz., carboxylic acids, esters, lactones, and anhydrides, have been investigated, and in this section, substituent effects on the ^{17}O chemical shifts of the single-bonded oxygen atoms in these systems will be discussed.

Carboxylic acids display only one ^{17}O resonance, suggesting the two oxygen atoms are equivalent due to fast proton exchange in dimeric or polymeric forms[49] (Scheme 22).

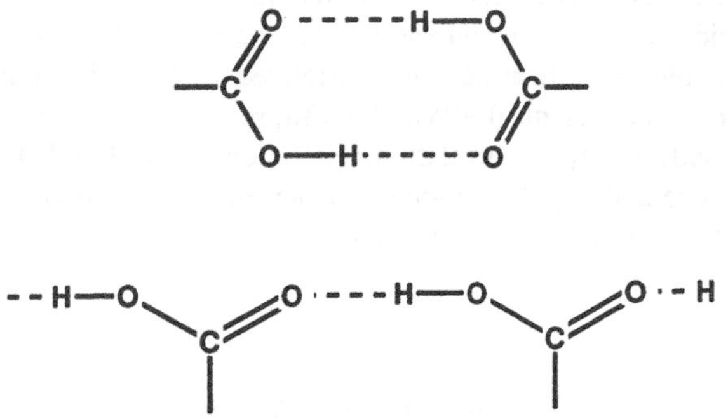

SCHEME 22

Substitution of hydrogen atoms in the acyl moiety by methyl groups results in small β-, γ-, and δ-effects on the ^{17}O shifts of carboxylic acids[49] (Table 32). While β- and γ-methyl groups produce slight shielding, the δ-methyl group deshields the oxygen by a small magnitude.

TABLE 32
β^{π}-, γ^{π}-, and δ^{π}-SCS(CH$_3$) Effects in ^{17}O NMR Spectra of Carboxylic Acids

No.	Compound	δ_{17O}[a]	SCS(CH$_3$)
			β^{π}-effect
276	HCOOH	253.5	
277	CH$_3$COOH	250.5	-3.0
			γ^{π}-effect
277	CH$_3$COOH	250.5	
278	CH$_3$CH$_2$COOH	244.0	-6.5
279	(CH$_3$)$_2$CHCOOH	242.0	-2.0
280	(CH$_3$)$_3$CCOOH	240.0	-2.0
			δ^{π}-effect
278	CH$_3$CH$_2$COOH	244.0	
279a	CH$_3$CH$_2$CH$_2$COOH	246.5	2.5
280a	(CH$_3$)$_2$CHCH$_2$COOH	248.5	2.0
281	(CH$_3$)$_3$CCH$_2$COOH	253.5	5.0

[a] Taken from Reference 49.

Boykin and co-workers[50] investigated the electronic effects of substituents on the ^{17}O chemical shifts of carboxylic acid oxygen atoms in a study of *p*-substituted benzoic and cinnamic acids (Table 33). Electron-donating substituents shielded the carboxylic oxygen atoms, while the electron-withdrawing ones provided deshielding. As expected from earlier work on substituent effects in these systems by other physical methods,[51,52] the transmission of effects in cinnamic acids was reduced relative to benzoic acids. The range of ^{17}O chemical shifts for benzoic acids was 11 ppm, whereas for cinnamic acids, it was only 5 ppm. The ^{17}O shift data for benzoic and cinnamic acids correlated with the Hammett-type substituent constant σ^+ (correlation coefficients 0.995 and 0.986, respectively), and in a dual substituent parameter analysis, with σ_I, σ_{R+} (correlation coefficients 0.999 and 0.994, respectively), indicating that resonance and π-polarization contributions, which affect the oxygen electron density, govern the ^{17}O shifts in these systems.[50]

TABLE 33
^{17}O NMR Chemical Shifts for
***p*-Substituted Benzoic and**
Cinnamic Acids

Substituent	$\delta_{^{17}O}$ [a]	$\delta_{^{17}O}$ [b]
NO$_2$	256.1	257.2
CN	255.1	256.8
CF$_3$	254.1	255.8
Cl	251.1	254.6
H	250.5	254.1
OAc	249.6	
F	249.6	253.6
CH$_3$	248.6	253.0
OCH$_3$	245.6	251.8

[a] Benzoic acids (from Reference 50).
[b] Cinnamic acids (from Reference 50).

For a limited series of substituted benzoic acids, Ricca and co-workers[53] observed that *ortho* substituents caused large deshielding of the carboxyl oxygen atoms while groups at the *meta* position had negligible effect. *Ortho* methoxy, chloro and nitro groups shifted the oxygen resonance downfield by 23 ppm, and the *meta* substituents produced little change. The authors attributed the origin of this effect to steric inhibition of resonance, and found a fair correlation between the ^{17}O SCS values and σ, calculated from the Yukawa and Tsuno equation[54] using terms to represent proximity polar effects and steric effects. The ortho substituents in phenylacetic acids, in which the carboxyl group is one carbon atom removed from the aromatic ring and devoid of steric interactions, show normal effects.[53] The electron-donating methoxy group shields (by 5 ppm) the carboxyl oxygen atoms, and the electron-attracting nitro substituent has the opposite effect (4 ppm). Correlation of the data with σ_p^o indicates that proximity polar effects are important in regulating the electron density of the oxygen atom, which determines the ^{17}O shifts of these compounds. Methoxy, chloro and nitro groups located at the *meta* position in phenylacetic acids produce deshielding effects (7 to 8 ppm) consistent with electron withdrawal through inductive interactions, and the *para* substituents result in smaller effects.[53]

A distinct ^{17}O resonance is observed for the single-bonded oxygen in carboxylic esters. Methyl substitution of the hydrogen atoms in the acyl group generates modest β^π, γ^π, and δ^π effects on the ^{17}O shifts of the single bonded oxygen.[49,55] As can be seen from Table 34, β^π and γ^π effects are shielding (3 to 4 ppm)[49,55] and δ^π effect is deshielding (1 to 5 ppm),[49]

TABLE 34
β^π-, γ^π-, and δ^π-SCS(CH$_3$) Effects on the Single-Bonded Oxygen in Carboxylic Esters due to Methyl Substitution in the Acyl Group

No.	Compound	δ(–O–)	SCS(CH$_3$)
			β^π-effect
272	HCOOCH$_3$	137.5[a]	
273	CH$_3$COOCH$_3$	134.0[a]	−3.5
274	HCOOC$_2$H$_5$	169.0[a]	
275	CH$_3$COOC$_2$H$_5$	165.0[a]	−4.0
			γ^π-effect
273	CH$_3$COOCH$_3$	134.0[a]	
282	CH$_3$CH$_2$COOCH$_3$	130.0[a]	−4.0
283	(CH$_3$)$_2$CHCOOCH$_3$	127.0[a]	−3.0
284	(CH$_3$)$_3$CCOOCH$_3$	124.0[a]	−3.0
275	CH$_3$COOC$_2$H$_5$	165.0[a]	
285	CH$_3$CH$_2$COOC$_2$H$_5$	161.0[a]	−4.0
286	(CH$_3$)$_2$CHCOOC$_2$H$_5$	158.5[a]	−2.5
287	(CH$_3$)$_3$CCOOC$_2$H$_5$	152.0[a]	−6.5
288	CH$_3$COOC(CH$_3$)$_3$	204.4[b]	
289	CH$_3$CH$_2$COOC(CH$_3$)$_3$	200.0[b]	−4.4
290	(CH$_3$)$_2$CHCOOC(CH$_3$)$_3$	198.9[b]	−1.1
			δ^π-effect
282	CH$_3$CH$_2$COOCH$_3$	130.0[a]	
283	CH$_3$CH$_2$CH$_2$COOCH$_3$	132.0[a]	2.0
284	(CH$_3$)$_2$CHCH$_2$COOCH$_3$	132.5[a]	0.5
285	(CH$_3$)$_3$CCH$_2$COOCH$_3$	137.5[a]	5.0

[a] Taken from Reference 49.
[b] Taken from Reference 55.

which parallel the corresponding shifts for carboxylic acids. In contrast to the effects shown in Table 34, substitution of β-hydrogen atoms in the alkyl fragment attached to the ester oxygen in esters by methyl substituents induces large downfield shifts on the single-bonded oxygen.[18,49,55] Table 35 exemplifies this effect; it is noted that the shift due to first methyl substitution is as large as 30 ppm and that successive substitution leads to an attenuation of the effect. These β-SCS(CH$_3$) effects are similar to those found for alcohols[10] and ethers,[11] but lower in magnitude, and may be attributed to smaller changes in the average excitation energy (ΔE term in Equation 1) upon methyl substitution. Introduction of a carbon-carbon double bond at the α,β position to the ester oxygen or an aryl group at the α position deshields the oxygen by 38 to 44 ppm[55] (Table 36). This observation is consistent with the

TABLE 35
β-SCS(CH₃) Effects on the Single-Bonded
Oxygen in Carboxylic Esters due to Methyl
Substitution in the Alkyl Group at the Ester
Oxygen

No.	Compound	δ_{17_O} (–O–)	β-SCS(CH₃)
272	HCOOCH₃	138.0[a]	
274	HCOOCH₂CH₃	169.0[a]	31.0
291	HCOOCH(CH₃)₂	200.0[b]	31.0
292	HCOOC(CH₃)₃	212.0[b]	12.0
273	CH₃COOCH₃	134.0[a]	
275	CH₃COOCH₂CH₃	165.0[a]	31.0
293	CH₃COOCH(CH₃)₂	196.0[c]	31.0
288	CH₃COOC(CH₃)₃	204.4[c]	8.4

[a] Taken from Reference 49.
[b] Taken from Reference 18.
[c] Taken from Reference 55.

TABLE 36
Effect of Carbon-Carbon Double Bond and
Aryl Groups on the ¹⁷O NMR Chemical
Shift of the Ester Oxygen in Esters

No.	Compound	δ_{17_O}(–O–)	Δδ[c]
275	CH₃COOCH₂CH₃	165.0[a]	
294	CH₃COOCH=CH₂	202.6[b]	37.6
295	CH₃COOC₆H₅	208.9[b]	43.9
296	CH₃COOCH₂CH=CH₂	165.4[b]	0.4
297	CH₃COOCH₂C₆H₅	171.1[b]	6.1

[a] Taken from Reference 49.
[b] Taken from Reference 55.
[c] Shift for vinyl or aryl ester minus shift for alkyl
     ester.

expectation that a reduction in electron density on the single-bonded oxygen, due to delocalization into the vinyl or aryl group, increases the $\langle r^{-3}\rangle_{2p}$ term in Equation 1 and produces deshielding. The downfield shifts (Δδ) seen in Table 36 are smaller relative to similar effects for dialkyl ethers (Table 25). Competition between the vinyl or aryl group and the carbonyl function for electron density on the single-bonded oxygen in esters is probably the cause for the attenuated effects seen in Table 36. When the double bond or the aryl group is one carbon atom removed from the ester oxygen, as in allyl or benzyl acetates, the resulting effect is, expectedly, small.

Electronic effects of substituents on the ^{17}O chemical shifts of the ester oxygen in p-substituted methyl benzoates and cinnamates were found to be similar to those obtained for the corresponding acids[50] (Table 37). The ^{17}O shift range was 6 ppm for benzoates and 3 ppm for cinnamates, revealing that the attenuation of transmission of effects across the carbon-carbon double bond in cinnamoyl systems is quite general. Dual substituent parameter analysis of the data for benzoates showed a good correlation with σ_{R+} (correlation coefficient, 0.995), indicating that electron density changes due to resonance interactions between $para$ substituents and the single-bonded oxygen determine the ^{17}O shifts in this system. The poor correlation for the cinnamates with σ_{R+} (correlation coefficient, 0.973) has been attributed to the small range of shifts in this series of compounds.[50]

TABLE 37
^{17}O **NMR Chemical Shifts for the Single-Bonded Oxygen in p-Substituted Methyl Benzoates, Methyl Cinnamates, and Benzyl Acetates**

Substituent	δ_{17O}		
X	p–X–C$_6$H$_4$COOCH$_3$[a]	p–X–C$_6$H$_4$CH=CHCOOCH$_3$[a]	p–X–C$_6$H$_4$CH$_2$OCOCCH$_3$[b]
NO$_2$	132.0	138.1	164.6
CN	131.3	137.0	165.1
CF$_3$	130.2	135.8	166.5
Cl	128.6	135.4	169.1
H	128.0	134.2	170.5
F	128.7	134.7	171.5
CH$_3$	126.9	134.0	171.6
OCH$_3$	125.7	134.8	174.3
N(CH$_3$)$_2$	—	—	177.6

[a] Taken from Reference 50.
[b] Taken from Reference 28a.

While the ester oxygen in p-substituted benzoates display normal substituent effects, the analogous oxygen in benzyl acetates respond in the opposite way toward $para$ substituents.[28a] As can be seen from Table 37, electron-withdrawing substituents in benzyl acetates shield the ester oxygen and electron-donating groups produce deshielding. These reversed substituent effects have also been observed in a comparison of ^{19}F chemical shifts of benzyl fluorides and benzoyl fluorides.[56]

An interesting example in which the ^{17}O shifts of the carbonyl oxygen and the ester oxygen show a large response to an equalization of the C=O and C–O bond orders in esters has been reported in the case of p-substituted (trifluorosilyl)methyl benzoates.[57] The data in Table 38 reveal that the carbonyl oxygen in the trifluorosilyl esters are shielded by 52 to 65 ppm relative to similar oxygen atoms in the trifluoroethyl esters, whereas the corresponding effect for the ester oxygen atom is deshielding by 18 to 25 ppm. These results

TABLE 38

^{17}O NMR Chemical Shifts for *p*-Substituted (Trifluorosilyl) Methyl Benzoates and 2,2,2-Trifluoroethyl Benzoates

Substituent (X)	$\delta_{17O}(-O-)^a$			$\delta_{17O}(C=O)^a$		
	p-x-C$_6$H$_4$COOCH$_2$SiF$_3$	p-x-C$_6$H$_4$COOCH$_2$CF$_3$	Δδ	p-x-C$_6$H$_4$COOCH$_2$SiF$_3$	p-x-C$_6$H$_4$COOCH$_2$CF$_3$	Δδ
OCH$_3$	148.2	125.3	22.9	265.2	329.7	−64.5
CH$_3$	148.9	125.9	23.0	275.8	337.6	−61.8
H	151.6	126.6	25.0	279.2	340.3	−61.1
F	149.4	124.9	24.5	280.2	339.3	−59.1
Cl	149.9	126.6	23.3	283.3	341.4	−58.1
Br	149.8	127.6	22.2	283.3	340.8	−57.5
NO$_2$	149.6	131.8	17.8	298.5	351.0	−46.5

^{a} Taken from Reference 57.

have been explained as the result of an intramolecular donor-acceptor bond between the carbonyl oxygen and the silicon atom in the silyl esters as shown in Scheme 23. The charge transfer from the carbonyl oxygen to the d-orbitals of silicon is accompanied by an increase in p,π-conjugation between the ester oxygen and the carbonyl group resulting in the observed effects.[57]

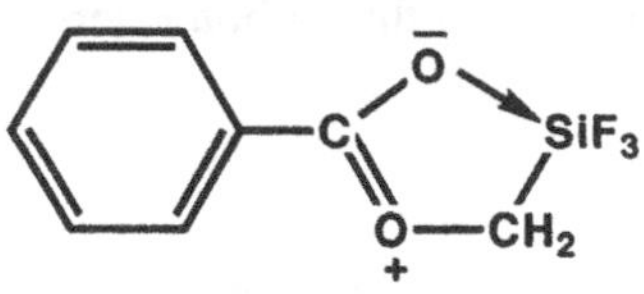

SCHEME 23

In a study of a series of sterically hindered aromatic methyl esters, in which electronic effects on ^{17}O chemical shifts were fairly constant, Baumstark et al.[58] found that the single-bonded ester oxygen experienced large deshielding as a result of increased steric interactions. The ^{17}O shifts of the esters listed in Table 39 correlated well with the torsion angle obtained by molecular mechanics calculations, defined as the difference angle between the planes of the aromatic ring and the carbonyl group (correlation coefficient 0.992). The authors[58] suggested this correlation stems from van der Waals interactions between the ester function and the adjacent substituents resulting in changes in torsion angle with substituents. This explanation is in accord with Chesnut's hypothesis[59,60] that repulsive van der Waals inter-actions produce contraction of orbitals in heavy-atom nuclei, which results in an increase in the $\langle r^{-3} \rangle_{2p}$ term in Equation 1 and a deshielding effect.

TABLE 39
^{17}O NMR Chemical Shifts for Aromatic Esters
$ArCOOCH_3$

Compound No.	Ar	δ_{17O} (–O–)[a]	Torsion angle[b]
298	C_6H_5	128.0	2
299	4-MeC_6H_4	127.0	2
300	2-naphthyl	129.0	2
301	2-MeC_6H_4	138.5	29
302	$2,3\text{-Me}_2C_6H_3$	141.0	29
303	1-Naphthyl	139.0	33
304	$2,6\text{-Me}_2C_6H_3$	150.0	54
305	$2,4,6\text{-Me}_3C_6H_2$	149.0	54
306	9-anthryl	154.0	67
307	$2,4,6\text{-tBu}_3C_6H_2$	162.0	76

[a] Taken from Reference 58.
[b] Difference angle between the plane of the aromatic ring and the plane of the carbonyl group.

Replacement of alkyl groups directly attached to the ester oxygen in carboxylic esters by analogous silicon substituents produces large upfield shifts for the single-bonded oxygen.[61] This shielding effect, exemplified in Table 40, ranges from -25 to -35 ppm, and is probably due to a higher excitation energy (ΔE) required to effect the $n \rightarrow \pi^*$ transition for the ester oxygen lone pair of electrons in the silicon esters. In these compounds, the p_π-d_π interaction between the oxygen lone pair of electrons and the silicon d-orbitals may be responsible for the higher ΔE relative to the carbon esters.

TABLE 40
A Comparison of the ¹⁷O NMR Chemical Shifts for the Single-Bonded Oxygen Atoms in Carboxylic Esters and Their Silicon Analogs, RCOOM(CH₃)₃; M = C, Si

R	δ_{17O}(-O-)		
	M = C[a]	M = Si[b]	$\Delta\delta$[c]
CH₃	204.4	179.0	-25.4
CH₃CH₂	200.0	174.0	-26.0
(CH₃)₂CH	198.9	169.0	-29.9
C₆H₅CH₂	203.6	176.0	-27.6
C₆H₅	198.0	163.0	-35.0

[a] Taken from Reference 55.
[b] Taken from Reference 61.
[c] Shift for the silicon compound minus shift for the carbon compound.

In carboxylic esters containing group IVA atoms attached to the ester oxygen, the silicon compounds display two clearly resolved ¹⁷O signals for the carbonyl and ester oxygen atoms, whereas the germanium and tin derivatives show only one signal. For instance, in silicon esters of the type $(CH_3)_3SiOC(O)R$ with R = H, CH₃, C₂H₅, (CH₃)₂CH, CH₃COCH₂CH₃, C₆H₅CH₂, C₆H₅, ClCH₂, Cl₂CH, Cl₃C, F₃C, NCCH₂, CH₃OCH₂, (CH₃)₃SiCH₂, and (CH₃)₃SiOCH₂, the ester and the carbonyl oxygen atoms resonate between 156 to 182 ppm and 370 to 396 ppm, respectively.[61] However, the germanium and tin compounds, $R_3GeOC(O)CH_3$ (R = n-C₄H₉, δ_{17O} = 277 ppm; R = C₆H₅, δ_{17O} = 290 ppm),[61] and (n-C₄H₉)₃SnOC(O)R (R = C₆H₅, δ_{17O} = 261 ppm; R = CH₂SC₆H₅, δ_{17O} = 276 ppm)[62] display only one ¹⁷O signal which appears approximately midway between the carbonyl and the ester oxygen resonances of the silicon esters. The equivalence of the two oxygen atoms has been ascribed to a fast intramolecular exchange of the group IVA metal substituent or a structure involving chelate bonding of the metal to the carboxy group as shown in Scheme 24.

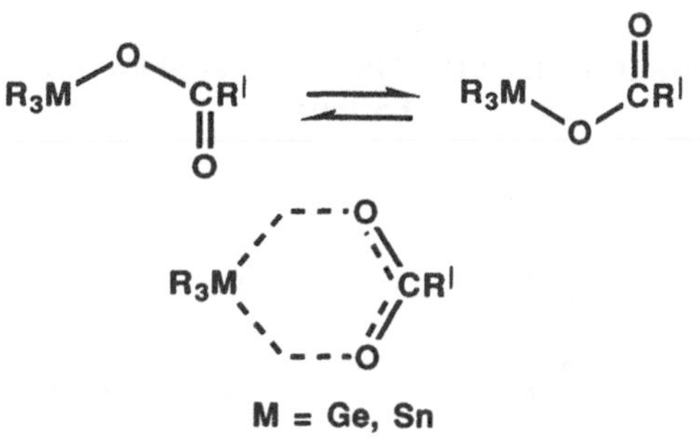

M = Ge, Sn

SCHEME 24

When the number of acyloxy groups attached to the metal atom increases to two, even the silicon esters display only one ^{17}O signal,[61] as is found for the compounds, $CH_3(XCH_2)Si(OCOCH_3)_2$ with $X = H$, $\delta_{17O} = 270$ ppm; $X = Cl$, $\delta_{17O} = 285$ ppm; $X = O-C(O)CH_3, \delta_{17O} = 284$ ppm. Tin derivatives containing two acyloxy groups also have a single peak in their ^{17}O spectra. For instance, esters of the type $(n-C_4H_9)_2Sn(OOCX)_2$ with $X = CH_3$, C_6H_5, $CH_2OC_6H_5$ and $CH_2SC_6H_5$ show an ^{17}O resonance at 291, 275, 249, and 272 ppm, respectively.[62] Chelate bonding of the two carboxy groups to the metal in a structure shown in Scheme 25 has been invoked to explain the appearance of a single ^{17}O peak for the diacetoxy esters. In the absence of such chelate bonding, geminal diacetates of the type $RCH(O(CO)CH_3)_2$ with $R = H$, CH_3, C_2H_5, $1-C_6H_{13}$, CCl_3, C_6H_5, $C_6H_5CH=CH$, 2-furyl exhibit two distinct ^{17}O resonances, with the single-bonded oxygen appearing at 185, 197, 203, 201, 188, 195, 219, and 201 ppm, respectively.[63] Even an increase in temperature over a range of 100°C did not produce equalization of the two signals.

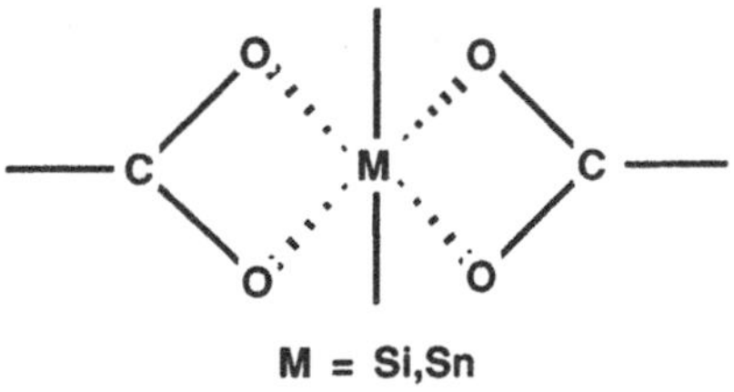

M = Si,Sn

SCHEME 25

When the single-bonded ester oxygen is situated in a ring system such as lactone, it undergoes substantial deshielding relative to its acyclic counterpart.[64,65a-c] This cyclization effect, $(\Delta\delta_{cyclization})$ after correcting for a β-alkyl effect of $+30$ ppm, shown in Table 41, depends strongly on the strain of the lactone ring which contains the ester oxygen. The downfield shift in lactones ranges from 1 to 77 ppm, and increases with increasing strain of the ring. Boykin et al.[65c] observed that the ^{17}O chemical shift of the ester oxygen in 4- to 6-membered lactones was strongly influenced by structural variations. The following effects were noted: deshielding (24 to 28 ppm) due to β-alkyl substitution, shielding (-3.7 ppm) from γ-alkyl substitution, and deshielding (32 ppm) by an α,β double bond. Due to extensive delocalization of its lone pair of electrons, the ester oxygen, in a 6-membered lactone containing a series of double bonds in conjugation with the carbonyl group, underwent large deshielding (79 ppm).[65c]

TABLE 41
Cyclization Effects on the 17O NMR Chemical Shifts of Ester Oxygen in Lactones

No.	Compound	δ_{17O}	$\delta_{corrected}$[a]	$\Delta\delta_{cyclization}$[b]
273		134.0[c]		
308		235.0[d]	205.0	71.0
		241.0[e]	211.0	77.0
282		130.0[c]		
309		174.0[d]	144.0	14.0
		178.5[e]	148.5	18.5
310		132.0[f]		
311		174.0[g]	144.0	12.0
		172.7[e]	142.7	10.7
283		132.0[c]		
312		163.0[d]	133.0	1.0
		167.0[e]	137.0	5.0

[a]   Shift for lactone minus shift due to β-alkyl effect taken to be +30 ppm.
[b]   Shift for lactone minus shift for the corresponding acyclic ester.
[c]   From Reference 49.
[d]   From Reference 64.
[e]   From Reference 65c.
[f]   From Reference 18.
[g]   From Reference 65a.

Cumulative deshielding effects are observed for the dicoordinated oxygen in going from dimethyl ether (-52.5 ppm)[11] to methyl acetate (134.0 ppm)[49] to acetic anhydride (273.0 ppm).[66] Cyclization also produces downfield shifts for the anhydride oxygen. As seen in Scheme 26, similar to the lactone oxygen, the cyclization effect increases in magnitude with greater ring strain.[66] The deshielding effect upon cyclization for the anhydride oxygen has been attributed to a greater overlap of lone pair orbitals with the carbonyl π system resulting in an increase in double bond character. Alkyl and aryl groups in the anhydride ring have small shielding effects on the single-bonded oxygen in succinic and glutaric anhydrides, as expected, since such a substitution involves the placement of these groups at the γ and δ positions to the anhydride oxygen.[66] Interestingly, these effects are similar to those observed for tetrahydropyran and 1,3-dioxanes.[38]

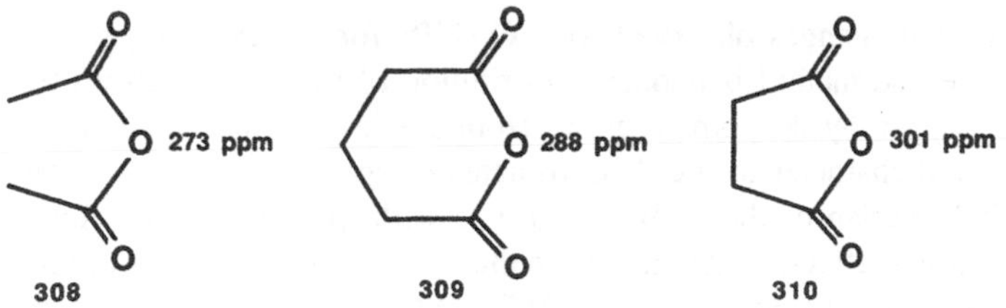

SCHEME 26

An endo double bond at the α,β position to the carbonyl group strongly shields the single-bonded oxygen in anhydrides, whereas an exo double bond has a less prominent effect.[66] This effect is exemplified in Scheme 27 by a comparison of succinic anhydride with maleic and itaconic anhydrides. In close similarity with succinic anhydrides, alkyl, aryl, and halogen substitutions in the unsaturated maleic anhydride ring system shield the anhydride oxygen by a small magnitude.[66]

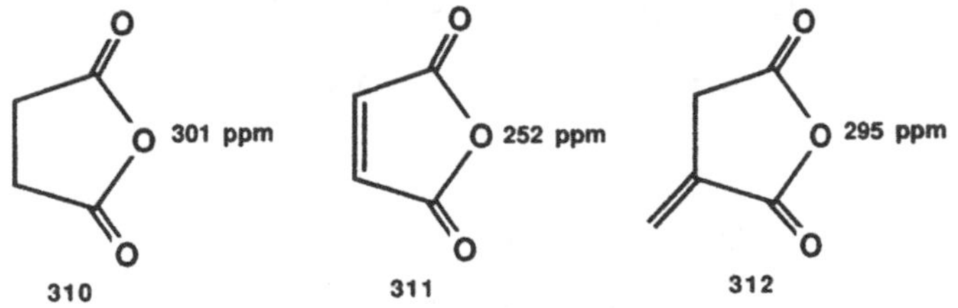

SCHEME 27

Conformational effects play a role in deshielding the single-bonded oxygen in *trans*-1,2-cyclohexane dicarboxylic anhydride relative to the *cis* derivative[66] (Scheme 28). Interaction of the double bond with the endo anhydride oxygen in the norbornene compound **315** produces a deshielding of 11 ppm with respect to succinic anhydride,[66] which is similar to shifts observed for ethers.[33] However, this effect disappears in the 7-oxanorbornene system.[66]

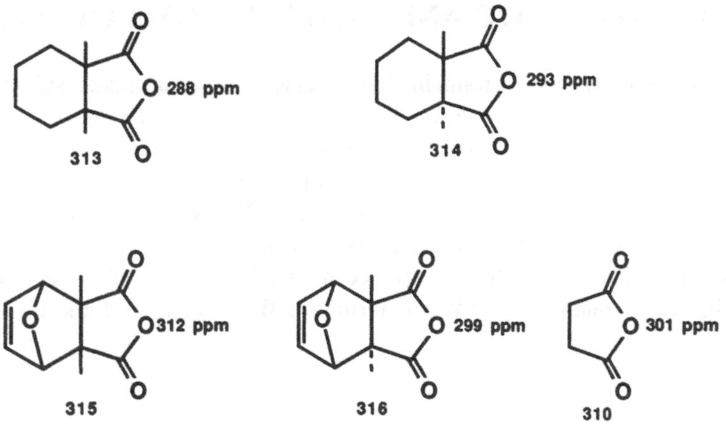

SCHEME 28

In analogy with changes observed for ¹⁷O shifts for the ester oxygen, in a comparison of methyl acetate and methyl benzoate, substitution of the methyl group by phenyl shields the anhydride oxygen by 20 ppm in going from acetic to benzoic anhydride.[66] A decrease in the double bond character of the dicoordinated oxygen in the aromatic anhydride, due to greater π-orbital overlap of the carbonyl groups with the phenyl ring, has been suggested to explain this upfield effect. Consistent with this explanation is the finding that the anhydride oxygen in fused aromatic systems such as **317** to **320** (Scheme 29), in which conformational restrictions promote greater π overlap with the carbonyl group relative to the aryl ring, the ¹⁷O shift of the anhydride oxygen is closer to that of acetic anhydride.[66,67]

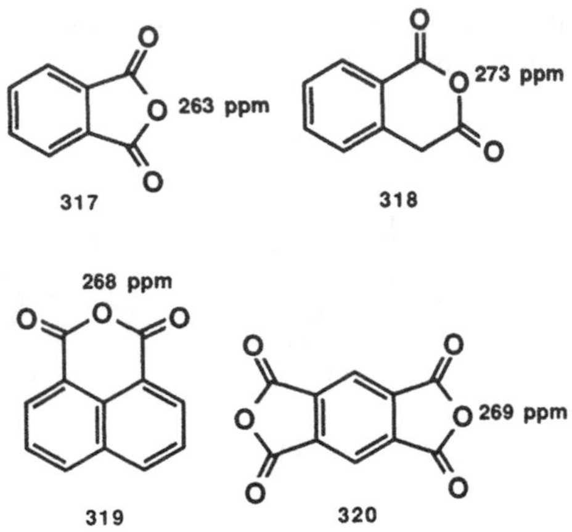

SCHEME 29

IX. AROMATIC AND HETEROAROMATIC ETHERS

Replacement of an alkyl group in dialkyl ethers by an aromatic substituent deshields the ether oxygen, as expected, due to delocalization of π-electron density from oxygen into the aromatic ring. In going from dimethyl ether to anisole, a 102 ppm downfield shift is observed.[26] By studying the electronic effects of substituents on ¹⁷O shifts of *p*-substituted anisoles, Iwamura et al.[26] have shown that the ¹⁷O shieldings in these compounds are dominated by the electron density on the oxygen atom, which affects its 2p orbital size, $\langle r^{-3} \rangle_{2p}$ term in Equation 1. The importance of the interaction of p-electrons of the ether oxygen with the aromatic π-system in determining the ¹⁷O shifts of alkyl phenyl ethers was demonstrated by Kalabin and co-workers.[27] These authors observed that the first two substitutions of the alkyl protons in anisole by methyl groups gave the expected downfield shifts due to β-CH$_3$ effects, 29 and 25 ppm, respectively (Table 42). However, the third methyl substitution produced no change in the ^{17}O shielding. Kalabin et al.[27] explained the origin of this effect as the increased localization of charge density on the ether oxygen in **323**, compared to **322** and **321**. Steric inhibition of coplanarity of the oxygen atom with the aromatic ring in **323**, caused by the bulky t-butyl group, produces a loss in p-π-interactions and an increase in the electronic charge on oxygen. By subtracting out the contribution of the alkyl groups (δ_{alkyl}, see footnote b in Table 42) in **321** to **323** from their ^{17}O chemical shifts, Kalabin and co-workers[27] showed that the resultant ^{17}O shifts, which should be a measure of p-π-interactions, are practically the same for anisole, phenetole (**321**), and phenyl isopropyl ether (**322**). However, this ^{17}O shift for phenyl t-butyl ether (**323**) is 10 ppm upfield of anisole, suggesting increased electron density on the oxygen atom or a change in the conformation of p-orbitals.

TABLE 42
^{17}O NMR Chemical Shifts of Aromatic Ethers

No.	Compound	$\delta_{^{17}O}$[a]	β-SCS(CH$_3$)	δ_{alkyl}[b]	$\delta_{corrected}$[c]
86	C$_6$H$_5$OCH$_3$	49			
321	C$_6$H$_5$OCH$_2$CH$_3$	78	29	29	49
322	C$_6$H$_5$OCH(CH$_3$)$_2$	103	25	51	52
323	C$_6$H$_5$OC(CH$_3$)$_3$	103	25	63	39

[a] Taken from Reference 27.
[b] Contribution to $\delta_{^{17}O}$ by alkyl groups (Et, i-Pr, t-Bu) calculated from the shift for ethyl alkyl ether minus the shift for ethyl methyl ether. Chemical shifts were taken from Reference 11.
[c] $\delta_{^{17}O}$ minus δ_{alkyl}

The response of the ^{17}O chemical shift of an aromatic ether oxygen to steric effects from substituents in the aromatic ring was clearly demonstrated in a study of *ortho*-substituted anisoles[68] (Scheme 30). While mono *ortho* substitution by alkyl groups provide negligible effects, even with the bulky t-butyl group, di-*ortho* substitution results in substantial upfield shifts of the ^{17}O resonances in **325** and **327** (38 and 21 ppm, respectively). Steric hindrance

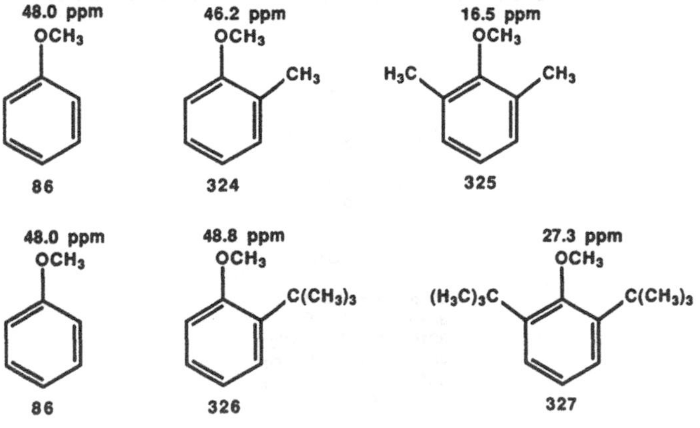

SCHEME 30

from the two *ortho* substituents for resonance interactions of the oxygen with the aromatic π system leading to increased electron density (decreased $\langle r^{-3}\rangle_{2p}$ term in Equation 1) has been suggested as the cause of the shielding effect.[68] However, this effect becomes smaller for the bulkier *t*-butyl substituent, 21 ppm in **327** as opposed to 32 ppm in **325**,[68] Scheme 30. Additional examples are shown in Scheme 31. As the size of the *ortho* alkyl substituents

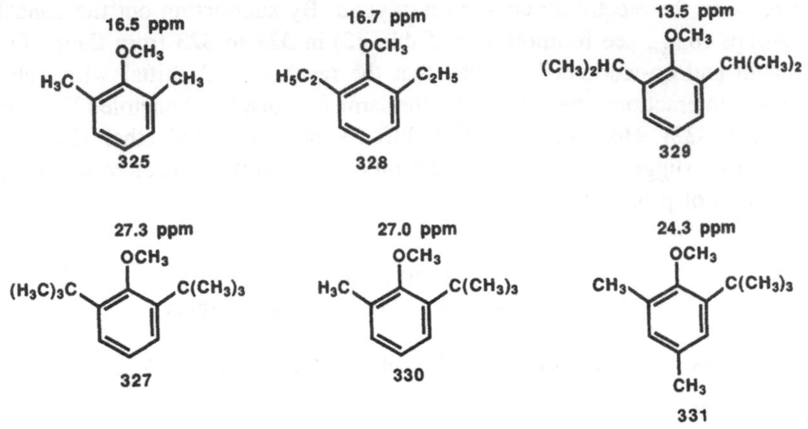

SCHEME 31

is increased from CH_3, C_2H_5 to i-Pr, shielding of the ether oxygen increases, but with two bulky *tert*-butyl groups, shielding decreases. This effect persists for cases (**330** and **331**) in which at least one of the two substituents is a large group. Repulsive van der Waals interactions between the substituents and the methoxy group have been invoked to explain the decrease in the upfield effect for the ^{17}O shifts in **327, 330,** and **331**.[68] By examining the orientation of the methoxy group in analogous 2,6-dialkylanisic acids (**331a** to **f**, Scheme 32) from their X-ray crystal structures, Schuster et al.[69] arrived at a similar conclusion. The authors suggested that in sterically congested anisoles, the nonbonded interactions between the methoxy oxygen lone pairs of electrons and the C-H bonds of the *ortho* substituents should localize electron density in the plane containing the oxygen, ipso, and *para* carbon atoms, thereby rendering the electron delocalization into the aromatic ring more feasible.

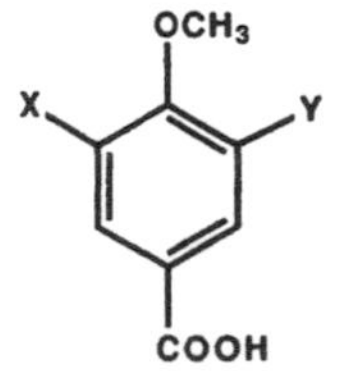

331a	X = Y = H
331b	X = Y = CH_3
331c	X = Y = C_2H_5
331d	X = Y = i-C_3H_7
331e	X = Y = t-C_4H_9
331f	X = t-C_4H_9 ; Y = H

SCHEME 32

In contrast to the effect produced by an *ortho* alkyl substituent, an *ortho* methoxy group induces a dramatic shielding of the aromatic ether oxygen (15 ppm) in 1,2-dimethoxybenzene (**332**) (Scheme 33). Wysocki et al[68] proposed that the placement of the methoxy group out

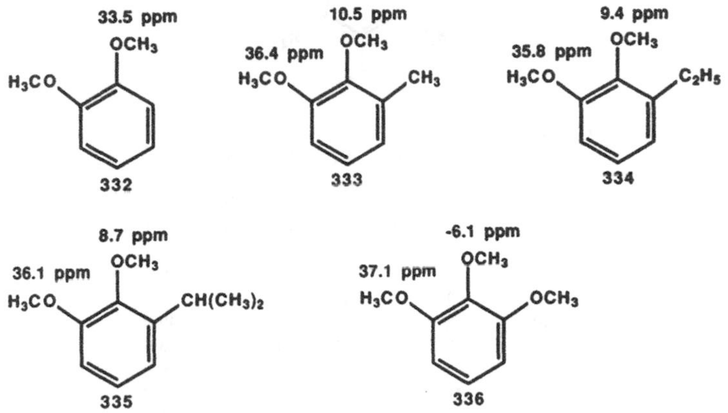

33.5 ppm — OCH$_3$, H$_3$CO **332**

36.4 ppm — H$_3$CO; 10.5 ppm — OCH$_3$; CH$_3$ **333**

35.8 ppm — H$_3$CO; 9.4 ppm — OCH$_3$; C$_2$H$_5$ **334**

36.1 ppm — H$_3$CO; 8.7 ppm — OCH$_3$; CH(CH$_3$)$_2$ **335**

37.1 ppm — H$_3$CO; -6.1 ppm — OCH$_3$; OCH$_3$ **336**

SCHEME 33

of conjugation with the aromatic ring is more pronounced in **332** than in **324**, due to repulsive interactions of the electron pairs in **332**. As a result, the increase in electron density (lowering of $\langle r^{-3} \rangle_{2p}$ term) on the ether oxygen leads to the upfield effect. The large shielding effect of 15 ppm in 1,2-dimethoxybenzene is actually due to a combination of two factors: rotation of the methoxy group from the plane of the aromatic ring and electronic effect of one methoxy group on the other. By considering the electronic effects in 1,2- and 1,4-dimethoxybenzenes to be the same (-8 ppm for the -OCH$_3$ group; compare 1,4-dimethoxybenzene, 40.0 ppm and anisole, 48.0 ppm), Wysocki et al.[68] estimated the shielding effect due to rotation of the methoxy group to be 6.5 ppm, which is quite large compared to the steric effects caused by mono alkyl substituents. The contrasting effects of *ortho* alkyl and methoxy substituents encountered for the ^{17}O shifts of anisoles are present in 1,2-dimethoxybenzenes also.[68] While alkyl substituents shield the ether oxygen by 23 to 25 ppm (**333** to **335**), the methoxy group produces a 40 ppm upfield effect (in **336**), again indicating that effects due to repulsive interactions of electrons are more dominant than simple steric effects. Chloro substituents also cause upfield shifts of 10 to 22 ppm for the methoxy oxygen in diortho and polysubstituted anisoles[70] (Table 43). However, the substituent-induced chemical shifts for the ether oxygen were not additive.

TABLE 43
^{17}O NMR Chemical Shifts of the Methoxy
Oxygen in Chlorinated Anisoles

Compound no.	R$_2$	R$_3$	R$_4$	R$_5$	R$_6$	δ$_{17O}$^{a}
86	H	H	H	H	H	47.7
337	Cl	H	H	H	H	50.1
338	H	Cl	H	H	H	52.1
87	H	H	Cl	H	H	49.1
339	Cl	Cl	H	H	H	58.6
340	Cl	H	Cl	H	H	49.9
341	Cl	H	H	Cl	H	53.9
342	Cl	H	H	H	Cl	26.0
343	H	Cl	Cl	H	H	52.8
344	H	Cl	H	Cl	H	56.7
345	Cl	Cl	Cl	H	H	56.4
346	Cl	Cl	H	Cl	H	64.9
347	Cl	H	Cl	Cl	H	58.9
348	Cl	Cl	H	H	Cl	32.7
349	Cl	H	Cl	H	Cl	26.0
350	H	Cl	Cl	Cl	H	60.5
351	Cl	Cl	Cl	Cl	H	61.7
352	Cl	Cl	Cl	H	Cl	32.2
353	Cl	Cl	H	Cl	Cl	37.5
354	Cl	Cl	Cl	Cl	Cl	36.3

^{a} Values from Reference 70 minus 3 ppm to correct for
D$_2$O as internal reference.

An interesting example of substantial shielding of an aromatic ether oxygen is provided by Duddeck et al[71] in the furocoumarin **355**, in which the methoxy oxygen at position 8 is shielded by 31 ppm relative to a similar substituent at position 5 in **356**, and by 36 ppm relative to anisole (Scheme 34). The endocyclic lactone oxygen in **355** is also shielded in comparison with its counterpart in **356** and **357** by 20 ppm, whereas the furan oxygen in **355** shows negligible effect. The authors attribute this large effect to through-space syn-periplanar interactions between the two γ oxygen atoms in **355**. Repulsion between the oxygen lone pairs of electrons, leading to deviations from planarity of the endocyclic lactone oxygen and the exocyclic ether oxygen, and thus, the shielding effect, cannot be excluded. These two interpretations, however, do not explain the lack of a reciprocal shielding effect on the furan oxygen in **355**. Additional examples of the effects[71] described above are presented in **358** and **359** in Scheme 34.

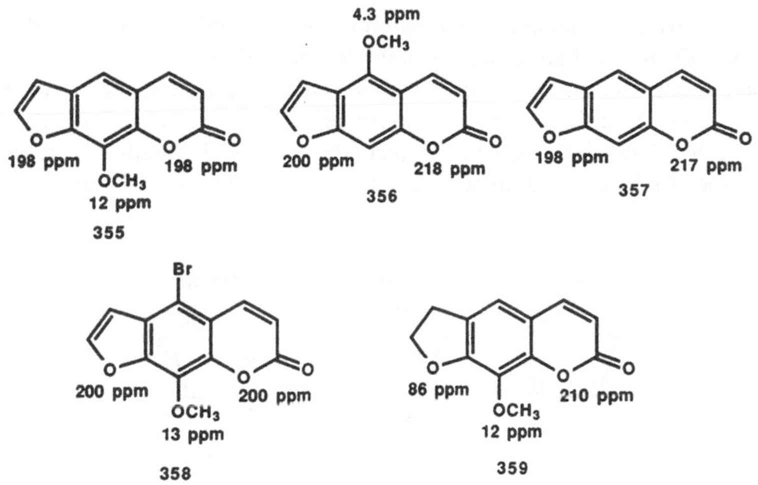

4.3 ppm
OCH$_3$

198 ppm   198 ppm   200 ppm   218 ppm   198 ppm   217 ppm

OCH$_3$
12 ppm

355   356   357

Br

200 ppm   200 ppm   86 ppm   210 ppm

OCH$_3$ OCH$_3$
13 ppm   12 ppm

358   359

SCHEME 34

The ^{17}O resonance of an oxygen atom in a heteroaromatic ring undergoes a dramatic downfield shift relative to the oxygen in the corresponding acyclic compound. For instance, the furan oxygen (360)[1] is deshielded by 111 ppm relative to the ether oxygen in divinyl ether,[27] Scheme 35. This effect is considerably large in comparison to the cyclization shifts (2 to 8 ppm) observed[33] for the conversion of aliphatic ethers to monocyclic ethers. The large deshielding of the heteroaromatic oxygen is probably due to the delocalization of electron density into the aromatic ring (increase in $\langle r^{-3}\rangle_{2p}$ term in Equation 1), which is not as extensive in the acyclic counterpart. Such a delocalization being totally absent in the aliphatic compounds, the effect in these cases is negligible. An additional example of the deshielding of the heteroaromatic oxygen is shown in dibenzofuran (362),[72] in which the ether oxygen resonates 43 ppm downfield of the oxygen atom in diphenyl ether (361).[72] The smaller effect in this case may be due to the possibility that oxygen electron density delocalization does not increase as much in going from diphenyl ether to dibenzofuran.

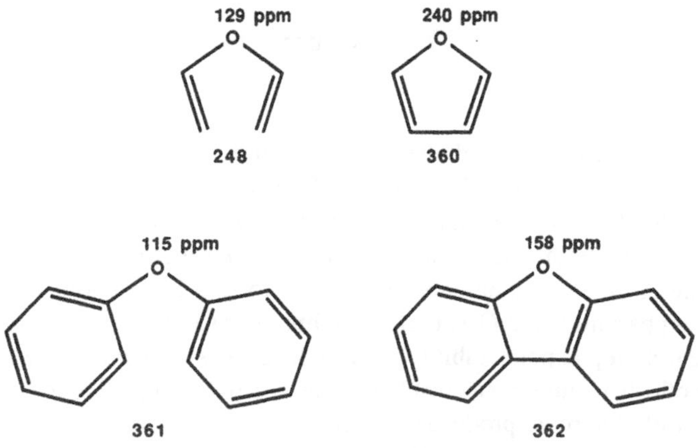

129 ppm   240 ppm
O   O

248   360

115 ppm   158 ppm
O   O

361   362

SCHEME 35

In a study of oxygen-containing heteroaromatic compounds, Chimichi and co-workers[73] found that the ¹⁷O chemical shifts were mainly determined by the π-electron density on the oxygen atom. As the sum of the electronegativities of the atoms attached to the ring oxygen increases, in going from furan (**360**) to isoxazole (**363**) to 3,4-dimethyl-1,2,5-oxadiazole (**364**),[73] the deshielding of the oxygen also increases, compatible with a decrease in electron density (Scheme 36). Downfield shifts of the ether oxygen due to β-CH₃ substitution observed

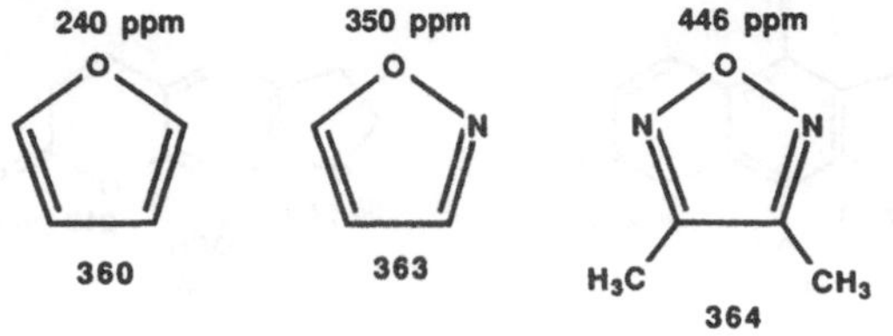

SCHEME 36

in alkyl vinyl ethers[27] are also present in furan and oxazole systems,[73] even though the effects in the latter are smaller. The β-SCS(CH₃) values range from 6 to 8 ppm and appear to be additive as seen for compounds **365** and **366**, relative to **360**, and **367** in comparison with **363**[73] (Scheme 37). These β-effects may be governed by electron density on the

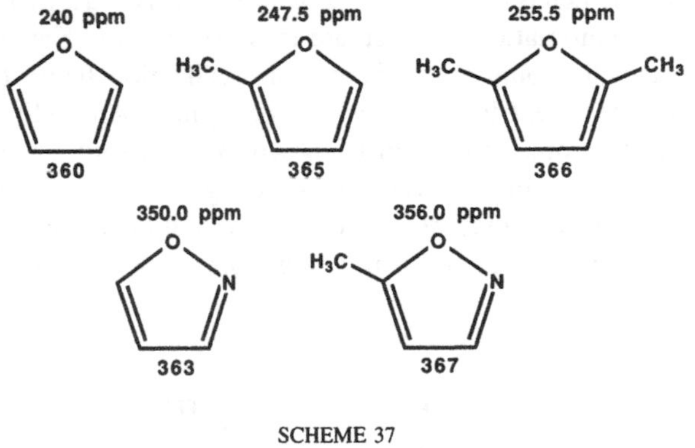

SCHEME 37

heterocyclic oxygen atom, even though Clennan and Mehrsheik-Mohammadi,[30] in their investigation of several 2- and 2,5-substituted furans, failed to find a correlation between the ¹⁷O shifts of the furan oxygen and its electron density. The authors, however, obtained a good correlation with the parameter Q, which is a function of the polarizability and ionization potential of the substituents. Exceptions to this correlation were the methyl and cyano groups. Apparently, the Q parameter is important only for those substituents which are large in size (relating to polarizability) or have lone pairs of electrons (relating to ionization potential), or both. It is quite conceivable that the methyl group, due to its smaller size and lack of lone pair of electrons, produces β-effect in furan and isoxazole which are dependent on the electron density on oxygen. γ-CH₃ effects in isoxazole[73] also parallel the corresponding effects in vinyl alkyl ethers.[27] Shielding of the ring oxygen in isoxazole, as a result of γ-CH$_3$ substitution, is 6 ppm and is found to be additive (Scheme 38). An interesting case, in which the two opposing β-CH$_3$ (downfield shifting) and γ-CH$_3$ (upfield shifting) effects of similar magnitude produce a net zero change in the ^{17}O shift, is illustrated in the isoxazole **370**.[73] The oxygen chemical shift in this disubstituted isoxazole is exactly the same as that observed for the unsubstituted derivative **363**.

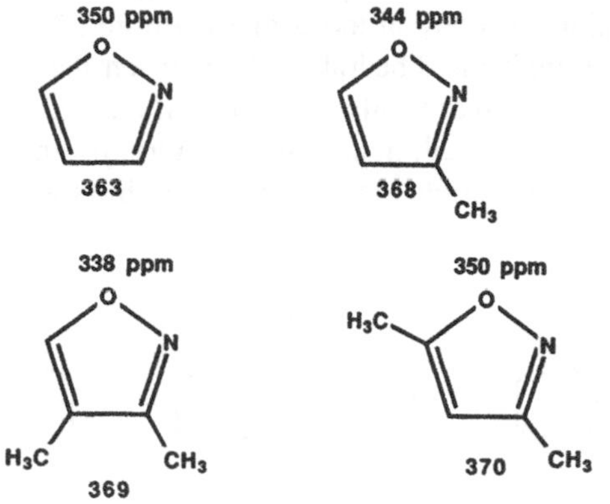

SCHEME 38

Annellation of a benzene ring to a heteroaromatic ether such as furan results in distinct shielding of the ether oxygen, and this effect is additive[71] (compare **371** with **360**, **362** with **371**, and **372** with **365** in Scheme 39). Chandrasekaran et al.[74] observed similar but smaller upfield shifts of the ether-type oxygen for the 4H-pyran-4-one series (**373** to **375**) in Scheme

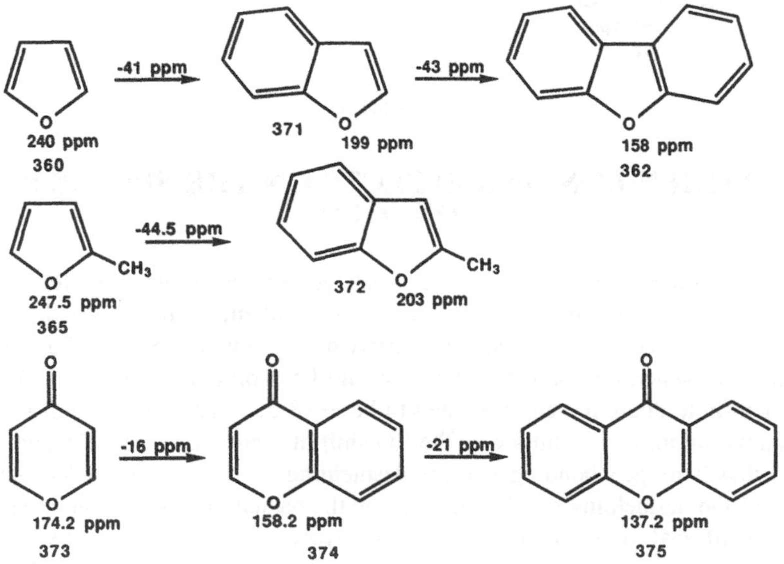

SCHEME 39

39. The effects in the pyranone series have been rationalized in terms of increasing electron density (373<374<375) on the dicoordinated oxygen, which is a consequence of cross-conjugation of the oxygen lone pair of electrons with the carbonyl group and the benzene rings.[74] However, shielding effects in the furan series are not readily explained in this way.

Boykin and Martin[75] found that the [17]O shift of the ether oxygen in dibenzopyran was quite sensitive to the replacement of the methylene function at the 4-position to the oxygen by groups of varying electronic properties (Scheme 40). Even though electron-withdrawing and donating groups produced deshielding and shielding, respectively, the large shifts observed could not be explained solely in terms of electronic effects. Based on a correlation of the C-O-C bond angle with the dihedral angle between the planes of the two aromatic rings in **378, 381,** and several other similar compounds, the authors[75] suggested that the [17]O shifts in **375** to **382** may be governed, in part, by structural parameters such as the molecular dihedral angle. However, no correlation between [17]O shifts and the dihedral angles was obtained.

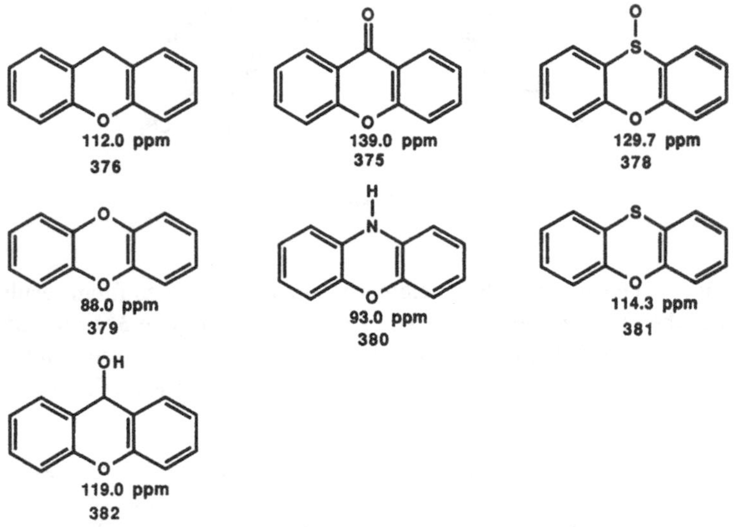

112.0 ppm
376

139.0 ppm
375

129.7 ppm
378

88.0 ppm
379

93.0 ppm
380

114.3 ppm
381

119.0 ppm
382

SCHEME 40

X. HYDROGEN-BONDING EFFECTS ON THE SINGLE-BONDED OXYGEN

One of the earliest investigations of the effects of hydrogen bonding on the [17]O chemical shifts of the single-bonded oxygen was reported by Reuben,[76] who recorded the [17]O shifts of water as a function of concentration in a variety of solvents. In solvents dioxane, acetone, and pyridine, in which the oxygen in water would be a proton donor, the [17]O resonances of water at infinite dilution were estimated to be -18.3, -12.1, and -8.3 ppm, respectively. Comparing these [17]O shifts with the [17]O shift of water vapor of -36 ppm, Reuben[76] concluded that hydrogen bonding induces deshielding of the single-bonded oxygen. The magnitude of the deshielding roughly depends on the basicity of the proton-accepting atom. From a detailed analysis of the two equilibria, involving one and two dioxane molecules per molecule of water in the water-dioxane system, Reuben[76] estimated the deshielding of the hydroxylic oxygen to be larger when it is a proton donor (12 ppm) than when it acts as a proton acceptor (6 ppm). These hydrogen-bonding-induced [17]O shifts of the hydroxylic oxygen have been explained[76] on the basis of valence bond resonance structures with the charge-transfer structure, $-O^-$ H$-^+$O playing a major role.

The first report of an ^{17}O NMR study of a hydrogen-bonded enol is found in the pioneering investigations by Christ and co-workers,[1,4] who found that the enol of acetylacetone exhibited a single resonance (269 ppm). Gorodetsky et al.[77] also investigated a large number of β-dicarbonyl compounds which exhibit keto-enol tautomerism, but were unable to observe a distinct signal for the enol oxygen due to fast exchange between tautomers. However, Lapachev and co-workers,[78] in a study of β-ketoesters of the type shown in Scheme 41, for which the enolization of the ester group is absent and the rate of keto-enol tautomerization is slow, identified discrete ^{17}O resonances for the hydrogen-bonded enol oxygen. The authors found that the enol oxygen was considerably deshielded as a result of hydrogen bonding with the carbonyl group. The observed range of shifts, 96 to 124 ppm, make the enol oxygen the most highly deshielded aliphatic hydroxyl oxygen.

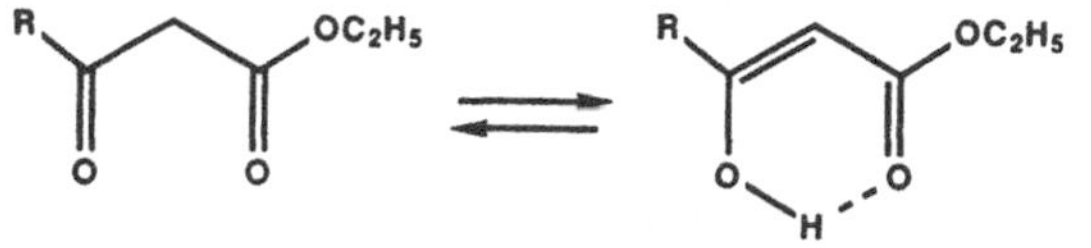

#	R	δ (O-H)
383	CH$_3$	124.0
384	C$_6$H$_5$	109.0
385	CF$_3$	96.0

SCHEME 41

Fiat and co-workers[79] observed that the aromatic hydroxyl oxygen was deshielded by 16 ppm in o-hydroxybenzaldehyde and 21 ppm in o-hydroxyacetophenone (relative to phenol) due to intramolecular hydrogen bonding with the adjacent carbonyl group. Chandrasekaran et al.[74] also observed deshielding of the phenol oxygen resonance in 1,4-dihydroxyanthra-quinone (28 ppm relative to 1,4-dihydroxynaphthalene, Scheme 42). However, the authors noted only one ^{17}O signal for 2,5-dihydroxybenzoquinone and 5,8-dihydroxynaphthoquinone (Scheme 42) due to fast intramolecular proton exchange. In a fairly extensive investigation of quinones and ketones containing intramolecular hydrogen bonds involving the hydroxy and carbonyl groups, Jaccard and Lauterwein[80] reported deshielding effects of the hydroxy group ranging from 3 to 27 ppm (Table 44).

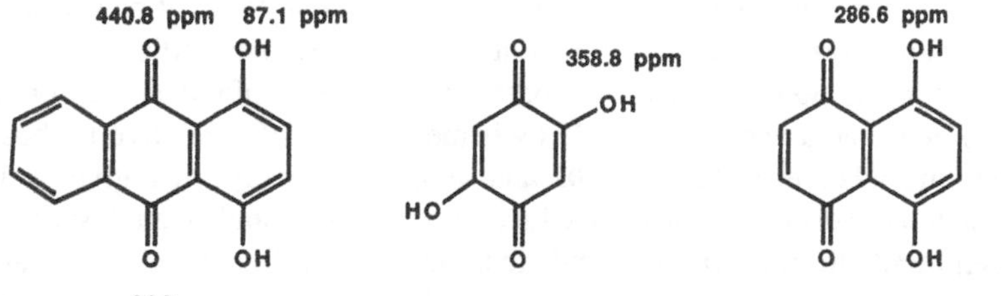

SCHEME 42

TABLE 44
Hydroxyl Oxygen Chemical Shifts in Compounds
Containing Intramolecular Hydrogen Bonds of the type
O–H--O=C

No.	Compound	δ_{OH}	$\Delta\delta^a$
387		96.4[b]	27.4
388		84.4[b]	15.4
389		79.5[b]	10.5
390		84.1[b]	15.1
391		84.4[b]	15.4
392		72.1[b]	3.1

[a] Hydrogen bonding effect calculated by subtracting the average shift for free aromatic-OH group (69 ppm) from δ_{OH}.
[b] Taken from Reference 80.

The [17]O chemical shifts for the single-bonded oxygen in hydrogen bonds of the type O–H---N, in which an aromatic nitrogen is the proton acceptor, fall into a narrow range of 90 to 97 ppm[78] (Scheme 43). This limited variation in the deshielding (21 to 28 ppm, relative to phenol) of the aromatic hydroxy oxygen is surprising, even though the basicity of the nitrogen moiety, and thus the hydrogen bond strength are probably not constant. However, as noted earlier,[78] the enols exhibit a larger range (96 to 124 ppm) of hydrogen-bond-induced [17]O chemical shifts.

In contrast to the deshielding of the hydroxylic oxygen, when it acts as a proton donor in hydrogen bond formation,[74,76,78-80] proton-accepting aromatic ether oxygen experiences shielding effects. Lauterwein and co-workers[81] observed upfield shifts for the methoxy oxygen in o-anisic acid and o-anisamide (Scheme 44) in CDCl_3 relative to those in CD_3CN or CD_3OD, and relative to those of **402** and **404**, respectively, in a given solvent. Based on an evaluation of net charges on the methoxy oxygen atoms and an analysis of the geometry of hydrogen bonds from *ab initio* calculations, the authors[81] concluded that an increase in π-charge (shielding effect) makes a higher contribution to the observed chemical shift than an increase in the σ charge (deshielding effect). However, it is quite likely that in the case of alcohols, for the charge transfer formalism of the hydrogen-bonded structure involving –OH as the proton donor and C=O or a basic nitrogen moiety as the acceptor, the decreased ionization potential (ΔE term in Equation 1) of the hydroxyl oxygen may determine the deshielding observed. For systems in which the proximity of the proton donor and acceptor results in torsion angle variations of the groups concerned, these effects should also be considered in the interpretation of hydrogen-bond-induced ^{17}O shifts.

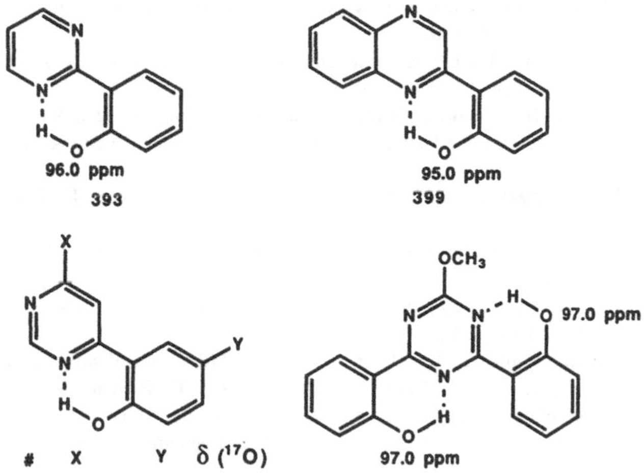

96.0 ppm
393

95.0 ppm
399

97.0 ppm

97.0 ppm
400

#	X	Y	$\delta\,(^{17}O)$
394	H	H	94.0
395	H	Br	95.0
396	H	OCH$_3$	90.0
397	Cl	H	95.0
398	N(C$_2$H$_5$)$_2$	H	97.0

SCHEME 43

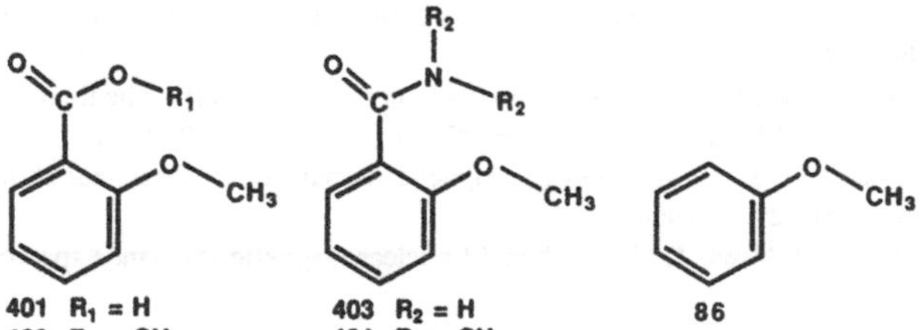

401 R$_1$ = H
402 R$_1$ = CH$_3$

403 R$_2$ = H
404 R$_2$ = CH$_3$

86

δ (OCH$_3$)

#	CDCl$_3$	CD$_3$CN	CD$_3$OD
401	47.0	51.6	52.0
402	53.0	53.2	53.4
403	50.5	52.1	52.7
404	52.5	52.1	52.5
86	47.7	47.6	47.9

SCHEME 44

REFERENCES

1. **Christ, H. A., Diehl, P., Schneider, H. R., and Dahn, H.**, Chemische Verschiebungen in der kernmagnetischen Resonanz von ^{17}O in organischen Verbindungen, *Helv. Chim. Acta,* 44, 865, 1961.
2. **Silver, B. L. and Luz, Z.**, Chemical applications of oxygen-17 nuclear and electron spin resonance, *Q. Rev. Chem. Soc.,* 21, 458, 1967.
3. **Greenzaid, P., Luz, Z., and Samuel, D.**, A nuclear magnetic resonance study of the reversible hydration of aliphatic aldehydes and ketones. I. Oxygen-17 and proton spectra and equilibrium constants, *J. Am. Chem. Soc.,* 89, 749 and 756, 1967.
4. **Christ, H. A. and Diehl, P.**, Lösungsmitteleinflüssee und Spinkopplungen in der kernmagnetischen resonanz von ^{17}O, *Helv. Phys. Acta,* 36, 170, 1963.
5. **Dahn, H., Schlunke, H. P., and Telmer, J.**, Chemische Verschiebungen bei ^{17}O-NMR und Hydrationsgeschwindigkeiten von cyclischen Ketonen mit transannularer Wechselwirkung, *Helv. Chim. Acta,* 55, 907, 1972.
6. **Kintzinger, J.-P.**, in *NMR of Newly Accessible Nuclei*, Vol. 2, Laszlo, P., Ed., Academic Press, New York, 1983, 79.
7. **Klemperer, W. G.**, in *The Multinuclear Approach to NMR Spectroscopy*, Lambert, J. B. and Riddell, F. G., Eds., Reidel, Dordrecht, Holland, 1983, 245.
8. **Kintzinger, J.-P.**, Oxygen-17 and silicon-29, in *NMR-17*, Diehl, P., Fluck, E., and Kosfeld, R., Eds., Springer-Verlag, New York, 1981.
9. (a) **Karplus, M. and Pople, J. A.**, Theory of carbon NMR chemical shifts in conjugated molecules, *J. Chem. Phys.,* 38, 2803, 1963; (b) **Pople, J. A. and Santry, D. P.** A molecular orbital theory of hydrocarbons. II. Ethane, ethylene and acetylene, *Mol. Phys.,* 9, 301, 1964.
10. **Crandall, J. K. and Centeno, M. A.**, Oxygen-17 nuclear magnetic resonance. 1. Alcohols, *J. Org. Chem.,* 44, 1183, 1979.
11. **Delseth, C. and Kintzinger, J.-P.**, Resonance magnetique nucleaire de ^{13}C et ^{17}O d'ethers aliphatiques. Effects γ entre les atomes d'oxygene et de carbone, *Helv. Chim. Acta,* 61, 1327, 1978.
12. **Eliel, E. L., Liu, K.-T., and Chandrasekaran, S.**, ^{17}O NMR spectra of equatorial and axial hydroxycyclohexanes and 5-hydroxy-1,3-dioxanes and their methyl ethers, *Org. Magn. Reson.,* 21, 179, 1983.
13. (a) **Sarneski, J. E., Suprenant, H. L., Molen, F. K., and Reilley, C. N.**, Chemical shifts and protonation shifts in carbon-13 nuclear magnetic resonance studies of aqueous amines, *Anal. Chem.* 47, 2116, 1975; (b) **Eliel, E. L. and Vierhapper, F. W.**, Carbon-13 nuclear magnetic resonance spectra of saturated heterocycles. IV. *trans*-decahydroquinolines, *J. Org. Chem.,* 44, 199, 1976.
14. (a) **Olah, G. A. and Donovan, D. J.**, ^{13}C nuclear magnetic resonance spectroscopic study of alkyl cations. The constancy of ^{13}C nuclear magnetic resonance methyl substituent effects and their application in the study of equilibrating carbocations and the mechanism of some rearrangements, *J. Am. Chem. Soc.,* 99, 5026, 1977; (b) **Bremser, W., Ernst, L., and Franke, B.**, *Carbon-13 NMR Spectral Data*, Verlag-Chemie, Weinheim, New York, 1979.
15. **Cheney, B. V. and Grant, D. M.**, Carbon-13 magnetic resonance. VIII. The theory of carbon-13 chemical shifts applied to saturated hydrocarbons, *J. Am. Chem. Soc.,* 89, 5319, 1967.
16. **Grant, D. M. and Paul, E. G.**, Carbon-13 magnetic resonance. II. Chemical shift data for the alkanes, *J. Am. Chem. Soc.,* 86, 2984, 1964.
17. **Lindeman, L. P. and Adams, J. Q.**, Carbon-13 nuclear magnetic resonance spectrometry, *Anal. Chem.,* 43, 1245, 1971.
18. **Sugawara, T., Kawada, Y., Katoh, M., and Iwamura, H.**, Oxygen-17 nuclear magnetic resonance. III. Oxygen atoms with a coordination number of two, *Bull. Chem. Soc. Jpn.,* 52, 3391, 1979.
19. **Takasuka, M.**, Relationship of the ^{17}O chemical shift to the stretching frequencies of the hydroxy-group in saturated alcohols, *J. Chem. Soc. Perkin Trans. 2*, p. 1558, 1981.
20. **Baird, N. C. and Teo, K. C.**, Effective excitation energies in ^{13}C NMR chemical shift calculations, *J. Magn. Reson.,* 24, 87, 1976.
21. **Arbuzov, B. A., Bredikhin, A. A., Enikeev, K. M., Ismaev, I. E., Il'yasov, A. V., and Vereshchagin, A. N.**, Investigation of several α-chloro ethers by ^{13}C and ^{17}O NMR, *Izv. Akad. Nauk SSSR, Ser. Khim.,* 2265, 1983.
22. **Manoharan, M. and Eliel, E. L.**, ^{17}O NMR spectra of tertiary alcohols, ethers, sulfoxides and sulfones in the cyclohexyl and 5-substituted 1,3-dioxanyl series and related compounds, *Magn. Reson. Chem.,* 23, 225, 1985.
23. **Stothers, J. B.**, *Carbon-13 NMR Spectroscopy*, Academic Press, New York, 1972, 64.
24. **Grover, S. H., Guthrie, J. P., Stothers, J. B., and Tan, C. T.**, The stereochemical dependence of δ-substituent effects in ^{13}C NMR spectra. Deshielding syn-axial interactions, *J. Magn. Reson.,* 10, 227, 1973.
25. **Levy, G. C., Lichter, R. L., and Nelson, G. L.**, *Carbon-13 Nuclear Magnetic Resonance Spectroscopy*, John S. Wiley & Sons, New York, 1980, 55.

26. **Katoh, M., Sugawara, T., Kawada, Y., and Iwamura, H.,** ^{17}O nuclear magnetic resonance studies. V. ^{17}O shieldings of some substituted anisoles, *Bull. Chem. Soc. Jpn.*, 52, 3475, 1979.

27. **Kalabin, G. A., Kushnarev, D. F., Valeyev, R. B., Trofimov, B. A., and Fedotov, M. A.,** ^{17}O NMR investigation of p,π-interactions in α,β-unsaturated and aromatic ethers, *Org. Magn. Reson.*, 18, 1, 1982.

28. (a) **Balakrishnan, P., Baumstark, A. L., and Boykin, D. W.,** ^{17}O NMR spectroscopy: unusual substituent effects in para-substituted benzyl alcohols and acetates, *Tetrahedron Lett.*, p. 169, 1984; (b) **Bromilow, J., Brownlee, R. T. C., and Page, A. V.,** The origin of fluorine-19 and carbon-13 substituent chemical shifts in substituted benzyl fluorides, *Tetrahedron Lett.*, p. 3055, 1976; (c) **Kusuyama, Y., Dyllick-Brenzinger, C., and Roberts, J. D.,** Carbon-13 nuclear magnetic resonance spectroscopy. Substituent-induced chemical shift effects on cyclopropyl carbons of 4-substituted cyclopropyl benzenes, *Org. Magn. Reson.*, 13, 372, 1980.

29. **Beraldin, M.-T., Vauthier, E., and Fliszar, S.,** Charge distributions and chemical effects. XXVI. Relationships between nuclear magnetic resonance shifts and atomic charges for ^{17}O nuclei in ethers and carbonyl compounds, *Can. J. Chem.*, 60, 106, 1982.

30. **Clennan, E. L. and Mehrsheik-Mohammadi, M. E.,** ^{17}O NMR spectra of 2-substituted and 2,5-disubstituted furans. The inapplicability of the Hammett LFER to correlate chemical shifts, *Magn. Reson. Chem.*, 23, 985, 1985.

31. **Hruska, F., Hutton, H. M., and Schaefer, T.,** Concerning the *ortho* shift in proton and fluorine magnetic resonance of some conjugated hydrocarbons, *Can. J. Chem.*, 43, 2392, 1965.

32. **Smith, W. B. and Proulx, T. W.,** A unified correlation of substituent effects with carbon, proton and fluorine chemical shifts in aromatic and olefinic systems, *Org. Magn. Reson.*, 8, 567, 1976.

33. **Nguyen, T. T.-T., Delseth, C., Kintzinger, J.-P., Carrupt, P.-A., and Vogel, P.,** Oxygen-17 nuclear magnetic resonance. The effects of remote unsaturation on ^{17}O chemical shifts in policyclic ethers, *Tetrahedron*, 36, 2793, 1980.

34. **Iwamura, H., Sugawara, T., Kawada, Y., Tori, K., Muneyuki, R., and Noyori, R.,** ^{17}O NMR chemical shifts versus structure relationships in oxiranes, *Tetrahedron Lett.*, p. 3449, 1979.

35. **Sauleau, A., Sauleau, J., Monti, J. P., and Faure, R.,** Oxygen-17 nuclear magnetic resonance study of some oxirane derivatives, *Org. Magn. Reson.*, 21, 403, 1983.

36. **Monti, J. P., Faure, R., Sauleau, A., and Sauleau, J.,** ^{13}C and ^{17}O NMR of some substituted oxiranes. Chemical shifts and quantitative correlations, *Magn. Reson. Chem.*, 24, 15, 1986.

37. **Kas'yan, L. I., Gnedenkov, L. Y., Stepanova, N. V., Sitnik, I. V., and Zefirov, N. S.,** The strain of cyclic epoxides and their characteristics, *Zh. Org. Khim.*, 22, 215, 1986.

38. **Eliel, E. L., Pietrusiewicz, K. M., and Jewell, L. M.,** O-17 NMR spectra of ring compounds. Correlation of ^{17}O and ^{13}C methyl substitution parameters, *Tetrahedron Lett.*, p. 3649, 1979.

39. **Kintzinger, J.-P., Delseth, C., and Nguyen, T. T.-T.,** ^{17}O nuclear magnetic resonance: mutual effect between two β-oxygen atoms and α,β-double bond effect on ^{17}O chemical shift, *Tetrahedron*, 36, 3431, 1980.

40. **Eliel, E. L., Chandrasekaran, S., Carpenter, L. E., II, and Verkade, J. G.,** ^{17}O NMR spectra of cyclic phosphites, phosphates, and thiophosphates, *J. Am. Chem. Soc.*, 108, 6651, 1986.

41. **Mattinen, J. and Pihlaja, K.,** ^{17}O NMR spectra of methyl-substituted 2-oxo-1,3,2-dioxathianes, *Magn. Reson. Chem.*, 25, 569, 1987.

42. **McKelvey, R. D., Kawada, Y., Sugawara, T., and Iwamura, H.,** Anomeric effect in 2-alkoxytetrahydropyrans studied by ^{13}C and ^{17}O NMR chemical shifts, *J. Org. Chem.*, 46, 4948, 1981.

43. **Kupce, E., Liepins, E., Zicmane, I., and Lukevics, E.,** Conformational dependence of ^{17}O chemical shifts in 1,3-dioxa-2-silacyclooctanes: anomeric effect, *Magn. Reson. Chem.*, 25, 1084, 1987.

44. **Kalabin, G. A. and Kushnarev, D. F.,** ^{17}O NMR spectra of unsaturated and aromatic ethers, *Dokl. Akad. Nauk SSSR*, 254, 1425, 1980.

45. **Afonin, A. V., Voronov, V. K., Enikeeva, E. I., and Andrayankov, M. A.,** ^{15}N and ^{17}O NMR spectra of vinyl ethers of pyridine and quinoline, *Izv. Akad. Nauk SSSR, Ser. Khim.*, p. 769, 1987.

46. **Stothers, J. B., Swenson, J. R., and Tan, C. T.,** ^{13}C nuclear magnetic resonance studies. XLV. The ^{13}C spectra of some β,γ-unsaturated polycyclic ketones, *Can. J. Chem.*, 53, 581, 1975.

47. **Quarroz, D., Sonney, J.-M., Chollet, A., Florey, A., and Vogel, P.,** Carbon-13 NMR spectra of 2,3-dimethylenebicyclo[2.2.1]heptanes and 2,3-dimethylene-7-oxabicyclo[2.2.1]heptanes, *Org. Magn. Reson.*, 9, 611, 1977.

48. **Kalabin, G. A., Kushnarev, D. F., Shostakovskii, S. M., and Voropayeva, T. K.,** Carbon-13 NMR spectra of substituted cyclopropanes. I. Chemical shifts of carbon-13 and the absence of σ_{π}-p$_{\pi}$ interaction in cyclopropyl alkyl ethers, *Izv. Akad. Nauk SSSR. Ser. Khim.*, p. 2459, 1975.

49. **Delseth, C., Nguyen, T. T.-T., and Kintzinger, J.-P.,** Oxygen-17 and carbon-13 nuclear magnetic resonance. Chemical shifts of unsaturated carbonyl compounds and acyl derivatives, *Helv. Chim. Acta*, 63, 498, 1980.

50. **Balakrishnan, P., Baumstark, A. L., and Boykin, D. W.,** ^{17}O NMR spectroscopy: effect of substituents on chemical shifts for p-substituted benzoic acids, methyl benzoates, cinnamic acids and methyl cinnamates, *Org. Magn. Reson.*, 22, 753, 1984.

51. **Silver, N. L. and Boykin, D. W.,** Substituent effects on the carbonyl stretching frequency of chalcones, *J. Org. Chem.,* 35, 759, 1970.

52. (a) **Caputo, J. A. and Fuchs, R.,** Synthesis and ionization constants of *meta-* and *para-*substituted *cis-*3-phenyl cyclobutanecarboxylic acids, *J. Org. Chem.,* 33, 1959, 1968; (b) **Thigben, A. B. and Fuchs, R.,** Synthesis and ionization constants of *meta-* and *para-*substituted 1-phenyl-2-methylcyclopropene-3-carboxylic acids, *J. Org. Chem.,* 34, 505, 1969.

53. **Monti, D., Orsini, F., and Ricca, G. S.,** Oxygen-17 NMR spectroscopy: effect of substituents on chemical shifts for *o- m- p-*substituted benzoic acids, phenylacetic and methyl benzoates, *Spectros. Lett.,* 19, 91, 1986.

54. **Fujita, T. and Nishioka, T.,** Progress in physical organic chemistry, 12, 49, 1976.

55. **Orsini, F. and Ricca, G. S.,** Oxygen-17 NMR chemical shifts of esters, *Org. Magn. Reson.,* 22, 653, 1984.

56. **Brownlee, R. T. C. and Craik, D. J.,** The relationship between ¹⁹F substituent chemical shifts and electron densities: *meta-* and *para-*substituted benzoyl fluorides, *Org. Magn. Reson.,* 14, 186, 1980.

57. **Liepin'sh, E. E., Zitsmane, I. A., Ignatovich, L. M., Lukevits, E., Guvanova, L. I., and Voronkov, M. G.,** ¹⁷O NMR chemical shifts of (trifluorosilyl)methyl *para-*substituted benzoates and their carbon analogs, *Zh. Obshch. Khim.,* 53, 1789, 1983.

58. **Baumstark, A. L., Balakrishnan, P., Dotrong, M., McCloskey, C. J., Oakley, M. G., and Boykin, D. W.,** ¹⁷O NMR spectroscopy: torsion angle relationships in aryl carboxylic esters, acids, and amides, *J. Am. Chem. Soc.,* 109, 1059, 1987.

59. **Li, S. and Chesnut, D. B.,** Intramolecular van der Waals interactions and chemical shifts: a model for β- and γ-effects, *Magn. Reson. Chem.,* 23, 625, 1985.

60. **Li, S. and Chesnut, D. B.,** Intramolecular van der Waals interactions and ¹³C chemical shifts: substituent effects in some cyclic and bicyclic systems, *Magn. Reson. Chem.,* 24, 93, 1986.

61. **Lycka, A., Holecek, J., Handlir, K., Pola, J., and Chvalovsky, V.,** ¹⁷O, ¹³C, and ²⁹Si NMR spectra of some acyloxy- and diacetoxysilanes and acetoxygermanes, *Coll. Czech. Chem. Comm.,* 51, 2582, 1986.

62. **Lycka, A. and Holecek, J.,** An ¹⁷O NMR study of some di- and tri-*n*-butyltin (IV) carboxylates, *J. Organomet. Chem.,* 294, 179, 1985.

63. **Lycka, A., Jirman, J., and Holecek, J.,** ¹⁷O and ¹³C NMR spectra of some geminal diacetates, *Coll. Czech. Chem. Comm.,* 53, 588, 1988.

64. **Kintzinger, J.-P.,** Oxygen-17 and silicon-29, in *NMR-17,* Diehl, P., Fluck, E., and Kosfeld, R., Eds., Springer-Verlag, New York, 1981, 28.

65. (a) **Canet, D., Goulon-Ginet, C., and Marchal, J. P.,** Accurate determination of parameters for ¹⁷O in natural abundance by fourier transform NMR, *J. Magn. Reson.,* 22, 537, 1976; (b) **Ebraheem, E. A. K. and Webb, G. A.,** Calculation of some ¹⁷O nuclear screening constants, *J. Magn. Reson.,* 25, 399, 1977; (c) **Boykin, D. W., Sullins, D. W., and Eisenbraum, E. J.,** ¹⁷O NMR spectroscopy of lactones, *Heterocycles,* 29, 301, 1989.

66. **Vasquez, P. C., Boykin, D. W., and Baumstark, A. L.,** ¹⁷O NMR spectroscopy (natural abundance) of heterocycles: anhydrides, *Org. Magn. Reson.,* 24, 409, 1986.

67. **Baumstark, A. L., Balakrishnan, P., and Boykin, D. W.,** ¹⁷O NMR spectroscopic study of steric hindrance in phthalic anhydrides and phthalides, *Tetrahedron Lett.,* p. 3079, 1986.

68. **Wysocki, M. A., Jardon, P. W., Mains, G. J., Eisenbraun, E. J., and Boykin, D. W.,** Steric effects in the ¹⁷O NMR spectroscopy of aromatic methyl ethers, *Magn. Reson. Chem.,* 25, 331, 1987.

69. **Schuster, I. I., Parvez, M., and Freyer, A. J.,** Enhancement of the resonance interaction of out-of-plane methoxy groups by ortho substituents in crowded anisoles, *J. Org. Chem.,* 53, 5819, 1988.

70. **Kolehmainen, E. and Knuutinen, J.,** ¹⁷O NMR study of chlorinated anisoles, *Magn. Reson. Chem.,* 26, 1112, 1988.

71. **Duddeck, H., Rosenbaum, D., Elgamal, M. H. A., and Shalaby, N. M. M.,** Natural abundance ¹⁷O NMR spectra of coumarins, furocoumarins and related compounds, *Magn. Reson. Chem.,* 25, 489, 1987.

72. **Kintzinger, J.-P.,** Oxygen-17 and silicon-29, in *NMR-17,* Diehl, P., Fluck, E., and Kosfeld, R., Eds., Springer-Verlag, New York, 1981, 24.

73. **Chimichi, S., Nesi, R., and DeSio, F.,** Oxygen-17 nuclear magnetic resonance study of some five-membered heterocyclic derivatives, *Org. Magn. Reson.,* 22, 55, 1984.

74. **Chandrasekaran, S., Wilson, W. D., and Boykin, D. W.,** ¹⁷O NMR studies on polycyclic quinones, hydroxyquinones and related cyclic ketones: models for anthracycline intercalators, *Org. Magn. Reson.,* 22, 757, 1984.

75. **Boykin, D. W. and Martin, G. E.,** ¹⁷O NMR investigation of cyclic aromatic ethers, *J. Heterocycl. Chem.,* 24, 365, 1987.

76. **Reuben, J.,** Hydrogen-bonding effects on oxygen-17 chemical shifts, *J. Am. Chem. Soc.,* 91, 5725, 1969.

77. **Gorodetsky, M., Luz, Z., and Mazur, Y.,** Oxygen-17 nuclear magnetic resonance studies of equilibria between enol forms of β-diketones, *J. Am. Chem. Soc.,* 89, 1183, 1967.

78. **Lapachev, V. V., Mainagashev, I. Y., Stekhova, S. A., Fedotov, M. A., Krivopalov, V. P., and Mamaev, V. P.,** ^{17}O NMR studies of enol and phenol compounds with intramolecular hydrogen bonds, *J. Chem. Soc., Chem. Commun.,* p. 494, 1985.

79. **St. Amour, T. E., Burgar, M. I., Valentine, B., and Fiat, D.,** ^{17}O NMR studies of substituent and hydrogen bonding effects in acetophenones and benzaldehydes, *J. Am. Chem. Soc.,* 103, 1128, 1981.

80. **Jaccard, G. and Lauterwein, J.,** Intramolecular hydrogen bonds of the C=O---H-O type as studied by ^{17}O-NMR, *Helv. Chim. Acta,* 69, 1469, 1986.

81. **Jaccard, G., Carrupt, P.-A., and Lauterwein, J.,** Study of intramolecular hydrogen bonding in *o*-anisic acid and *o*-anisamide by ^{17}O NMR and *ab initio* MO calculations, *Magn. Reson. Chem.,* 26, 239, 1988.

Chapter 8

17O NMR SPECTROSCOPIC DATA FOR CARBONYL COMPOUNDS:
I. ALDEHYDES AND KETONES
II. CARBOXYLIC ACIDS AND DERIVATIVES

David W. Boykin* and Alfons L. Baumstark*

TABLE OF CONTENTS

I. INTRODUCTION AND SCOPE

Heteronuclear nuclear magnetic resonance (NMR) chemical shift data are dependent upon the paramagnetic (deshielding) σ_p^p and the diamagnetic (shielding) σ_0^d screening constants.[1-3] ^{17}O NMR chemical shift data are considered to be dependent essentially upon the paramagnetic term and are usually described by the Karplus-Pople[4] equation. For a detailed discussion of ^{17}O NMR chemical shift theory, the reader is referred to Chapter 1 as well as several reviews.[1-3] ^{17}O NMR spectroscopy is presently receiving considerable attention.[1-2] Earlier studies were limited in number due to the low natural abundance (0.037%) and to the quadrupolar properties of the [17]O nucleus.[1] These difficulties, broad lines and low signal-to-noise ratios encountered in early stages of the field, have been significantly reduced by use of FT NMR spectrometers, by carrying out measurements at higher temperatures, and by use of low viscosity solvents.[1] The applications of ^{17}O NMR spectroscopy to organic chemistry problems have been greatly facilitated and are shown to provide insights into molecular structure.[5]

The focus of this chapter is on the ^{17}O NMR spectroscopic properties of carbonyl functional groups. This important family of functional groups is among the most extensively studied and the one for which chemical shift-structure relationships are perhaps the best understood. ^{17}O NMR signals for carbon oxygen double bonds cover a wide chemical shift range (Figure 1) and are highly sensitive to electronic effects and to steric interactions. This topic has been divided into two general sections: the first deals with aldehydes and ketones and the second covers carboxylic acids and related derivatives. An attempt has been made to include data from the major papers in this field which have appeared within the last ten years. The reader is referred to the original literature for the pioneering work of Christ et al.[6] In certain cases only selected compounds are listed, when a series showed little variation with structure or when large error limits on the chemical shifts were reported. While we attempted to be thorough, we have not been encyclopedic; isolated reports may not be included. ^{17}O NMR spectroscopic studies on 1,2- and 1,3-diketones (and related hydrogen bonding data) have been omitted since these topics are treated in earlier chapters. In addition, ^{17}O NMR spectroscopic studies of biological systems[7] (peptides, etc.) are beyond the scope of this review. However, selected studies on amino acids and related compounds have been included. Since the signals for these compounds are pH and solvent dependent, the reader is referred to the original literature for details of these effects (see Section III.A).[7]

II. ALDEHYDES AND KETONES

A. ALDEHYDES

The ^{17}O NMR chemical shift values[1,8-12] for the carbonyl oxygen for the aldehyde functional group are presented in Table 1. The observed chemical shifts for this functional group range from approximately 500 to 600 ppm. The chemical shift data for aliphatic aldehydes reflect conventional β and γ shielding and δ deshielding effects.[1] The data for the substituted benzaldehydes show a large dependence on the electronic character of the substituent; electron donors cause shielding of the signal whereas electron-attracting groups result in deshielding.[9,10] Shifts for sterically hindered aryl aldehydes are sensitive to changes in the torsion angle between the formyl group and the aryl ring (see Chapter 3 for detailed discussion of this effect).[11] Intramolecular hydrogen bonding results in large upfield shifts of the aldehyde signal. In general, hydrogen-bonding-induced shifts are larger than other substituent effects (see Chapter 5 for a detailed discussion of hydrogen-bonding effects).[9,10] The effect of α,β-unsaturation on the relative aldehyde [17]O chemical shift tends to be minimal.[1]

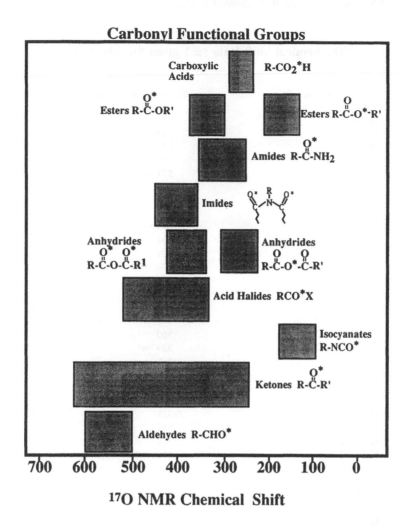

Carbonyl Functional Groups

^{17}O NMR Chemical Shift

FIGURE 1. General chemical shift ranges of carbonyl-containing functional groups.

B. ALIPHATIC ACYCLIC KETONES

The observed chemical shift range for aliphatic acyclic ketones is small (540 to 570). The data are summarized in Table 2. The variations in the chemical shift data of aliphatic acyclic ketones are generally interpretable by employing standard β, γ, and δ effect arguments (β and γ effects are shielding while δ effects are deshielding). The ^{17}O NMR chemical shift data for ketone carbonyls are sensitive to hydrogen bonding; see results for acetone **42** in water solution and Chapter 5. Data obtained from neat liquids and in dry acetonitrile as solvent seem comparable. Note in chloroform solution, intermolecular hydrogen-bonding influences are expected. Cyclopropyl substitution (**50a** and **50b**) causes large shielding shifts and offers the potential to study cyclopropane-carbonyl group interactions.

C. ALIPHATIC CYCLIC KETONES

The ^{17}O data for aliphatic cyclic ketones occur over a relatively small range and are listed in Table 3.[1,8,13-15] The influence of increasing ring size for cyclic ketones on ^{17}O NMR chemical shift data results in deshielding (cf.**67,68,69** and **99**). The effects on ^{17}O NMR data of varying substituents and for conformational changes in cyclohexanones have been extensively studied by Crandall, and empirical relationships were developed.[14] Other substituent effects are consistent with the acyclic ketone data.

TABLE 1
¹⁷O Chemical Shift Data (±1 ppm) for Aldehydes

Compound No.	Structure/name	δ(C=O) ppm	δ(R)	Conditions[a]	Ref.
1	MeCHO	592	—	A	1,8
		550	—	B	1,8
2	EtCHO	579.5	—	A	1,8
		538	—	B	1,8
3	i-PrCHO	574.5	—	A	1,8
4	t-BuCHO	564	—	A	1,8
5	n-PrCHO	589	—	A	1,8
6	n-BuCHO	587	—	A	1,8
7	i-BuCHO	593.5	—	A	1,8
8	Me₃CCH₂CHO	604	—	A	1,8
9	Et₂CHCHO	589	—	A	1,8
10	CH₂=CHCHO	579	—	A	1,9
11	CH₃CH=CHCHO	563	—	A	1,9
12	XC₆H₄CHO; X = H	569	—	A	1,9
		562	—	C	10
13	12; X = 2-MeO	555	—	C	10
14	12; X = 3-MeO	560	—	C	10
15	12; X = 4-MeO	539	—	C	10
16	12; X = 2-HO	505	85	C	10
		502	78	A	1,9
17	12; X = 3-HO	559	92	C	10
		532.5	68	A	1,9
18	12; X = 4-HO	528	108	C	10
		517	82	A	1,9
19	12; X = 2-Br	573	—	C	10
20	12; X = 3-Br	566	—	C	10
21	12; X = 4-Br	563	—	C	10
22	12; X = 2-NO₂	576	—	C	10
23	12; X = 3-NO₂	526	—	C	10
24	12; X = 4-NO₂	580	—	C	10
25	12; X = 2-Me	575	—	D	11
26	12; X = 4-Me	557	—	D	11
27	2,4,6-Me₃C₆H₂CHO	585	—	D	11
28	1-Naphthaldehyde	575	—	D	11
29	2-Naphthaldehyde	564	—	D	11
30	9-Anthrylaldehyde	597	—	D	11
31	2,3,4-Trimethoxybenzaldehyde	545	—	C	10
32	2,3,4-Trimethoxybenzaldehyde	538	—	C	10
33	2,4,5-Trimethoxybenzaldehyde	565	—	C	10
34	Furfural	524	234.5	A	1,9
		536.4	238.1	D	12
35	5-Nitrofurfural	574	225	D	12
36	3-Furaldehyde	557.3	246.1	D	12
37	2-Thiophenecarboxaldehyde	538	—	D	12
38	5-Nitrothiophenecarboxaldehyde	567	—	D	12
39	3-Thiophenecarboxaldehyde	552.8	—	D	12
40	2-Pyrrolecarboxaldehyde	496	—	D	12
41	N-Methyl-2-pyrrolecarboxaldehyde	501.8	—	D	12

[a] A = Neat liquids at RT; B = Aqueous solution at RT; C = Dioxane solvent at 30°; D = Acetonitrile solvent at 75°.

D. α,β-UNSATURATED KETONES

The ¹⁷O NMR data for α,β-unsaturated ketones are summarized in Table 4.[1,9,17-19] In general, based upon charge density considerations, a relative upfield shift of the carbonyl

TABLE 2
[17]O Chemical Shift Data (±1 ppm) for Aliphatic, Acyclic Ketones

Compound no.	Structure/name	δ(C=O)ppm	Conditions[a]	Ref.
42	Me$_2$CO	569	A	1,8
		523	B	1,8
		571	C	11
43	EtMeCO	557.5	A	1,8
		558	C	11
44	n-PrMeCO	563	A,C	1,8,11
45	i-PrMeCO	557	A,C	1,8,11
46	t-BuMeCO	561	A	1,8
		560	C	11
47	i-BuMeCO	568	A	1,8
48	n-BuMeCO	563	A	1,8
49	Me$_3$CCH$_2$MeCO	575	A	1,8
50a	CyclopropylMeCO	521	C	11
50b	DicyclopropylCO	472	A	1,8
51	CyclohexylMeCO	560	C	11
52	Et$_2$CO	547	A	1,8
53	n-PrEtCO	550	A	1,8
54	i-PrEtCO	543.5	A	1,8
55	t-BuEtCO	547.5	A	1,8
56	i-BuEtCO	552.5	A	1,8
57	Et$_3$CEtCO	566	A	1,8
58	n-Pr$_2$CO	555	A	1,8
59	n-PrtBuCO	553.5	A	1,8
60	i-Pr$_2$CO	535	A	1,8
61	i-Pri-BuCO	548	A	1,8
62	i-Prt-BuCO	538.5	A	1,8
63	i-Bu$_2$CO	564	A	1,8
64	t-Bu$_2$CO	564.5	A	1,8
65	ClCH$_2$MeCO	528	B	1,8
66	(ClCH$_2$)$_2$CO	521	B	1,8

[a] A = Neat liquids at RT; B = Aqueous solution at RT; C = Acetronitrile solution at 75°.

^{17}O NMR signal would be expected upon conjugation; however, very little difference is noted between the ^{17}O NMR chemical shift of the simple unsaturated ketones (e.g., 101 and 102) and their saturated analogs (43 and 52).[1,9] A similar trend is noted for analogous aldehydes (see Section II.A). However, a greater difference in ^{17}O NMR chemical shifts has been noted between cyclic unsaturated ketones and their saturated cyclic analogs (cf.106 and 112 to 68 and 69). The effect of the α-methyl substitution of the [17]O chemical shift data for the cyclic α,β-unsaturated ketones is less than the noted for β-substitution.[17] Interestingly, substituent effects (shielding) for the cyclohexenone system are significantly less than those for the cyclopentenones. In contrast, for a corresponding acyclic α,β-unsaturated ketone, α-methyl substitution results in 4 ppm deshielding (cf. 101 and 104). Interpretation of the effect of substitution in acyclic systems is complicated by the presence of cisoid and transoid conformers of approximately equal energy. Additional investigations are required to more fully understand the competing factors which determine the [17]O chemical shifts in these simple acyclic and cyclic α,β-unsaturated systems.

Substituent effects for the aryl α,β-unsaturated chalcone system show excellent correlations with Hammett-Brown constants.[19] The effect of 4′-substituents (123 to 129) is equivalent to that observed for 4-substituted acetophenones — whereas, the results for 4-substituted chalcones (116 to 122) show a reduced sensitivity as expected, due to the insulating effect

TABLE 3
17O Chemical Shift Data (±1 ppm) for Aliphatic, Cyclic Ketones

Compound no.	Structure/name	δ(C=O)ppm	Conditions[a]	Ref.
67	Cyclobutanone	545	A	1,8
68	Cyclopentanone	543	A	1,8
		545	C	13
69	Cyclohexanone	558	A	1,8
		559.9	B	14
		572.5	E	16
70	2-Methycyclohexanone	548.8	B	14
71	3-Methylcyclohexanone	561.8	B	14
72	4-Methylcyclohexanone	560.1	B	14
73	2-Ethylcyclohexanone	553.3	B	14
74	2-Isopropylcyclohexanone	558.7	B	14
75	2-*t*-Butylcyclohexanone	561.9	B	14
76	3-*t*-Butycyclohexanone	561.4	B	14
77	4-*t*-Butylcyclohexanone	557.4	B	14
78	2-*n*-Propylcyclohexanone	552.2	B	14
79	2,2-Dimethylcyclohexanone	554.9	B	14
80	3,3-Dimethylcyclohexanone	572.3	B	14
81	4,4-Dimethylcyclohexanone	561.4	B	14
82	*cis*-2,6-Dimethylcyclohexanone	540.8	B	14
83	*trans*-2,6-Dimethylcyclohexanone	550.8	B	14
84	*cis*-3,5-Dimethylcyclohexanone	561.2	B	14
85	*trans*-3,5-Dimethylcyclohexanone	571.3	B	14
86	*cis*-2-Me-4-*t*-Butylcyclohexanone	549.3	B	14
87	3,3,5-Trimethylcyclohexanone	570	A	14
88	3,3,5,5-Tetramethylcyclohexanone	576.5	B	14
89	Bicyclo[2.2.1]heptan-2-one	524.3	B	14
90	1-Methylbicyclo[2.2.1]heptan-2-one	514.2	B	14
91	*exo*-3-Methylbicyclo[2.2.1]heptan-2-one	517.7	B	14
92	*endo*-3-Methylbicyclo[2.2.1]heptan-2-one	509.0	B	14
93	3,3-Dimethylbicyclo[2.2.1]heptan-2-one	503.4	B	14
94	1,3,3-Trimethylbicyclo[2.2.1]heptan-2-one	495.3	B	14
95	1,7,7-Trimethylbicyclo[2.2.1]heptan-2-one	520.1	B	14
96	Bicyclo[2.2.2]octan-2-one	545.9	B	14
97	3-Methylbicyclo[2.2.2]octan-2-one	539.4	B	14
98	3,3-Dimethylbicyclo[2.2.2]octan-2-one	531.6	B	14
99a	Cycloheptanone	566	A	1,8
99b	Cyclooctanone	578.8	E	16
99c	Cyclodecanone	575.7	E	16
99d	Cyclododecanone	581.4	E	16
100	Camphor	514	D	1,15

[a] A = Neat liquids at RT; B = Dioxane solution at RT with drop of H_2O^{17}; C = Acetonitrile solution at 75°; D = Benzene solution; E = Cyclohexane solution; referenced to external D_2O.

of the conjugated carbon-carbon double bond. These results can be considered normal electronic effects which reflect the relative charge distribution on the carbonyl oxygen.

E. ARYL KETONES

The *17O* NMR chemical shift data for aryl ketones are reasonably well understood. Representative chemical shift data are listed in Table 5.[10,11,18,21-25] A study of substituent effects on the *17O* chemical shift data of unhindered acetophenones has shown that the chemical shifts are dependent on the π electron density of the carbonyl oxygen.[21] The *17O* chemical shift data for hindered aryl ketones have been correlated with the magnitude of

TABLE 4
[17]O Chemical Shift Data (± 1 ppm) for α,β-Unsaturated Ketones

Compound no.	Structure/name	$\delta(C=O)$ ppm	$\delta(R)$	Conditions[a]	Ref.
101	Methylvinyl ketone	561	—	A	1,9
102	Ethylvinylketone	543.5	—	A	1,9
103	Methylpropenylketone	553	—	A	1,9
104	Methylisopropenylketone	565	—	A	1,9
105	Mesityl oxide	549	—	A	1,9
106	2-Cyclohexen-1-one	546.0	—	B	17
		565.4	—	C	18
107	2-Methyl-2-cyclohexen-1-one	539.5	—	B	17
108	3-Methyl-2-cyclohexen-1-one	531.4	—	B	17
109	3-Phenyl-2-cyclohexen-1-one	543.2	—	B	17
110	2-Ethoxy-1-cyclohexen-1-one	535.3	70.5	B	17
111	3-Ethoxy-2-cyclohexen-1-one	509.5	122.1	B	17
112	2-Cyclopenten-1-one	508.2	—	B	17
113	2-Methyl-2-cyclopenten-1-one	497.5	—	B	17
114	3-Methyl-2-cyclopenten-1-one	495.3	—	B	17
115	X-C_6H_4-CH=CH-CO-Ph; X = H	524.6	—	C	19
116	115; X = NMe_2	509.3	—	C	19
117	115; X = MeO	519.3	—	C	19
118	115; X = Me	522.9	—	C	19
119	115; X = F	523.1	—	C	19
120	115; X = Cl	525.1	—	C	19
121	115; X = CN	533.3	—	C	19
122	115; X = NO_2	535.4	—	C	19
123	PhCH=CH-CO-C_6H_4-Y; Y = MeO	508.6	—	C	19
124	123; Y = Me	518.3	—	C	19
125	123; Y = F	520.5	—	C	19
126	123; Y = Cl	525.4	—	C	19
127	123; Y = Br	526.6	—	C	19
128	123; Y = CN	537.1	—	C	19
129	123; Y = NO_2	541.0	—	C	19

[a] A = Neat liquids at RT; B = Acetonitrile solvent at 75°; C = Toluene solution at 95°.

the torsion angle between the aryl ring and the carbonyl group (see Chapter 3 for a detailed discussion of this effect).[22] The trend in ^{17}O NMR chemical shifts for the relatively unhindered alkyl aryl ketones is an increase in shielding with increasing size of the alkyl group.[11] However, when steric hindrance becomes important, torsion angle variation occurs and deshielding is observed. The ^{17}O NMR chemical shift for phenylcyclopropylketone shows a large shielding effect compared to its isopropyl analog.[11]

The ^{17}O NMR spectral data for 1-indanone (**180**), 1-tetralone (**189b**), and 1-benzosuberone (**190**) show significant sensitivity to ring size.[11] The result of benzene ring fusion on the cyclic ketones is an upfield shift of their signals compared to their acyclic counterparts. The downfield trend for the data of **180**, **189b**, and **190** is explicable, by changes in charge density as a consequence of torsion angle difference.[11] The effect of substituents on the [17]O chemical shifts of indanones has been extensively studied; Table 5 contains only a representative selection of the data for over 30 indanones reported.[24] The data for the saturated ring substituted indanones show normal β and γ effects, and hindered indanones show evidence for repulsive van der Waals interactions (compressional effects, see Chapter 4).

506 PPM   **528 PPM**   **562 PPM**

180   **189b**   **190**

F. QUINONES

Surprisingly, the ^{17}O chemical shift data for quinones appear downfield relative to α,β-unsaturated ketones. The ^{17}O NMR data for quinones are presented in Table 6.[17,18,25,26] In general, for the unsymmetrical quinones, two signals are observed.[17] The effect of α-methyl substitution (shielding) is greater on the signal of the 1-carbonyl group than on that for the 4-carbonyl group on the benzoquinones in contrast to that observed for the cyclic enones (Table 4). The large shielding effects observed for most α-substitutions in the benzoquinone system appear to be the result of a combination of electronic and attractive van der Waals effects.[17] Intramolecular hydrogen bonding is apparent from the ^{17}O NMR data for the hydroxyquinones.[18,25] In fact, only a single ^{17}O NMR resonance has been observed for 2,5-dihydroxybenzoquinone (**201**) and 5,8-dihydroxynaphthoquinone (**206**), indicating rapid proton exchange between the oxygen atoms of these compounds.[18] However, 1,4-dihydrox-yanthraquinone (**213**) shows two discrete ^{17}O signals indicating that proton exchange between oxygen atoms is slow in this compound.[18] The ^{17}O NMR signal for the hindered carbonyl group in rigid planar quinones showed large downfield shifts which were correlated with local repulsive van der Waals energies (for a detailed discussion of this effect see Chapter 4.)[26]

G. HETEROCYCLIC AND MISCELLANEOUS KETONES

Table 7 contains ^{17}O NMR data for selected heterocyclic ketones, cyclopropenones, and other related compounds.[12,15,18,26-30] The data for simple furan, thiophene, and pyrrole ketones (**226** to **237**) qualitatively correlated with the electronic character of the heterocyclic rings.[12] Based upon the available data, it appears that electronic effects (cf. **234** and **235**) and torsion angle effects for hindered heterocyclic ketones (**230** and **233**) play important roles in determining the relative ^{17}O NMR chemical shift value. In rigid, planar heterocyclic ketones (cf. **219,220,224,** and **225**) the influence of repulsive van der Waals interactions on the ^{17}O NMR chemical shift is apparent.[26]

The ^{17}O NMR signal for two cyclopropenones, **238** and **239**, show remarkable upfield shifts for a carbonyl group, suggesting that in these cases the functional group has substantial single bond character.[29] Interestingly, the ^{17}O NMR chemical shift value for tropone (**240**) is similar to that of the enones. In contrast, the data for 4H-pyran-4-one (**215**) reflects an increase in single bond character for the carbonyl group.[18] The upfield chemical shift of certain heterocyclic ketones has been presented as evidence for a transannular interaction of the lone pair of the heteroatom with the carbonyl oxygen.[16,30] Representative data for this effect are present in the Table (**243** to **245**); additional data are available in the original reports.

TABLE 5
^{17}O Chemical Shift Data (± ppm) for Aryl Ketones (ArCOR)

Compound no.	Structure/name	δ(C=O) ppm	δ(R)	Conditions[a]	Ref.
130	XC$_6$H$_4$COCH$_3$; X = H	554	—	A	20
		548.6	—	B	21
		552	—	C	22
		552	—	D	10
		550	—	E	10
131	130; X = 4-MeO	548	—	A	20
		534.5	60.3	B	21
		535	—	E	10
132	130; X = 3-MeO	551	—	E	10
		548	60	D	10
133	130; X = 2-MeO	561	—	E	10
		563	56	D	10
134	130; X = 4-NH$_2$	513.9	—	B	21
		518.7	—	C	23
		511	—	E	10
		507	—	F	10
135	130; X = 3-NH$_2$	548	—	E	10
		536	—	F	10
136	130; X = 2-NH$_2$	512	—	D	10
		524.1	—	C	23
		508	—	E	10
137	130; X = 4-OH	513	—	E	10
138	130; X = 3-OH	546	—	E	10
139	130; X = 2-OH	488	—	E	10
140	130; X = 4-Me	555	—	A	20
		543.5	—	B	21
		543	—	E	10
		546	—	C	22
141	130; X = 3-Me	548	—	E	10
142	130; X = 2-Me	586	—	A	20
		579	—	E	10
		582	—	C	22
143	130; X = 4-F	545.8	—	B	21
		545	—	E	10
144	130; X = 2-F	562	—	E	10
145	130; X = 4-Cl	551.1	—	B	21
		550	—	E	10
146	130; X = 2-Cl	586	—	E	10
147	130; X = 4-Br	552	—	A	20
		551	—	E	10
		548	—	D	10
148	130; X = 3-Br	556	—	E	10
		555	—	D	10
149	130; X = 2-Br	589	—	E	10
		586	—	D	10
150	130; X = 4-COCH$_3$	559	559	B	21
151	130; X = 4-CN	562.8	—	B	21
		561	—	E	10
152	130; X = 4-NO$_2$	567	—	A	20
		564.3	578.5	B	21
		562	—	D	10
		567	—	E	10
153	130; X = 3-NO$_2$	556	—	E	10
		566	—	D	10
154	130; X = 2-NO$_2$	573	—	E	10

TABLE 5 (continued)
¹⁷O Chemical Shift Data (± ppm) for Aryl Ketones (ArCOR)

Compound no.	Structure/name	δ(C=O) ppm	δ(R)	Conditions[a]	Ref.
155	2,4-Dimethylacetophenone	576	—	C	22
156	2,5-Dimethylacetophenone	582	—	C	22
157	3,4-Dimethylacetophenone	545	—	C	22
158	2,6-Dimethylacetophenone	608	—	A	20
159	2,4,6-Trimethylacetophenone	593	—	A	20
160	2,4,6-Trimethylacetophenone	601	—	C	22
161	2,4,5-Trimethylacetophenone	575	—	C	22
162	2,4,6-Tri-isopropylacetophenone	607	—	C	22
163	2,3,5,6-Tetramethylacetophenone	596	—	C	22
164	1-Acetynaphthalene	585	—	C	22
165	2-Acetylnaphthalene	553	—	C	22
166	9-Acetylanthracene	613	—	C	22
167	PhCOR; R = Et	540	—	C	11
168	**167**; R = n-Pr	543	—	C	11
169	**167**; R = i-Pr	535	—	C	11
170	**167**; R = t-Bu	565	—	C	11
171	**167**; R = Cyclopropyl	495	—	C	11
172	**167**; R = Cyclobutyl	530	—	C	11
173	**167**; R = Cyclopentyl	529	—	C	11
174	**167**; R = Cyclohexyl	538	—	C	11
175	**167**; R = Ph	552	—	C	11
176	**130**; X = 2-NHCOCH₃	535.4	373.3	C	23
177	**130**; X = 4-NHCOCH₃	539.5	366.3	C	23
178	**130**; X = 2-NHCOCF₃	531.0	343.3	C	23
179	**130**; X = 4-NHCOCF₃	551.0	339.1	C	23
180	Indanone	505.3,506	—	C	24
181	2-Methylindanone	499.8	—	C	24
182	3-Methylindanone	507.7	—	C	24
183	7-Methylindanone	520.2	—	C	24
184	2,2-Dimethylindanone	494.0	—	C	24
185	3,3-Dimethylindanone	510.9	—	C	24
186	3,3-Dimethyl-7-*t*-butylindanone	541	—	C	24
187	Fluorenone	510	—	F	25
188	α-Hydroxyfluorenone	476.4	—	F	25
189a	β-Tetralone	556	—	A	1
189b	α-Tetralone	539.4	—	G	18
		528	—	C	11
190	Benzosuberone	562	—	C	11
191	9-Anthranone	491	—	G	18

[a] A = In acetone solution, RT; B = Acetonitrile at 60°; C = Acetonitrile at 75°; D = Neat liquid; E = Dioxane solution at 30°; F = In CHCl₃ solution; G = Toluene solution.

TABLE 6
^{17}O Chemical Shift Data (± 1 ppm) for Quinones

Compound no.	Structure/Name	δ(C=O)ppm	δ(R)	Conditions[a]	Ref.
192	Benzoquinone	635.1	—	A	18
		624	—	B	17
193	Methylbenzoquinone	609,614.5	—	B	17
194	2,3-Dimethylbenzoquinone	598	—	B	17
195	2,5-Dimethylbenzoquinone	601	—	B	17
196	2,6-Dimethylbenzoquinone	593,604	—	B	17
197	Tetramethylbenzoquinone	578	—	B	17
198	t-Butylbenzoquinone	614,621	—	B	17
199	2,3-di-t-Butylbenzoquinone	611	—	B	17
200	2,6-di-t-Butylbenzoquinone	605,612	—	B	17
201	2,5-Dihydroxybenzoquinone	357.7	—	A	18
202	Naphthoquinone	580.7	—	A	18
		571.8	—	B	17
		568.7	—	C	25
203	2-Methylnaphthoquinone	558.2	—	C	25
204	5-Hydroxynaphthoquinone	570.4,498.3	—	C	25
205	2-Methyl-5-hydroxynaphthoquinone	560.3,488.5	—	C	25
206	5,8-Dihydroxynaphthoquinone	282.6	—	A	18
207	5-Ethoxynaphthoquinone	560.2	—	C	25
208	Anthraquinone	531.4	—	A	18
		524	—	B	26
209	2-Methylanthraquinone	524,519	—	B	26
210	1-Methylanthraquinone	552,521	—	B	26
211	2-t-Butylanthraquinone	523,523	—	B	26
212	1-t-Butylanthraquinone	572,522	—	B	26
213	1,4-Dihydroanthraquinone	440.8	87.1	A	18
214	1,4-Chrysenequinone	602	—	B	26

[a] A = Toluene solution at 95°; B = Acetonitrile solution at 75°; C = Chloroform solution at 40°.

TABLE 7
^{17}O Chemical Shift Data ($\pm$ ppm) for Selected Heterocyclic and Miscellaneous Ketones

Compound no.	Structure/name	$\delta(C=O)$ ppm	$\delta(R)$	Conditions[a]	Ref.
215	4H-pyran-4-one	463.6	174.2	A	18
216	1-Benzopyran-4-one	450.8	158.2	A	18
		446	162	B	27
217	Xanth-9-enone	454.8	137.1	A	18
		443.6	139.0	C	28
218	Flavone	438	158	C	26
		426	160	B	27
219	7,8-Benzoflavone	433	—	C	26
220	5,6-Benzoflavone	451	—	C	26
221	4-Chromanone	515	70	B	27
222	2-Phenyl-4-chromanone	518	93	B	27
223	2,2-Dimethyl-4-chromanone	518	117	B	27
224	5-Methyl-7-methoxy-2,2-Dimethyl-4-chromanone	525	120	B	27
225	7-Methoxy-2,2-Dimethyl-4-chromanone	500	118	B	27
226	2-Acetylfuran	525.5	241.6	C	12
227	5-Methyl-2-acetylfuran	516.3	249.6	C	12
228	3-Acetylfuran	545	244	C	12
229	2-Acetylpyrrole	490.8	—	C	12
230	N-Methyl-2-acetylpyrrole	505.4	—	C	12
231	N-Methyl-3-acetylpyrrole	509.4	—	C	12
232	2-Acetylthiophene	528.1	—	C	12
233	3-Methyl-2-acetylthiophene	535.8	—	C	12
234	5-Methyl-2-acetylthiophene	523.3	—	C	12
235	5-Chloro-2-acetylthiophene	531.0	—	C	12
236	3-Acetylthiophene	539.1	—	C	12
237	2,5-Dimethyl-3-acetylthiophene	546.1	—	C	12
238	Diphenylcyclopropenone	248	—	D	29
239	Dimethylcyclopropenone	233	—	D	29
240	Tropone	502	—	E	29
241	2,2-Di-t-butylcyclopropanone	524	—	B	29
242	Tropolone	250	—	D	15
243	N-Et-4-piperidone	561	—	F	30
244	Tetrahydro-4H-pyran-4-one	568	—	F	30
245	Tetrahydrothiopyran-4-one	569	—	F	30

[a] A = Toluene solution at 95°; B = Chloroform solution; C = Acetonitrile solution at 75°; D = Benzene solution; E = Neat liquid; F = Dioxane solution.

III. CARBOXYLIC ACIDS AND THEIR DERIVATIVES

A. CARBOXYLIC ACIDS

The ^{17}O NMR spectra for carboxylic acids exhibit only one signal, which appears at approximately 250 ppm; the data are shown in Table 8.[1,9,31-34] The equivalence of both oxygens for the carboxylic acids has been attributed to fast proton exchange in dimeric or higher aggregates.[1,9] The chemical shift value for the signal for the carboxylic acid group is roughly at the average of the value of the carbonyl signal and the single-bonded oxygen signal observed for methyl esters.[1,9] Aliphatic carboxylic acids show only small variation of chemical shift value. In addition, the ^{17}O NMR data for carboxylic acids in water solution are pH dependent.[31] ^{17}O NMR signals for aromatic carboxylic acids are sensitive to electronic effects (Hammett relationships).[32,34] The rho value from the Hammett correlation of benzoic acid data is approximately twice that for an analogous series of cinnamic acids, consistent with previous studies on transmission of substituent effects by the carbon-carbon double bond. A study of a limited number of p-substituted phenylacetic acids suggests that the ^{17}O NMR signals show similar electronic effects, but of lower magnitude.[34] Torsion angle variation induced by steric hindrance has been shown to produce downfield chemical shifts for the aryl carboxylic acids.[33] These effects are discussed in detail in Chapter 3.

B. ESTERS

Table 9 contains ^{17}O NMR chemical shift data for aliphatic and aromatic esters.[1,9,35-40] The ^{17}O NMR signals for both the carbonyl oxygen and the single-bonded oxygen of esters are sensitive to electronic effects and steric effects. For aryl esters the ^{17}O NMR data for the double-bonded oxygens are approximately twice as responsive to electronic effects as those of the single-bonded oxygens. These effects have been correlated by the Hammett approach.[32] Quantitative relationships between ^{17}O chemical shifts and torsion angles (carboxylate-ring) for aromatic esters have also been reported (see Chapter 3 for details).[33] Substitution on the group attached to the carbonyl function causes large changes in the carbonyl ^{17}O chemical shift with smaller relative shifts noted for the single-bonded oxygen. Change in structure of the group attached to the single-bonded oxygen (α-effects) results in large shifts of resonance for that oxygen and smaller relative shifts for the carbonyl oxygen. It is difficult to determine if the signal for the carbonyl group of α,β-unsaturated esters is shielded relative to the saturated analogs, because the data for α,β-unsaturated esters have been obtained in chloroform, a solvent which will produce upfield shifts due to hydrogen bonding. However, data obtained on lactones (*vide infra*) suggest that shielding of the carbonyl signal of α,β-unsaturated esters occurs.

C. AMIDES

The ^{17}O NMR spectra of amides are of considerable interest because of the importance of this functional group to biochemistry. The focus of this review is on structural influences on ^{17}O chemical shifts and, consequently, deals with studies which are carried out primarily in organic solvents (Table 10).[1,9,33,38,41,42] A number of reports have appeared which examine the ^{17}O NMR properties of amides, amino acid amides, and peptides in aqueous solution or mixed solvents; some discussion of this field may be found in Chapter 6, and key references are noted.[7] Excluding studies carried out in water and mixed solvents, relatively few investigations of the ^{17}O NMR properties of simple amides have appeared. Variation of structure on the carbonyl portion of the amide unit causes changes in the carbonyl ^{17}O NMR signal similar to that noted for the analogous change in structure for the carbonyl group of esters. Substitution on nitrogen reduces (monosubstitution) or eliminates (disubstitution) self-association by hydrogen bonding, and generally, for N,N-dialkyl substituted amides, the ^{17}O NMR signals are downfield of the analogous simple amides and N-alkylamides. The ^{17}O chemical shifts of hindered arylamides and hindered N-phenylacetamides have been found to depend on torsion angles (see Chapter 3).[33,38]

TABLE 8

^{17}O NMR Chemical Shift Data (± 1 ppm) for Selected Carboxylic Acids

Compound no.	Structure/name	$\delta(CO_2H)$	Conditions[a]	Ref.
246	HCO_2H	253	A	1,9
		263.4	E	31
247	$MeCO_2H$	251	A	1,9
		257.6	E	31
248	$EtCO_2H$	244	A	1,9
		251.9	E	31
249	i-$PrCO_2H$	242	A	1,9
		247.5	E	31
250	t-$BuCO_2H$	240	A	1,9
251a	p–X–$C_6H_4CO_2H$; X = H	250.5	B,C	32,33
		250	D	34
251b	251a; X = NO_2	256.1	B	32
251c	251a; X = CN	255.1	B	32
251d	251a; X = CF_3	254.1	B	32
251e	251a; X = Cl	251.1	B	32
251f	251a; X = OAc	249.6	B	32
251g	251a; X = F	249.6	B	32
251h	251a: X = CH_3	248.6	B,C	32,33
251i	251a; X = OCH_3	245.6	B	32
252a	o-X-$C_6H_4CO_2H$; X = MeO	272.5	D	34
252b	252a; X = Cl	273	D	34
252c	252a; X = NO_2	271.9	D	34
253a	m-X-$C_6H_4CO_2H$; X = MeO	251	D	34
253b	253a; X = Cl	250	D	34
253c	253a; X = NO_2	249	D	34
254	$PhCH_2CO_2H$	266	D	34
255a	o-X-C_6H_4-CH_2CO_2H; X = MeO	261	D	34
255b	255a; X = Cl	266	D	34
255c	255a; X = NO_2	270.3	D	34
256a	m-X-C_6H_4-CH_2CO_2H; X = MeO	272.1	D	34
256b	256a; X = Cl	273.8	D	34
256c	256a; X = NO_2	272	D	34
257a	p-X-C_6H_4-CH_2CO_2H; X = MeO	266	D	34
257b	257a; X = Cl	271.1	D	34
257c	257a; X = NO_2	277	D	34
258a	p-X-$C_6H_4CH{=}CHCO_2H$; X = H	254.1	B	32
258b	258a; X = NO_2	257.2	B	32
258c	258a; X = CN	256.8	B	32
258d	258a; X = CF_3	255.8	B	32
258e	258a; X = Cl	254.6	B	32
258f	258a; X = F	253.6	B	32
258g	258a; X = CH_3	253.0	B	32
258h	258a; X = OCH_3	251.8	B	32
259	2-$MeC_6H_4CO_2H$	265	C	33
260	2,3-$Me_2C_6H_3CO_2H$	269	C	33
261	2,6-$Me_2C_6H_3CO_2H$	280	C	33
262	2,4,6-$Me_3C_6H_2CO_2H$	280	C	33
263	1-Naphthyl CO_2H	267	C	33
264	2-Naphthyl CO_2H	251.5	C	33
265	9-Anthryl CO_2H	287	C	33
266	Gly[b]	254.5	E	31
269	Ala[b]	251.3	E	31
270	Val[b]	256.6	E	31
271	Ile[b]	257.1	E	31
272	Leu[b]	252.9	E	31
273	Ser[b]	255.8	E	31

TABLE 8 (continued)
^{17}O NMR Chemical Shift Data (± ppm) for Selected Carboxylic Acids

Compound no.	Structure/name	δ(CO$_2$H)	Conditions[a]	Ref.
274	Thr[b]	255.9	E	31
275	Pro[b]	251.7	E	31
276	Lys[b]	254.2	E	31
277	Arg[b]	254.2	E	31
278	Met[b]	253.9	E	31
279	Trp[b]	255.1	E	31
280	Phe[b]	255.8	E	31
281	Tyr[b]	255.3	E	31
282	His[b]	256.8	E	31

[a] A = Neat liquids at RT; B = Acetone at 40°C; C = Acetonitrile solution at 75°; D = DMSO; E = H$_2$O; pH ≅ 0.5.

[b] Amino acids.

D. LACTONES

The ^{17}O NMR data for lactones are listed in Table 11.[1,44] Both signals for the oxygens of lactones are sensitive to ring size. Alkyl substitution effects are consistent with the effects of similar substitution noted for simple esters and ketones. Interestingly, the carbonyl signal for thiolactones is significantly deshielded relative to their oxygen analogs.[44] Introduction of a double bond in conjugation with the carbonyl group results in shielding of the carbonyl signal, whereas placing a double bond in conjugation with the single-bonded oxygen results in deshielding of the carbonyl resonance.[44] In rigid, planar lactones (e.g., **448** and **449**) large downfield shifts are noted which are attributable to repulsive van der Waals interactions (see Chapter 4.)[46]

E. LACTAMS, IMIDES, AND RELATED COMPOUNDS

The ^{17}O NMR carbonyl data for lactams, imides, and related compounds are shown in Table 12.[48-53] Substituent effects for simple lactams show trends similar to those for the corresponding lactones. Caution should be used in interpreting small chemical shift differences for N-H lactams since self-association (hydrogen bonding) often occurs at concentrations normally used for ^{17}O NMR investigations. As noted for lactones, introduction of groups which can interact with the amide group by conjugation can cause either shielding or deshielding of the carbonyl signal depending upon whether the group is in direct conjugation with the nitrogen or the carbonyl group.[48]

For the imides the ^{17}O NMR data show large deshielding effects as the steric bulk of the N-substituent was increased; introduction of a 3-substituent in the phthalimide series also yields deshielding.[51] The deshielding effects of 3-substitution and N-substitution in the phthalimides have been shown to be additive. In these cases, the ^{17}O NMR data have been correlated with in-plane bond angle distortions. The results for N-arylphthalimides show that the aromatic ring is rotated out of the plane of the phthalimide ring. The deshielding effects noted for the rigid, planar, lactams and imides are correlated with repulsive van der Waals energies (see Chapter 4.)[51]

A study of 5-substituted uracils shows that the chemical shift range between the 5-methoxy and the 5-nitro compound for oxygen 2 to be 40 ppm; and the range for oxygen 4 for the same compounds is 20 ppm.[53] The ^{17}O data for oxygen 2 gives a good correlation with Hammett and DSP treatments, whereas the data for oxygen 4 do not.

F. ANHYDRIDES

Prior to 1986, ^{17}O NMR data had been reported only for acetic and propionic anhydrides; subsequently, several studies have appeared and the data are summarized in Table 13. The

TABLE 9
17O Chemical Shift Data (±1 ppm) for Esters

Compound no.	Structure/name	δ(C=O) ppm	δ(–O–) ppm	Conditions[a]	Ref.
283	HCO$_2$Me	359,364	137,143	A	1,9,35
284	MeCO$_2$Me	357,355	134, 148	A	1,9,35
		361	141.2	B	36
		355	148	C	37
285	EtCO$_2$Me	350	130	A	1,9
		353.5	136.7	B	36
2856a	*i*-PrCO$_2$Me	347	127	A	1,9
		348.3	134.7	B	36
286b	*t*-BuCO$_2$Me	347	124	A	1,9
		347.6	131.1	B	36
287	HCO$_2$Et	364,360.5	173,109	A	1,9,35
288	HCO$_2$iPr	364	200	A	1,9
289	HCO$_2$-*t*-Bu	380	212	A	1,9
290	MeCO$_2$Et	359.5,363	165,169	A	1,9,35
		363	169	C	37
291	MeCO$_2$-*i*-Pr	363	196	A	1,9
		363	196	C	37
292	MeCO$_2$-*t*-Bu	375	207	A	1,9
		368.7	204.4	C	37
293	EtCO$_2$Et	351.5	161	A	1,9
294	*i*-PrCO$_2$Et	346	158.5	A	1,9
295	*t*-BuCO$_2$Et	348	152	A	1,9
296	*n*-PrCO$_2$Me	353	132	A	1,9
		355.5	138.3	B	36
297	*i*-BuCO$_2$Me	355	132.5	A	1,9
298	neo-PenCO$_2$Me	361	137.5	A	1,9
299	CH$_2$ClCO$_2$Me	357.1	135.8	B	36
300	CHCl$_2$CO$_2$Me	351.0	133.8	B	36
301	CCl$_3$CO$_2$Me	345.5	128.6	B	36
302	CCl$_2$FCO$_2$Me	346.5	130.5	B	36
303	CF$_3$CO$_2$Me	352.8	133.0	B	36
304	CH$_2$BrCO$_2$Me	361.0	138.1	B	36
305	CH$_2$OCH$_3$CO$_2$Me	352.5	131.7	B	36
306	CH(OCH$_3$)$_2$CO$_2$Me	354.1	134.8	B	36
307	CH$_2$CNCO$_2$Me	362.0	140.7	B	36
308	CH$_2$(CO$_2$CH$_3$)CO$_2$Me	363.1	142.3	B	36
309	CH$_2$[CH(OCH$_3$)$_2$]CO$_2$Me	359.1	141.3	B	36
310	CH$_2$[CH$_2$N(CH$_3$)$_2$]CO$_2$Me	356.3	139.1	B	36
311	MeCO$_2$CH$_2$CH=CH$_2$	360	165.4	C	37
312	MeCO$_2$CH$_2$Ph	361.2	171.1	C	37
313	MeCO$_2$CH=CH$_2$	363.4	202.6	C	37
314	MeCO$_2$C$_6$H$_4$-X; X = H	370.0	201.3	B	38
		366.6	208.9	C	37
315	314; X = 4-NO$_2$	373	203	B	38
316	314; X = 4-MeO	369.2	197.3	B	38
317	314; X = 2-Me	371	199.6	B	38
318	MeCO$_2$-2,6-Me$_2$C$_6$H$_3$	371	196.6	B	38
319	MeCO$_2$-1-Naphthyl	371.6	195.3	B	38
320	MeCO$_2$-2-Naphthyl	371.3	201.3	B	38
321	X-C$_6$H$_4$-CO$_2$Me; X = H	341.3	128	E	32
		356	147	D	34
322	321; X = p-NO$_2$	350	132	E	32
		336.1	134.3	D	34
323	321; X = p-CN	348.3	131.3	E	32
324	321; X = p-CF$_3$	346.8	130.2	E	32

TABLE 9 (continued)
^{17}O Chemical Shift Data (± 1 ppm) for Esters

Compound no.	Structure/name	δ(C=O) ppm	δ(-O-) ppm	Conditions[a]	Ref.
325	321; X = p-Cl	342.3	128.6	E	32
		331.5	129.2	D	34
326	321; X = p-F	340.6	128.7	E	32
327	321; X = p-CH$_3$	338.9	126.9	E	32
328	321; X = p-MeO	335.0	125.7	E	32
		328.4	135.1	D	34
329	321; X = o-MeO	354.8	139.3	D	34
330	321; X = m-MeO	329.1	132.7	D	34
331	321; X = o-Cl	358.9	143.6	D	34
332	321; X = m-Cl	333.9	127.2	D	34
333	321; X = o-NO$_2$	354.3	138.9	D	34
334	321; X = m-NO$_2$	329.9	129.8	D	34
335	p-X-C$_6$H$_6$-CH=CH-CO$_2$Me; X = H	339.9	134.2	E	32
336	335; X = MeO	336.6	134.8	E	32
337	335; X = Me	339.0	134.0	E	32
338	335; X = F	341.1	134.7	E	32
339	335; X = Cl	341.3	135.4	E	32
340	335; X = CF$_3$	343.5	135.8	E	32
341	335; X = CN	345.3	137.0	E	32
342	335; X = NO$_2$	345.6	138.1	E	32
343	EtCO$_2$t-Bu	356.8	200.0	C	37
344	i-PrCO$_2$t-Bu	352.5	198.9	C	37
345	PhCH$_2$CO$_2$t-Bu	364.2	203.6	C	37
346a	Ph-CO$_2$t-Bu	345.6	198.0	C	37
346b	Ph-CO$_2$Ph	334	193	C	37
347	CH$_3$CH=CH-CO$_2$tBu	335.5	208.7	C	37
348	BrCH$_2$-CO$_2$t-Bu	365.5	202.4	C	37
349	CH$_3$CHBrCO$_2$t-Bu	325.2	201.1	C	37
350	Me$_2$CBrCO$_2$t-Bu	348.2	194.6	C	37
351	PhCHBrCO$_2$t-Bu	359.3	202.5	C	37
352	Me-CH=CH-CO$_2$Me	327.2	129.5	C	37
353	Et-CH=CH-CO$_2$Me	332	130	C	37
354	n-PrCH=CHCO$_2$Me	332.8	128.7	C	37
355	Me$_2$C=CHCO$_2$Me	341.5	139.2	C	37
356	E-MeCH=CMeCO$_2$Me	335	125.2	C	37
357	BrCH$_2$CH=CH-CO$_2$Me	346.9	134.1	C	37
358	CH$_3$CHBrCH=CH-CO$_2$Me	337.8	131.3	C	37
359	E-MeCH$_2$BrC=CH-CO$_2$Me	349.0	142.7	C	37
360	Z-MeCH$_2$BrC=CH-CO$_2$Me	342.5	141.6	C	37
361	2-MeC$_6$H$_4$CO$_2$Me	359	138.5	B	33
362	2,3-Me$_2$C$_6$H$_3$CO$_2$Me	363	141	B	33
363	2,6-Me$_2$C$_6$H$_3$CO$_2$Me	377	150	B	33
364	2,4,6-Me$_3$C$_6$H$_2$CO$_2$Me	376	149	B	33
365	1-naphthyl CO$_2$Me	361	139	B	33
366	2-naphthyl CO$_2$Me	341	129	B	33
367	9-anthryl CO$_2$Me	385	154	B	33
368	2,4,6-t-BuC$_6$H$_2$CO$_2$Me	392	162	B	33
369	p-X-C$_6$H$_4$-CO$_2$-CH$_2$CF$_3$; X = H	340.3	126.6	C	39
370	369; X = MeO	329.7	125.3	C	39
371	369; X = Me	337.6	125.9	C	39
372	369; X = F	339.3	124.9	C	39
373	369; X = Br	340.8	127.6	C	39
374	369; X = NO$_2$	351.0	131.8	C	39
375	p-X-C$_6$H$_4$-CO$_2$-CH$_2$SiF$_3$; X = H	279.2	151.6	C	39
376a	(CH$_3$CO$_2$)CHR; R = H	369	185	A	40

TABLE 9 (continued)
^{17}O Chemical Shift Data (± 1 ppm) for Esters

Compound no.	Structure/name	δ(C=O) ppm	δ(–O–) ppm	Conditions[a]	Ref.
376b	376a; R = Me	372	197	A	40
376c	376a; R = Et	377	203	A	40
376d	376a; R = n-Hex	369	201	A	40
376e	376a; R = Cl₃C	379	188	A	40
376f	376a; R = Ph	373	195	A	40
376g	376a; R = PhCH=CH	381	219	A	40
376h	376a; R = 2-Furyl	377	201	A	40

[a]   A = Neat liquids at RT; B = Acetonitrile solution at 75°; C = Chloroform solution; D = DMSO solution; E = Acetone solution at 40°.

^{17}O NMR data for simple anhydrides in dry acetonitrile at 75°C have been found to be highly sensitive to ring size and conformation [acetic anhydride, δ 411 ppm (C=O), δ 273; glutaric anhydride, δ 396 ppm (C=O), δ 288; succinic anhydride, δ 376 ppm (C=O), δ 301; maleic anhydride, δ 401 ppm (C=O), δ 252; phthalic anhydride, δ 374 ppm (C=O), δ 263].[54] Substituent effects on a series of maleic anhydrides are consistent with electronic effects. In contrast, the ^{17}O NMR data for polycyclic succinic anhydrides show that the chemical shifts of the carbonyl oxygens are roughly constant, while those of the central oxygens are sensitive to structure. The ^{17}O NMR data for homophthalic anhydride 549 [δ 397 (C=O), δ 372 (C=O)] are consistent with those for **522** and **523**; and those for 3,4-dihydronaphthalene-1,2-dicarboxylic anhydride [δ 392 (C=O), δ 387 (C=O), δ 252] have been explained by compressional and electronic effects.[54]

^{17}O NMR data for sterically hindered 3-substituted phthalic anhydrides shown unusual deshielding effects which have been correlated with in-plane bond angle distortions, indicative of van der Waals repulsions.[46,47] Steric interactions of substituents *ortho* to the carbonyl groups result in deshielding effects (9 to 22 ppm) relative to parent compounds regardless of the electronic character of the substituents. The relationships between ^{17}O chemical shifts and regiochemistry of the phthalic anhydrides have been discussed.[46,47]

G. ACID CHLORIDES, ISOCYANATES, HYDROXAMIC ACIDS, AND MISCELLANEOUS COMPOUNDS

Data for acid chlorides, isocyanates, and hydroxamic acids are summarized in Table 14.[1,9,55-58] ^{17}O NMR studies of acid halides have been limited in number and additional study is needed.[1,9] From a recent study of aryl acid halides, it is apparent that the carbonyl chemical shift parallels the electron donating ability of halogen, e.g., F>Cl>Br. A wide range in chemical shift, 350 to 521 ppm, is noted for the compounds **567a** to **567c**.[55]

The ^{17}O NMR signals for the isocyanates **569** to **584** show interesting variation with structure (Table 14). The signals for the simple alkyl isocyanates appear between 83 and 93 ppm, upfield by about 15 ppm from the general range noted for aryl isocyanates (100 to 120 ppm).[56] The chemical shift noted for methyl isocyanate (84 ppm) is upfield of that of phenyl isocyanate (111 ppm) by 27 ppm. Thus, the chemical shift of methyl isocyanate has the most upfield value for any carbon-oxygen double-bond system yet reported.[56] The substantial difference in chemical shifts between methyl and phenyl isocyanate may arise from a combination of the shielding effect of the methyl group (σ inductive) and the deshielding effect (σ and π inductive and δ effect) of the phenyl group. Increasing the length for the straight-chain alkyl groups shows a modest deshielding effect on the ^{17}O NMR signal relative to **569**. The deshielding effect is presumably a result of δ effects as described

TABLE 10
^{17}O Chemical Shift Data (± 1 ppm) for Amides

Compound no.	Structure/name	δ(C=O) ppm	Conditions[a]	Ref.
377	HCONH$_2$	310,303	A	1,9,41
		283	B	1
378	CH$_3$CONH$_2$	313.5	A	1,9
		286	B	1
		300	D	41
379	EtCONH$_2$	308	A	1,9
380	i-PrCONH$_2$	306	A	1,9
381	t-BuCONH$_2$	314.5	A	1,9
382	PhCONH$_2$	329	C	33
383	4-MeC$_6$H$_4$CONH$_2$	327	C	33
384	2-MeC$_6$H$_4$CONH$_2$	350	C	33
385	2,6-Me$_2$C$_6$H$_3$CONH$_2$	353	C	33
386	1-Naphthyl CONH$_2$	359	C	33
387	2-Naphthyl CONH$_2$	331	C	33
388	9-Anthryl CONH$_2$	365	C	33
389	HCONHMe	304	A	41
390	HCONMe$_2$	323	A	41
391	CH$_3$CONMe$_2$	341	A	41
		348	E	42
392	PhCONMe$_2$	348	C	33
393	1-Naphthyl CONMe$_2$	352	C	33
394	9-Anthryl CONMe$_2$	357	C	33
395	CH$_3$CONEt$_2$	342	A	1,9
		348	E	42
396	CH$_3$CONHMe	309	A	1,9
397	HCONHt-Bu (trans)	323	A	7b
		286	B	7b
398	HCONHt-Bu (cis)	267	B	7b
399	CH$_3$CONHAr; Ar = Ph	355.3	C	38
400	399; Ar = 4-NO$_2$C$_6$H$_4$	372.0	C	38
401	399; Ar = 4-NO$_2$-2-MeC$_6$H$_3$	367.0	C	38
402	399; Ar = 4-MeOC$_6$H$_4$	348.6	C	38
403	399; Ar = 2-MeOC$_6$H$_4$	359.1	C	38
404	399; Ar = 4-EtOC$_6$H$_4$	349.1	C	38
405	399; Ar = 4-MeO-2MeC$_6$H$_3$	343.5	C	38
406	399; Ar = 4-MeC$_6$H$_4$	352.0	C	38
407	399; Ar = 2-MeC$_6$H$_4$	349.0	C	38
408	399; Ar = 2,6-Me$_2$C$_6$H$_3$	343.0	C	38
409	399; Ar = 4-HOC$_6$H$_4$	342.0	C	38
410	399; Ar = 2-HOC$_6$H$_4$	327.8	C	38
411	399; Ar = 4-ClC$_6$H$_4$	359.0	C	38
412	399; Ar = 4-BrC$_6$H$_4$	358.0	C	38
413	399; Ar = 1-Naphthyl	350.5	C	38
414	399; Ar = 2-Naphthyl	358.1	C	38
415	CH$_3$CONMePh	350	C	38
416	CH$_3$CONMe(4-MeOC$_6$H$_4$)	347.2	C	38
417	CH$_3$CONMe(4-NO$_2$C$_6$H$_4$)	363.0	C	38
418a	N-Acetylpiperidine	347.1	E	42
418b	N-Acetylpyrrolidine	349.5	E	42
419	N-Acetyl-L-proline	298.4	B	43
		314	F	43

[a] A = Neat liquids; B = Aqueous solution; C = Acetonitrile solution at 75°; D = MeOH
solution; E = Acetone solution at 25°C; F = Chloroform solution

TABLE 11
^{17}O Chemical Shift Data (± 1 ppm) for Lactones

Compound no.	Structure/name	δ(C=O) ppm	δ(–O–) ppm	Conditions[a]	Ref.
420	β-Propiolactone	340	235	A	1
		348.5	241	B	44
421	γ-Butyrolactone	333,335	174,176	A	1
		340.5	178.5	B	44
422	δ-Valerolactone	360	163	A	1
		367	167	B	44
423	ε-Caprolactone	377	175.4	B	44
424	2(5H)-Furanone	323	174	A	1
		326.2	172.7	B	44
425	α-Angelicalactone	349	237	A	1
		351.3	239	B	44
426	β-Butyrolactone	348.8	268.5	B	44
427	γ-Valerolactone	340.3	206.6	B	44
428	α-Bromo-γ-butyrolactone	340.3	170.8	B	44
429	α-Bromo-γ-valerolactone	339.1	199.1	B	44
430	α-Methyl-γ-butyrolactone	344.4	174.8	B	44
431	γ-n-Pentyl-γ-butyrolactone	340.1	202.8	B	44
432	γ-n-Heptyl-γ-butyrolactone	340.0	202.5	B	44
433	δ-n-Hexyl-γ-valerolactone	365.3	189.5	B	44
434	γ-Butyrothiolactone	494.3	—	B	44
435	2(5H)Thiophenone	462.7	—	B	44
436	3-Methoxy-2(5H)thiophenone	445.5	—	B	44
437	2-Coumaranone	357.3	224.5	B	44
438	Dihydrocoumarin	375.8	204.0	B	44
439	Coumarin	347.7	217.6	B	44
		351	219	C	45
440	2H-Pyran-2-one	333.3	245.6	B	44
441	5,6-Dihydro-2H-pyran-2-one	361.5	142.5	B	44
442	3,4-Dihydro-4,4-dimethyl-2H-pyran-2-one	303.1	198.1	B	44
443	3,6-Dihydro-4,6,6-trimethyl-2H-pyran-2-one	361.1	206.5	B	44
444	4-Methoxy-6-methyl-2H-pyran-2-one	315.5	247.5	B	44
445	Methyl coumate	338.9,343.8	248.8,127.1	B	44
446	Phthalide	320.0	170.0	B	46,47
447	7-R-Phthalide; R = MeO	333	168	B	47
448	7-R-Phthalide; R = Me	332	170	B	46,47
449	7-R-Phthalide; R = t-Bu	346	168	B	46,47
450	7-R-Phthalide; R = F	334	171	B	47
451	7-R-Phthalide; R = Cl	335	170	B	47
452	7-R-Phthalide; R = Br	334	170	B	47
453	7-R-Phthalide; R = I	330	171	B	47
454	7-R-Phthalide; R = NO$_2$	337	172	B	47
455	4-R-Phthalide; R = MeO	332	170	B	47
456	4-R-Phthalide; R = t-Bu	319	173	B	46,47
457	4-R-Phthalide; R = F	325	169	B	47
458	4-R-Phthalide; R = Cl	327	170	B	47
459	4-R-Phthalide; R = Br	327	171	B	47
460	4-R-Phthalide; R = I	327	171	B	47
461	4-R-Phthalide; R = NO$_2$	325	174	B	47
462	6-Methylcoumarin	351	220	C	45
463	7-Methylcoumarin	352	221	C	45
464	7-Methoxycoumarin	347	225	C	45
465	Furocoumarin (psoralen)	352	219	C	45
466	5-Methoxyfurocoumarin (bergapten)	351	220	C	45
467	8-Methoxyfurocoumarin (xanthotoxin)	350	200 ± 5	C	45

TABLE 11 (continued)
^{17}O Chemical Shift Data (± ppm) for Lactones

Compound no.	Structure/name	δ(C=O) ppm	δ(–O–) ppm	Conditions[a]	Ref.
468	5,8-Dimethoxyfurocoumarin (isopimpinellin)	351	200 ± 7	C	45

[a] A = Neat liquids; B = Acetonitrile solution at 75°; C = 1,2-Dibromoethane solution at 70°C.

TABLE 12
^{17}O NMR Chemical Shift Data (± 1 ppm) for Lactams, Imides, and Related Compounds

Compound no.	Structure/name	δ(C=O) ppm	Conditions[a]	Ref.
469	2-Azetidinone	314.3	A	48
470	2-Pyrrolidinone	301.1	A	48
471	δ-Valerolactam	320.6	A	48
472	ε-Caprolactam	336.1	A	48
473	2-Azacyclooctanone	328.0	A	48
474	2-Azacyclononanone	335.5,327.8	A	48
475	2-Azacyclotridecanone	323.5	A	48
476	5-Methyl-2-pyrrolidinone	302.9	A	48
477	1-Methyl-2-pyrrolidinone	297.6	A	48
		294	B	49
		289	C	49
478	1-Ethyl-2-pyrrolidinone	300	B	49
		289	C	49
479	1-Vinyl-2-pyrrolidinone	321.3	A	48
480	1-Phenyl-2-pyrrolidinone	328.8	A	48
481	N-Methyloxindole	323.9	A	48
482	Phthalamide	282	A	48
483	N-Methylphthalamide	281	A	48
484	N-t-Butylphthalamide	300	A	48
485	N-Methyl-δ-valerolactam	326.1	A	48
486	2-(1H)Pyridone	269.5	A	48
487	6-Methyl-2(1H)pyridone	266.1	A	48
488	1-Methyl-2-pyridone	294.0	A	48
489	1-Methyl-2-quinolinone	317.8	A	48
490	3,4-Dihydro-1-methyl-2-quinolinone	354.0	A	48
491	Pyrrolizin-3-one	386	C	50
492	Succinimide	373.5	A	51
493	N-Methylsuccinimide	371	A	51
494	N-t-Butylsuccinimide	392	A	51
495	N-Phenylsuccinimide	376	A	51
496	Maleimide	411	A	51
497	N-Methylmaleimide	407	A	51
498	N-t-Butylmaleimide	426	A	51
499	N-Phenylmaleimide	412	A	51
500	Phthalimide	379	A	51
501	N-Methylphthalimide	374	A	51
502	N-i-Propylphthalimide	383	A	51
503	N-t-Butylphthalimide	394	A	51
504	3-t-Butylphthalimide	407.3,370.6	A	51
505	N-t-Butyl-3-t-butylphthalimide	423.3,385.3	A	51
506	N-Phenylphthalimide	378.3	A	51
507	N-(2-Methylphenyl)phthalimide	381.3	A	51
508	N-(2,6-Dimethylphenyl)phthalimide	384.0	A	51
509	N-(4-Methoxyphenyl)phthalimide	377.5	A	51

TABLE 12 (continued)
^{17}O NMR Chemical Shift Data ($\pm$ ppm) for Lactams, Imides, and Related Compounds

Compound no.	Structure/name	δ(C=O) ppm	Conditions[a]	Ref.
510	*N*-(2-Methyl-4-methoxyphenyl)phthalimide	380.0	A	51
511	1-Methyl-2-thiohydantoin	348.0	A	52
512	1-Phenyl-2-thiohydantoin	352.0	A	52
513	Uracil	232,301	D	53
514	5-Methoxyuracil	220,296	D	53
515	5-Methyluracil	224,301	D	53
516	5-Fluorouracil	230,294	D	53
517	5-Chlorouracil	234,306	D	53
518	5-Bromouracil	235,311	D	53
519	5-Trifluoromethyluracil	247,312	D	53
520	5-Nitrouracil	261,317	D	53

[a]   A = Acetonitrile solution, 75°C; B = Neat liquids; C = Chloroform solution; D = Water solution at 95°C.

previously. Similar results have been noted for the ^{15}N chemical shifts of alkyl isocyanates.[59] The natural abundance ^{17}O NMR spectra of aryl isocyanates show electronic effects (Hammett plot) consistent with previous studies. However, the steric effects of *ortho*-alkyl groups show shielding effects analogous to *N*-arylacetamides.[57]

The ^{17}O NMR chemical shifts of the carbonyl oxygens have been reported for benzo-hydroxamic acid (**602**) and its *N*-methyl (**603**), *O*-methyl (**604**), and *N,O*-dimethyl (**605**) derivatives.[58] The chemical shifts were determined in several solvents as well as at different pHs (only data in dioxane solvent are shown in the table).[58] The observed chemical shift values (330 to 350 ppm downfield in dioxane from external H$_2$O) are in the range characteristic of benzamides, supporting the amide structure. Upfield shifts of about 31 to 57 ppm are observed for the *N*-methyl and NH compounds, respectively, in methanol solutions, indicative of strong hydrogen bonding with the solvent molecules in the NH compounds. ^{17}O NMR chemical shift titration curves for these compounds, including pK$_a$ values, may be found in the original work.[58]

TABLE 13
^{17}O Chemical Shift Data (± 1 ppm) for Anhydrides

Compound no.	Structure/name	δ(C=O) ppm	δ(–O–) ppm	Conditions[a]	Ref.
521	Acetic anhydride	411	273	A	54
		393	259	B	55
522	Glutaric anhydride	396	288	A	54
523	2-Phenylglutaric anhydride	397	288	A	54
524	3-Methylglutaric anhydride	398	287	A	54
525	3,3-Dimethylglutaric anhydride	401	284	A	54
526	3,3-Tetramethyleneglutaric anhydride	401	286	A	54
527	Diglycolic anhydride	391	283, −20	A	54
528	Succinic anhydride	376	301	A	54
529	Methylsuccinic anhydride	376,372	300	A	54
530	2,2-Dimethylsuccinic anhydride	379,367	294	A	54
531	Propionic anhydride	390	248	A	54
532	cis-1,2-Cyclohexanedicarboxylic anhydride	374	288	A	54
533	trans-1,2-Cyclohexanedicarboxylic anhydride	374	293	A	54
534	cis-1,2,3,6-Tetrahydrophthalic anhydride	371	301	A	54
535	cis-1,2-Cyclobutanedicarboxylic anhydride	373	300	A	54
536	cis-5-Norbornene-endo-2,3-dicarboxylic anhydride	374	312	A	54
537	endo-Bicyclo[2.2.2]oct-5-ene-2,3-dicarboxylic anhydride	370	306	A	54
538	Itaconic anhydride	376,350	295	A	54
539	Maleic anhydride	401	252	A	54
540	2,3-Dimethylmaleic anhydride	381	246	A	54
541	2,3-Tetramethylenemaleic anhydride	382	249	A	54
542	2-Phenylmaleic anhydride	398,395	254	A	54
543	2-Bromomaleic anhydride	400,388	253	A	54
544	2,3-Dichloromaleic anhydride	388	249	A	54
545	3,4-Dihydronaphthalene-1,2-dicarboxylic anhydride	392,387	252	A	54
546	1,8-Naphthalenedicarboxylic anhydride	370	268	A	54
547	1,2,4,5-Benzenetetracarboxylic anhydride	384	269	A	54
548	2,3-Pyridinedicarboxylic anhydride	383,380	263	A	54
549	Homophthalic anhydride	397,372	273	A	54
550	Isatoic anhydride	354,245	262	A	54
551	Benzoic anhydride	389	242	A	54
		384	237	C	55
552	Phthalic anhydride	374	263	A	47
553	3-R-Phthalic anhydride; R = EtO	384,370	264	A	47
554	3-R-Phthalic anhydride; R = MeO	383,371	264	A	47
555	3-R-Phthalic anhydride; R = Me	383,372	264	A	46,47
556	3-R-Phthalic anhydride; R = t-Bu	396,367	262	A	46,47
557	3-R-Phthalic anhydride; R = F	385,376	264	A	47
558	3-R-Phthalic anhydride; R = Cl	386,375	264	A	47
559	3-R-Phthalic anhydride; R = Br	385,374	264	A	47
560	3-R-Phthalic anhydride; R = I	382,373	263	A	47
561	3-R-Phthalic anhydride; R = NO$_2$	395,377	262	A	47
562	4-NO$_2$-Phthalic anhydride	384,381	268	A	47

[a] A = Acetonitrile solution at 75°; B = Neat liquid; C = Chloroform solution.

TABLE 14
¹⁷O Chemical Shift Data (±1 ppm) for Acid Chlorides, Isocyanates, Hydroxamic Acids, and Related Compounds

Compound no.	Structure/name	δ(C=O) ppm	δ(R)	Conditions[a]	Ref.
563a	MeCOCl	502	—	A	1,9
563b	MeCOBr	536	—	A	1
564a	EtCOCl	495	—	A	1,9
564b	EtCOBr	526	—	A	1
565a	n-PrCOCl	498	—	A	9
565b	i-PrCOCl	494	—	A	1,9
566a	i-BuCOCl	501	—	A	9
566b	t-BuCOCl	497	—	A	1,9
566c	neoPenCOCl	504.5	—	A	9
567a	PhCOF	350	—	A	55
567b	PhCOCl	452	—	A	55
567c	PhCOBr	521	—	A	55
568	Benzoyl cyanide	559	—	C	55
569	MeNCO	83.6	—	B	56
570	EtNCO	86.8	—	B	56
571	ClCH₂NCO	116.3	—	B	56
572	n-PrNCO	84.8	—	B	56
573	i-PrNCO	88.8	—	B	56
574	n-BuNCO	85.0	—	B	56
575	t-BuNCO	92.9	—	B	56
576	Cyclohexyl NCO	88.1	—	B	56
577	1,6-Dicyanatohexane	86.0	—	B	56
578	PhCH₂NCO	91.7	—	B	56
579	Ph₃CNCO	103.5	—	B	56
580	EtO₂CCH₂NCO	101.7	344.6,158.1	B	56
581	ClCH₂CONCO	146.8	467.8	B	56
582	CCl₃CONCO	158.7	452.1	B	56
583	EtO₂CNCO	143.6	309.3,153.6	B	56
584	PhSO₂NCO	158.6	180.8	B	56
585	ArNCO; Ar=Ph	111.3	—	B	57
586	ArNCO; Ar=4-NO₂C₆H₄	122.8	572.2	B	57
587	ArNCO; Ar=4-EtO₂CC₆H₄	118.4	338.9,159.2	B	57
588	ArNCO; Ar=4-CF₃C₆H₄	118.7	—	B	57
589	ArNCO; Ar=4-ClC₆H₄	114.7	—	B	57
590	ArNCO; Ar=4-CH₃C₆H₄	109.5	—	B	57
591	ArNCO; Ar=2-CH₃C₆H₄	106.2	—	B	57
592	AfNCO; Ar=2,6-DiCH₃C₆H₃	101.0	—	B	57
593	ArNCO; Ar=4-C₂H₅C₆H₄	110.7	—	B	57
594	ArNCO; Ar=2-C₂H₅C₆H₄	106.3	—	B	57
595	ArNCO; Ar=4-i-PrC₆H₄	110.2	—	B	57
596	ArNCO; Ar=2-i-PrC₆H₄	101.0	—	B	57
597	ArNCO; Ar=4-Br-2,6-Me₂C₆H₂	105.5	—	B	57
598	ArNCO; Ar=2,6-Di-i-PrC₆H₃	99.4	—	B	57
599	ArNCO; Ar=2-iPr-6-CH₃C₆H₃	101.2	—	B	57
600	ArNCO; Ar=4-CH₃OC₆H₄	108.5	48.7	B	57
601	ArNCO; Ar=1-naphthyl	110.8	—	B	57
602	Ph-CO-NHOH	333	—	D	58
		302	79	B	13
603	Ph-CO-NMeOH	330	—	D	58
604	Ph-CO-NH-OMe	341	—	D	58
605	Ph-CO-NMe-OMe	342	—	D	58

[a]   A=Neat liquids; B=Acetonitrile solution at 75° (35); C=Chloroform solution; D=Dioxane solution.

ACKNOWLEDGMENTS

Acknowledgment is made to the Donors of the Petroleum Research Fund, administered by the American Chemical Society, for partial support of this research, to NSF (CHE-8506665), and to the Georgia State University Research Fund.

REFERENCES

1. (a)**Kintzinger, J.-P.,** Oxygen NMR characteristic parameters, in *NMR-17. Oxygen-17 and Silicon-29,* Diehl, P., Fluck, E., and Kosfeld, R., Eds., Springer-Verlag, New York, 1981, 1; (b) **Kintzinger, J.-P.,** Oxygen-17 NMR, in *NMR of Newly Accessible Nuclei,* Vol. 2, Laszlo, P., Ed., Academic Press, New York, 1983, 79.

2. **Klemperer, W. G.,** Application of ^{17}O NMR spectroscopy to structural problems, in *The Multinuclear Approach to NMR Spectroscopy,* Proc. NATO Adv. Study Inst. on The Multinuclear Approach to NMR Spectroscopy, Lambert, J. D. and Riddell, F. G., Eds., Reidel, Dordrecht, Holland, 1983, 245.

3. **Butler, L. G.,** The NMR parameters for oxygen-17, in *^{17}O NMR Spectroscopy in Organic Chemistry,* Boykin, D. W., Ed., CRC Press, Boca Raton, Florida, 1990.

4. **Karplus, M. and Pople, J. A.,** Theory of carbon NMR chemical shifts in conjugated molecules, *J. Chem. Phys.,* 38, 2803, 1963.

5. **Baumstark, A. L. and Boykin, D. W.,** 17-O NMR spectroscopy: assessment of steric perturbation of structure in organic compounds, *Tetrahedron,* 45, 3613, 1989.

6. **Christ, H. A., Diehl, P., Schneider, H. R., and Dahn, H.,** Chemische Verschiebungen in der Kernmagnetischen Resonanz von ^{17}O in Organische Verbindungen, *Helv. Chim. Acta,* 44, 865, 1961.

7. (a) **Steinschneider, A. and Fiat, A.,** Carbonyl-^{17}O NMR of amino acid and peptide carboxamide and methyl ester derivatives, *Int. J Pept. Protein Res.,* 23, 591, 1984. (b) **Valentine, B., Steinschneider, A., Dhawan, D., Burgar, M. I., St. Amour, T., and Fiat, D.,** Oxygen-17 NMR of peptides, *Int. J. Pept. Protein Res.,*25, 56, 1985; (c) **Burgar, M. I., Dhawan, D., and Fiat, D.,** ^{17}O and ^{14}N spectroscopy of ^{17}O-labeled nucleic acid bases, *Org. Magn. Reson.,* 20, 184, 1982; (d) **Sakarellos, C., Gerothanassis, I. P., Birlirakis, N., Karayannis, T., Sakarellos-Daitsiotis, M., and Marraud, M.,** ^{17}O-NMR studies of the conformational and dynamic properties of enkephalins in aqueous and organic solutions using selectively labeled analogues, *Biopolymers,* 28, 15, 1989, and references cited therein.

8. **Delseth, C., and Kintzinger, J.-P.,** Résonance magnétique nucléaire de ^{17}O. Aldéhydes et cétones aliphatiques: additivite des effects de substitution at corrélation avec la ^{13}C-RMN, *Helv. Chim. Acta,* 59, 466, 1976.

9. **Delseth, C., Nguyên, T. T.-T., and Kintzinger, J.-P.,** Oxygen-17 and carbon-13 NMR. Chemical shifts of unsaturated carbonyl compounds and acyl derivatives, *Helv. Chim. Acta,* 63, 498, 1980.

10. **St. Amour, T. E., Burger, M. I., Valentine, B., and Fiat, D.,** 17-O NMR studies of substituent and hydrogen-bonding effects in substituted acetophenones and benzaldehydes, *J. Am. Chem. Soc.,* 103, 1128, 1981.

11. **Boykin, D. W., Balakrishnan, P., and Baumstark, A. L.,** 17-O NMR studies in aryl-alkyl ketones and aromatic aldehydes, *Magn. Reson. Chem.,* 25, 248, 1987.

12. **Boykin, D. W. and Rhodes, R. A.,** ^{17}O NMR investigation of acetyl and formyl thiophenes, furans and pyrroles, *J. Heterocycl. Chem.,* 25, 643, 1988.

13. **Boykin, D. W.,** unpublished results.

14. **Crandall, J. K., Centeno, M. A., and Borresen, S.,** Oxygen-17 NMR. 2. Cyclohexanones, *J. Org. Chem.,* 44, 1184, 1979.

15. **Gorodetsky, M., Luz, Z., and Mazur, Y.,** Oxygen-17 nuclear magnetic resonance studies of equilibria between enol forms of β-diketones, *J. Am. Chem. Soc.,* 89, 1183, 1967.

16. **Spanka, G., Rademacher, P., and Duddeck, H.,** Transannular interactions in difunctional medium rings. 3. ^{13}C and ^{17}O NMR studies on cyclic amino ketones, *J. Chem. Soc. Perkin Trans. 2,* p. 2119, 1988.

17. **Boykin, D. W., Baumstark, A. L., Mehdizadeh, A., and Venkatramanan, M. K.,** ^{17}O NMR spectroscopic study of substituted benzoquinones and α,β-unsaturated cyclic ketones, *Magn. Reson. Chem.,* 23, 305, 1990.

18. **Chandrasekaran, S., Wilson, W. D., and Boykin, D. W.,** ^{17}O NMR studies on polycyclic quinones and related cyclic ketones: models for anthracycline intercalators, *Org. Magn. Reson.,* 22, 757, 1984.

19. **Boykin, D. W., Baumstark, A. L., Perjéssy, A., and Hrnciar, P.,** ^{17}O NMR studies on 4- and 4'-substituted chalcone and p-substituted b-nitrostyrenes, *Spectrochim. Acta,* 40A, 887, 1984.

20. **Sardella, D. J., and Stothers, J. B.**, Nuclear magnetic resonance studies. XVI. Oxygen-17 shieldings of some substituted acetophenones, *Can. J. Chem.*, 47, 3089, 1969.
21. **Brownlee, R. T. C., Sadek, M., and Craik, D. J.**, Ab initio MO calculations and 17-O NMR at natural abundance of *p*-substituted acetophenones, *Org. Magn. Reson.*, 21, 616, 1983.
22. **Oakley, M. G. and Boykin, D. W.**, Relationship of torsion angle to 17-O NMR data for aryl ketones, *J. Chem. Soc. Chem. Commun.*, p. 439, 1986.
23. **Baumstark, A. L., Graham, S. S., and Boykin, D. W.**, ¹⁷O NMR spectroscopy: intramolecular hydrogen bonding in 2-amino, 2-acetamido and 2-trifluoroacetamidoacetophenone, *J. Chem. Soc. Chem. Commun.*, p. 767, 1989.
24. **Boykin, D. W., Eisenbraun, E. J., Delphon, J. K., and Hertzler, R. L.**, ¹⁷O NMR studies on alkyl indanones: steric effects, *J. Org. Chem.*, 54, 1418, 1989.
25. **Jaccard, G. and Lauterwein, J.**, Intramolecular hydrogen bonds of the C=O . . . H–O type as studied by ¹⁷O-NMR, *Helv. Chim. Acta*, 69, 149, 1986.
26. **Baumstark, A. L., Dotrong, M., Stark, R. R. and Boykin, D. W.**, ¹⁷O spectroscopy: origin of deshielding effect in rigid, planar molecules, *Tetrahedron Lett.*, p. 2143, 1988.
27. **Duddeck, H., Jaszberenyi, J. C., Levai, A., Timar, T., and Elgamal, M. H. A.**, Natural abundance ¹⁷O NMR spectra of chromanones, chromones, flavanones and flavones, *Magn. Reson. Chem.*, 27, 170, 1989.
28. **Boykin, D. W. and Martin, G. E.**, ¹⁷O NMR studies of cyclic aromatic ethers, *J. Heterocyl. Chem.*, 24, 365, 1987.
29. **Dahn, H. and Ung-Truong, M.-N.**, ¹⁷O NMR spectra of cyclopropenones and tropone. Oxygen exchange with water, *Helv. Chim. Acta*, 70, 2130, 1987.
30. **Dahn, H., Schlumbe, H. P., and Temler, J.**, Chemische verschiebungen bei ¹⁷O-NMR und hydrationsgeschwindigkeiten von cyclischen ketonen mit transannularer wechselwirkung, *Helv. Chim. Acta*, 55, 907, 1972.
31. **Gerothanassis, I. P., Hunston, R. N., and Lauterwein, J.**, ¹⁷O NMR chemical shifts of the twenty protein amino acids in aqueous solution, *Magn. Reson. Chem.*, 23, 659, 1985.
32. **Balakrishnan, P., Baumstark, A. L., and Boykin, D. W.**, 17-O NMR spectroscopy: effect of substituents on chemical shifts for *p*-substituted benzoic acids, methyl benzoates, cinnamic acids, and methyl cinnamates, *Org. Magn. Reson.*, 22, 753, 1984.
33. **Baumstark, A. L., Balakrishnan, P., Dotrong, M., McCloskey, C. J., Oakley, M. G., and Boykin, D. W.**, 17-O NMR spectroscopy: torsion angle relationships in aryl carboxylic esters, acids and amides, *J. Am. Chem. Soc.*, 109, 1059, 1987.
34. **Monti, D., Orsini, F., and Ricca, G. S.**, Oxygen-17 NMR spectrocopy. Effect of substituents on chemical shifts for substituted benzoic acids, phenylacetic and methyl benzoates, *Spectrosc. Lett.*, 19, 91, 1986.
35. **Sugawara, T., Kawada, Y., and Iwamura, H.**, Oxygen-17 NMR chemical shifts of alcohols, ethers and esters, *Chem. Lett.*, p. 1371, 1978.
36. **Boykin, D. W., Subramanian, T. S., and Baumstark, A. L.**, Natural abundance ¹⁷O NMR study of α-substituted methyl acetates, *Spectrochim. Acta*, 45A, 335, 1989.
37. **Orsini, F. and Ricca, G. S.**, Oxygen-17 NMR chemical shifts of esters, *Org. Magn. Reson*, 22, 653, 1984.
38. **Boykin, D. W., Deadwyler, G. H., and Baumstark, A. L.**, 17-O NMR studies on substituted *N*-arylacetamides and aryl acetates: torsion angle and electronic effects, *Magn. Reson. Chem.*, 26, 19, 1988.
39. **Liepinish, É. É., Zitsmane, I. A., Ignatovich, L. M., Lukevits, É., Guvanova, L. I., and Voronkov, M. G.**, ¹⁷O NMR chemical shifts of (trifluorosilyl) methyl *p*-substituted benzoates and their carbon analogs, *Zh. Obshch. Khim.*, 53, 1789, 1984.
40. **Lycka, A., Jirman, J., and Holecek, J.**, ¹⁷O and ¹³C NMR spectra of some geminal diacetates, *Collect. Czech. Chem. Commun.*, 53, 588, 1988.
41. **Burger, M. I., St. Amour, T. E., and Fiat, D.**, ¹⁷O and ¹⁴N NMR studies of amide systems, *J. Phys. Chem.*, 85, 502, 1981.
42. **Pinto, B. M., Grindley, T. B., and Szark, W. A.**, Effects of substitution on nitrogen on barriers to rotation of amides. III. Effect of variation of ring size of cyclic substituents, *Magn. Reson. Chem.*, 24, 323, 1986.
43. **Hunston, R. N., Gerothanassis, I. P., and Lauterwein, J.**, A study of L-proline, sarcosine, and the *cis trans* isomers of *N*-acetyl-L-proline and *N*-acetylsarcosine in aqueous and organic solution by ¹⁷O NMR, *J. Am. Chem. Soc.*, 107, 2654, 1985.
44. **Boykin, D. W., Sullins, D. W., and Eisenbraun, E. J.**, ¹⁷O NMR spectroscopy of lactones, *Heterocycles*, 29, 301, 1989.
45. **Duddeck, H., Rosenbaum, D., Elgamal, M. H. A., and Shalaby, N. M. M.**, Natural abundance ¹⁷O NMR spectra of coumarins, furocoumarins and related compounds, *Magn. Reson. Chem.*, 25, 489, 1987.
46. **Baumstark, A. L., Balakrishnan, P., and Boykin, D. W.**, ¹⁷O NMR spectroscopy as a probe of steric hindrance in phthalic anhydrides and phthalides, *Tetrahedron Lett.*, p. 3079, 1986.

47. **Boykin, D. W., Baumstark, A. L., Kayser, M. M., and Soucy, C. M.,** ^{17}O NMR spectroscopic study of substituted phthalic anhydrides and phthalides, *Can. J. Chem.*, 65, 1214, 1987.

48. **Boykin, D. W., Sullins, D. W., Pourahmady, N., and Eisenbraun, E. J.,** ^{17}O NMR spectroscopy of lactams, *Heterocycles*, 29, 307, 1989.

49. **Ruostesuo, P., Hakkinen, A. M., and Peltola, K.,** Carbon-13, nitrogen-15 and oxygen-17 NMR chemical shifts of 1-ethyl-2-pyrrolidinone and 1-methyl-2-pyrrolidinone in some solvents, *Spectrochim. Acta,* 41A, 739, 1985.

50. **McNab, H.,** An analysis of the NMR spectra of pyrrolizin-3-one-pyrrolo[1,2-c]imidazol-5-one, pyrrolo[1,2-a]imidazol-5-one and pyrrolo[1,2-b]pyrazol-6-one, *J. Chem. Soc. Perkin Trans. 1,* p. 657, 1987.

51. **Baumstark, A. L., Dotrong, M., Oakley, M. G., Stark, R., and Boykin, D. W.,** ^{17}O NMR study of steric interactions in hindered *N*-substituted imides, *J. Org. Chem.*, 52, 3640, 1987.

52. **Boykin, D. W.,** A natural abundance ^{17}O NMR investigation of substituted 1-methyl and 1-phenyl-2-thiohydantoins, *Heterocycles*, 26, 773, 1987.

53. **Chandrasekaran, S., Wilson, W. D., and Boykin, D. W.,** ^{17}O NMR spectroscopy of 5-substituted uracils, *J. Org. Chem.*, 50, 829, 1985.

54. **Vasquez, P. C., Boykin, D. W., and Baumstark, A. L.,** ^{17}O NMR spectroscopy (natural abundance) of heterocycles: anhydrides, *Magn. Reson. Chem.*, 24, 409, 1986.

55. **Cheng, C. P., Lin, S. C., and Shaw, G.-S.,** Correlation of ^{17}O NMR and ^{17}NQR data for some aromatic carbonyl compounds, *J. Magn. Reson.*, 69, 58, 1986.

56. **Boykin, D. W.,** ^{17}O NMR investigation of alkyl isocyanates: electronic and steric effects, *J. Chem. Res.*, p. 338, 1987.

57. **Boykin, D. W.,** ^{17}O NMR investigation of aryl isocyanates: electronic and steric effects, *Spectros. Lett.*, 20, 4154, 1987.

58. **Lipczynska-Kochany, E. and Iwamura, H.,** Oxygen-17 nuclear magnetic resonance studies of the structures of benzohydroxamic acids and benzohydroxamate ions in solution, *J. Org. Chem.*, 47, 5277, 1982.

59. **Yavari, I., Staral, J. S., and Roberts, J. D.,** Nitrogen-15 nuclear magnetic resonance spectra of isocyanates, isothiocyanates and *N*-sulfinylamines, *Org. Magn. Reson.*, 12, 340, 1979.

Chapter 9

I. OXYGEN BOUND TO NITROGEN
II. OXYGEN BOUND TO OXYGEN

David W. Boykin* and Alfons L. Baumstark*

TABLE OF CONTENTS

INTRODUCTION AND SCOPE

This chapter provides a review of the literature of the ^{17}O nuclear magnetic resonance (NMR) spectroscopy of compounds containing oxygen bound to nitrogen and oxygen bound to oxygen. Section I deals with the relationships between structure and ^{17}O chemical shift for organic compounds with oxygen bound to nitrogen. Section II describes such relationships for organic compounds containing oxygen bound to oxygen. For convenience each part is self-contained with respect to references and compound numbers. Throughout this chapter it will be noted that for many functional groups only limited studies have been carried out to date, and, consequently, this subfield of ^{17}O NMR studies represents an area where considerable future developments are expected.

I. OXYGEN BOUND TO NITROGEN*

A. NITRO COMPOUNDS
1. Aliphatic Systems
A limited number of aliphatic nitro compounds have been studied by ^{17}O NMR spectroscopy.[1] Most of the data for aliphatic compounds which have been reported are on neat liquids, were acquired on earlier generations of instrumentation, and probably should have their chemical shifts checked prior to a detailed analysis of shift differences. The reported chemical shift range for simple aliphatic nitro compounds was between 587 and 605 ppm (Table 1). The shift variation with changing alkyl group was qualitatively analogous to those noted for alkyl aldehydes and ketones.[2] Introduction of halogens alpha to the nitro group causes shielding of the signal (**5** to **7**). Similar effects on the ^{17}O signal for ketones,[2] esters,[3] and isocyanates[4] were noted. The origin of this shielding effect is not clear; however, similar observations have been made for ^{13}C signals of halogenated ketones.[5] New insights into this effect may be forthcoming.[6] Introduction of additional nitro groups does not produce large shifts; compare **1** with **8, 9,** and **10**.

The ^{17}O resonance for the nitromethane anion has been reported, and it appears approximately 400 ppm upfield from nitromethane, reflecting considerable localization of charge on the oxygen atoms.[1] The signals for the anion of dinitromethane and trinitromethane show a net shielding of approximately 180 and 80 ppm, respectively, reflecting the charge delocalization over the oxygen atoms of the additional nitro groups. It has been concluded[7] from study of a wide range of nitrocompounds, including some simple aliphatic ones, that the nitro resonance position is most influenced by the ΔE^{-1} term of the Poplé-Karplus equation, with the r^{-3} and Q terms playing a lesser role.

2. Aromatic Systems
Aromatic nitro compounds have been studied more extensively by ^{17}O NMR spectroscopy than their aliphatic counterparts, and chemical shifts for a large number of systems have been reported (Tables 2, 3, and 4). The effect of substituents on the chemical shift of substituted nitrobenzenes has been examined in detail (Table 2).[8-10] In contrast to an earlier report,[8] Stothers[9] and co-workers and Craik and co-workers[10] have shown a wide chemical shift range for the ^{17}O chemical shifts of substituted nitrobenzenes and predictable shift variation. A typical spectrum for an aromatic nitro compound is shown in Figure 1. Craik and co-workers[10] studied a wide range of substituents in both the *meta* and *para* positions by both ^{17}O and ^{15}N NMR spectroscopy. As is generally the case, the chemical range for ^{17}O shifts is substantially larger than the range for the ^{15}N shifts; a range of 40 ppm vs. 6 ppm was observed for the *para* substituted nitrobenzenes. Craik and co-workers[10] analyzed

---

* References for Sections I and II appear at the end of this chapter.

TABLE 1
^{17}O Chemical Shift Data (ppm) of Aliphatic Nitro Compounds[a]

Compound no.	Structure	Chemical shift
1	CH_3NO_2	605
2	$CH_3CH_2NO_2$	597
3	$(CH_3)_2CHNO_2$	591
4	$(CH_3)_3CNO_2$	587
5	CCl_3NO_2	559
6	$CH_3CHClNO_2$	588
7	$CH_3CCl_2NO_2$	577
8	$CH_2(NO_2)_2$	610
9	$HC(NO_2)_3$	605
10	$C(NO_2)_4$	600
11	$CH_3C(NO_2)_3$	598
12	$FC(NO_2)_3$	598
13	$ClC(NO_2)_3$	587
14	$BrC(NO_2)_3$	592

[a] Reference 1, p. 29.

TABLE 2
^{17}O Chemical Shift Data (ppm) of Monosubstituted Nitrobenzenes[a]

Compound no.	X	δ (ortho)	δ (para)	δ (meta)
15p	NEt_2	—	542.6[b]	—
16o, p, m	NH_2	558[c]	549.4	575.8
17o, p	OCH_3	601[c]	565.2	—
18o, p	OH	555[c]	568.3	—
19p, m	F	—	573.7	577.9
20o, p	Cl	610[c]	572.5	—
21p, m	Br	—	572.9	579
22o, p, m	CH_3	602[d]	572.7	574.6
23	H	575[d]	575[d]	575[d]
24p, m	CHO	—	583.1	576.9
25p	$COCH_3$	—	578.3[b]	—
26p	CO_2CH_3	—	582.2[b]	—
27p, m	CF_3	—	582.2	577.7
28p, m	CN	—	582.5	578.5
29o, p, m	NO_2	609	586.5	580

[a] Unless otherwise noted SCS values from Reference 10 were added to 575 ppm, the value for nitrobenzene in Reference 12.
[b] SCS values from Reference 9 added to 575 ppm as in [a].
[c] From Christ, H. A. and Diehl, P., *Helv. Phys. Acta*, 36, 170, 1963.
[d] From Reference 12.

TABLE 3
^{17}O Chemical Shift Data (ppm) for Monosubstituted Nitrostyrenes (p-X-C$_6$H$_4$-CH=CH-NO$_2$)[a]

Compound no.	p-X	δ^{17}O (NO$_2$)
30	NMe$_2$	562.4
31	OH	572.0
32	OMe	573.0
33	Me	577.1
34	F	578.4
35	H	578.8
36	Cl	579.5
37	Br	579.6
38	CF$_3$	582.1
39	CN	583.5
40	NO$_2$	586.2

[a] From Reference 11.

TABLE 4
^{17}O Chemical Shift Data (ppm) for Heteroaromatic Nitro Compounds[a]

Compound no.	Name	Chemical shift (ppm)[a]
48	2-Nitrothiophene	569
49	5-Nitrothiophene-2-carboxaldehyde	578
50[b]	5-Nitrofuran-2-carboxaldehyde	574
51[c]	2-Nitropyrrole	520
52[c]	1-Methyl-2-nitropyrrole	536
53	6-Nitroquinoline	577
54	5-Nitroquinoline	602
55	8-Nitroquinoline	626
56	7-Methyl-8-nitroquinoline	633
57	5-Nitroisoquinoline	600
58	5-Nitroindole	562
59	7-Nitroindole	568
60	6-Nitroindole	566
61	4-Nitroindole	578
62	5-Nitroindoline	546
63	6-Nitroindoline	572
64	5-Nitroindazole	571
65	6-Nitroindazole	575

[a] Unless otherwise noted from Reference 14.
[b] From Reference 16.
[c] From Reference 15.

both the ^{17}O and ^{15}N data by the dual substituent parameter (DSP) approach. It was found for the *para* substituted nitrobenzenes that the ^{17}O data was best correlated with σ_R^+, indicative of enhanced interaction of donor substituents. The magnitude of the coefficients from Equation 1 indicates that the resonance term was the most important.

$$^{17}\text{O SCS}_{para} = 13.5\sigma_I + 15.6\sigma_{R^+} \tag{1}$$

A similar treatment of the ^{15}N data gave the best correlation using σ_R^0 and σ_I was the dominant term, indicating the importance of π polarization on the ^{15}N chemical shifts. The

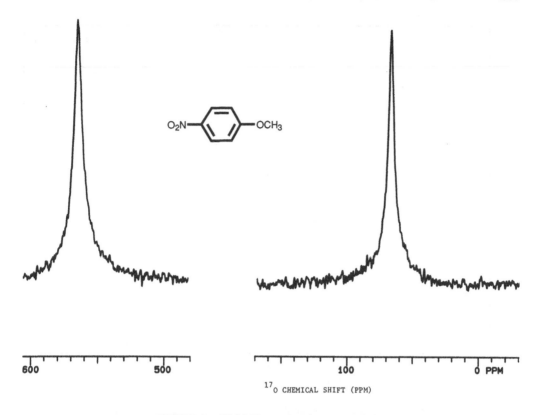

FIGURE 1. ^{17}O NMR spectrum for 4-nitroanisole.

range of chemical shifts for the *meta* substituted series was significantly compressed compared to the *para* system (6 ppm vs. 36 ppm). The DSP correlation of the ^{17}O chemical shift data for the *meta* series showed a clear dominance by the σ_I term (Equation 2).

$$^{17}O \text{ SCS}_{meta} = 6.5\sigma_I + 0.6\sigma_{R-} \tag{2}$$

The ^{17}O chemical shift data for the *para*-substituted nitrobenzene gave a reasonable correlation with calculated (STO-3G) electron density at the oxygen atoms, which suggests that the change in chemical shifts with substituents is primarily controlled by electronic effects.

The sensitivity of the nitro group ^{17}O chemical shift to electronic effects in substituted β-nitrostyrenes has been reported (Table 3)[11]. The effect of the intervening double bond resulted in a reduction of the chemical shift range by approximately one third compared to the nitrobenzenes. A reasonable correlation was found using the DSP approach, and, as was the case for nitrobenzenes, the best correlation was found with σ_R^+. The resonance term was found to be the more significant one (Equation 3).

$$^{17}O \text{ SCS} = 7.82\sigma_I + 9.04\sigma_{R+} \tag{3}$$

A plot of the ^{17}O chemical shift data for the nitrobenzenes vs. the nitrostyrenes gave a good correlation (Figure 2) and indicated that the chemical shifts of the nitrostyrenes were also highly dependent upon the electronic charge on the nitro group oxygen atoms.

The influence of steric congestion on the ^{17}O chemical shift of aromatic nitro compounds has been discussed in detail in Chapter 3. A quantitative relationship for the torsion angle between the aromatic ring and the nitro group has been developed; as the torsion angle increases, the ^{17}O chemical shift increases. Representative results[12] are shown with structures **41** to **43**, and for additional examples the reader is referred to Table 1 in Chapter 3.

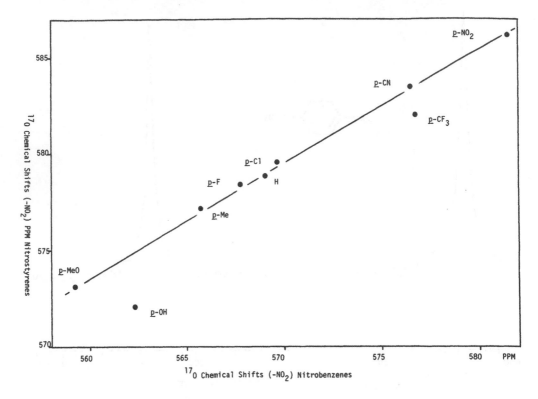

FIGURE 2.   Plot of nitro group ^{17}O chemical shifts for β-nitrostyrenes vs. those for nitrobenzenes.

575 PPM	605 PPM	637 PPM
NO₂	NO₂	NO₂
41	**42**	**43**

^{17}O chemical shift data for a limited number of polyfluoronated nitrobenzenes (**44** to **47**) were reported.[13] The chemical shift values for these compounds are downfield of those for their hydrocarbon analogs (compare **44** to **23**). This result is consistent with the electron attraction by multiple fluorine atoms. However, interpretation of the data for this series is not straightforward since substitution in the 4-position by electron donors (compare **45** with **46** and **47**) is reported to result in additional deshielding.

627 ppm	631 ppm	644 ppm	633 ppm
$C_6F_5NO_2$	$4\text{-}HC_6F_4NO_2$	$4\text{-}Et_2C_6F_4NO_2$	$4\text{-}CH_3C_6F_4NO_2$
44	**45**	**46**	**47**

^{17}O chemical shift data for several nitro-substituted heterocyclic systems were reported.[14] The chemical shifts of a few isolated examples of nitro-substituted five-membered heterocyclics[14-16] (**48** to **52**) are included in Table 4. Despite the limited amount of data on the thiophene system, it is apparent that the nitro group on the nitrothiophene ring is slightly shielded compared to that for a nitro group on the benzene ring. Compare the shifts of **48** and **23** and **49** and **24**. The one result for a nitrofuran **50** suggests that the furan ring also causes shielding (compare to nitrobenzene **23**). The most significant shielding is noted for the two nitropyrroles **51** and **52**. The signal for **51** is 55 ppm upfield from nitrobenzene. While there may be some contribution to the upfield shift from hydrogen bonding, the value for **52** for which hydrogen bonding is not possible shows clearly that considerable electron donation results from the pyrrole ring. Interestingly, comparison of the shifts of **51** and **52** suggests that torsion angle rotation arises from the *ortho*-like *N*-methyl group (see Chapter 3). A similar effect was observed for *N*-methyl 2-acyl pyrroles.

The influence of fusion of the electron-attracting pyridine ring to nitrobenzene and the similarity of the chemical shifts of the quinolines **53, 54,** and **57** to their naphthalene counterparts have been discussed in Chapter 3. It is also noteworthy that the torsion angle for the 8-nitro group and the quinoline ring for 8-nitroquinoline has been estimated to be 69°. Apparently the nitro group lone pair repulsive van der Waals interaction is greater than the analogous *peri*-hydrogen-nitro group interaction (Chapter 3).

Nitro groups on benzene rings fused to electron-rich heterocycles were expected to have more single bond character for the nitrogen-oxygen bonds; thus their ^{17}O chemical shift was expected to be upfield, compared to the value for nitrobenzene. The ^{17}O chemical shift of 5-nitroindole (**58**) is 562 ppm. This corresponds to approximately a 15 ppm upfield shift compared to nitrobenzene, confirming expectations. The ^{17}O chemical shift for 7-nitroindole (**59**) which should be electronically equivalent to its isomer **58**, is slightly downfield from **58** at 568 ppm. This small downfield shift is in the correct direction for a *peri*-type compressional effect; however, hydrogen bonding would be expected to cause shielding. A significant contribution from the latter effect seemed unlikely, since it has been reported that nitro groups form weak hydrogen bonds. The two nitroindoles **60** and **61** represenet electronically equivalent systems, and, consequently, their differing ^{17}O chemical shift of 566 to 578 ppm, respectively, indicated that *peri* type interactions, although small compared to quinolines and napthalenes, were also operative for the indoles. The ^{17}O chemical shift for 5-nitroindazole **64** (571 ppm) falls between that of the quinoline **53** (577 ppm) and the indole **58** (562 ppm) which was consistent with expectations, since the indazole ring is more electron-rich than quinoline and more electron-deficient than indole. Similarly, the chemical shift of 6-nitroindazole (**65**) (575 ppm) was slightly downfield with respect to the chemical shift value for **60**.

562 PPM

568 PPM

58

59

566 PPM

578 PPM

60

61

TABLE 5
^{17}O Chemical Shift Data (ppm) for Miscellaneous Nitroso Compounds

Compound no.	Structure	Chemical shift
66a[a]	$(CH_3)_2N-NO$	660
66b[b]	$(CH_3CH_2)_2N-NO$	682
66c[b]	$(CH_2)_4N-N=O$	693
66d[b]	$(CH_2)_5N-N=O$	668
66e[c]	$(NCCH_2)(CH_3)N-N=O$ (isomers)	696, 705
66f[d]	C_6F_5NO	699
67[d]	C_6H_5NO	620

[a]   Reference 7.
[b]   Data from 0.5 M acetonitrile solutions by authors.
[c]   Reference 42.
[d]   Reference 13.

The two indolines 5-nitroindoline (**62**) and 6-nitroindoline (**63**) contain saturated heterocyclic ring systems and give ^{17}O NMR results similar to those reported for nitroanilines. The chemical shift of 546 ppm for **62** was substantially upfield to that of **58**, which was consistent with electron donation by a *para* amino function, and the shift of 572 ppm for **63** reflects the smaller shift expected from a *meta* amino substituent.

B. NITROSO COMPOUNDS

Only a very limited number of organic nitroso compounds have been studied by ^{17}O NMR methods.[7,13] The chemical shift for this group appears downfield from other systems with oxygen bound to nitrogen. It has been suggested that the downfield position of absorbance for this functional group can be attributed substantially to enhanced excitation energies of the nitroso group. However, further investigation seems to be necessary in order to develop a clear understanding of the factors that contribute to the chemical shift of this functional group.

Table 5 contains ^{17}O chemical shift data for miscellaneous organic nitroso compounds found in the literature and for a few for which data were obtained for this review in our laboratories. Figure 3 shows a typical ^{17}O NMR spectrum for an organic nitroso compound.

C. N-OXIDES

The unusual chemical reactivity of heteroraromatic N-oxides has been attributed to the ability of the N-oxide group to function both as a π electron donor or as a π electron acceptor.[17-20] Earlier examinations of the dual properties of the N-oxide function have included ¹H,[21] ¹⁵N,[22,23] and ¹³C[22,24] NMR investigations, IR studies,[25] and polarographic reduction investigations.[26] The sensitivity of ^{17}O NMR chemical shifts to changes in electron density makes ^{17}O NMR spectroscopy an important method for studying the electronic properties of N-oxides. This section deals with recent reports[27-29] which have demonstrated the influence of electronic factors, steric factors, and hydrogen bonding on the ^{17}O NMR chemical shift data for heteroaromatic N-oxides.

1. Electronic Effects

The ^{17}O chemical shift data for 4-subsituted pyridine N-oxides in acetonitrile are listed in Table 6. The chemical shift for the N-oxide function is very sensitive to substituents;[27] the range of the shift from 4-methoxy to 4-nitro substituent is 102 ppm (see Figure 4). This result can be compared to that for similar substituents for 4-substituted anisoles and 4-substituted acetophenones which gave ranges of 29 ppm and 50 ppm, respectively. The

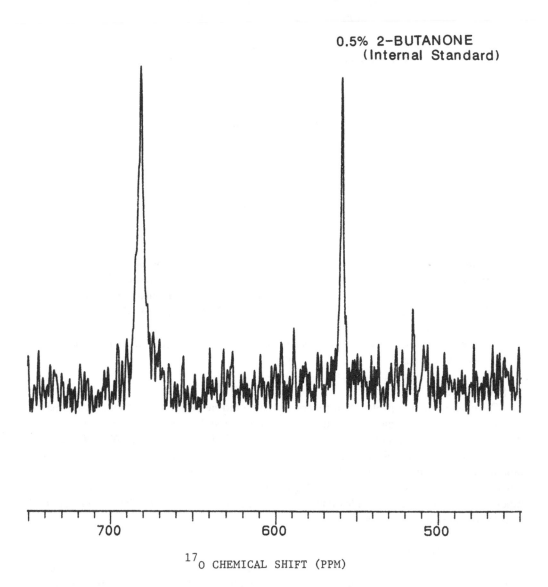

FIGURE 3. ^{17}O NMR spectrum for nitrosodiethylamine.

greater sensitivity to substituent effects is illustrated by plotting the pyridine N-oxide vs. anisole ^{17}O NMR data (see Figure 5), which gives a good correlation with a slope of 3.7. The chemical shift data for the pyridine N-oxides gave a reasonable correlation with the Hammett expression; sigma plus gave a slightly better correlation than either sigma or sigma minus (Table 7). Treatment of the data by the dual substituent parameter (DSP) approach gives a better correlation than that obtained by the single parameter approach (see Table 7). The DSP method allows, by comparison of the magnitude of the regression coefficients, the assessment of the contribution of the mesomeric and polar effects to the correlation. It was found that the mesomeric contribution was large. These results were in accord with the considerable double-bond character for the N-oxide group. Recently, similar ^{17}O NMR results have been found for 4-substituted pyridine N-oxides in dimethylsulfoxide solution.[30] Anisole

TABLE 6
^{17}O Chemical Shift Data (ppm) for 4-Substituted Pyridine N-Oxides[a]

Compound no.	R	δ(ppm) (N⁺–O⁻)	δ(ppm)R
68	CH_3O	312	56
69	C_2H_5O	317	86
70	PhO	320	116
71	CH_3	336	
72	H	349	
73	Ph	351	
74	Cl	352	
75	Ac	389	554
76	CN	404	
77	NO_2	414	573

[a]   From Reference 27.

^{17}O chemical shift data[31] were correlated with calculated π-electron densities and π-bond order of the oxygen atom. The correlation of anisole and pyridine N-oxide data indicated a strong dependence of the N-oxide ^{17}O chemical shift on π-electron density and π-bond order term of the Karplus-Pople equation.[32] A recent ^{17}O NQR study of 4-pyridine N-oxides also demonstrated that their chemical shifts varied with bond order and charge density. The oxygen chemical shift of N-oxides is deshielded as the bond order increases and as the charge density decreases.

2. Steric Effects

The change in chemical shift for sterically hindered heteroaromatic N-oxides compared to unhindered isomers has been found to be large enough to use ^{17}O chemical shifts to distinguish between structural isomers.[22] The chemical shifts of hindered N-oxides are deshielded compared to their unhindered isomers. The influence of steric factors on heteroaromatic N-oxide and the origin of the change in ^{17}O chemical shift with changing steric environment is discussed in detail in Chapter 4. A summary of all the chemical shifts for hindered heteroaromatic N-oxides and their unhindered isomers which have been reported to date are shown in Table 8.

3. Hydrogen Bonding

In view of the sensitivity of pyridine N-oxide ^1H NMR chemical shifts[21-23] to protic solvent and the large ^{17}O chemical shift change noted for acetone[33] on addition of water, the influence of hydrogen bonding on the pyridine ^{17}O chemical shift was studied. The ^{17}O chemical shift of pyridine N-oxide was markedly changed in the presence of water (see Figure 6). The chemical shift difference between pyridine N-oxide in acetonitrile and water of 75 ppm is larger than the noted difference (52 ppm) for acetone neat and acetone at infinite dilution in water. These large upfield shifts have been attributed to hydrogen bonding. The role of protic solvents on the substituent-induced chemical shift for pyridine N-oxides was evaluated by comparing the chemical shift values for 4-nitro- and 4-methoxy-pyridine N-oxide (**68** and **77**) measured in aqueous solutions (Table 9). On comparison of the chemical shift range in the two solvents (water and acetonitrile), it was found that substituent effects were suppressed from 102 to 76 ppm in water solution. Comparing the chemical shifts in the two solvents ($\delta_{CH_3CN}-\delta_{H_2O}$) showed a delta value of 59 ppm for the methoxy compound **68**, 75 ppm for the parent compound **72**, and 85 ppm for the nitro compound **77**. The greater

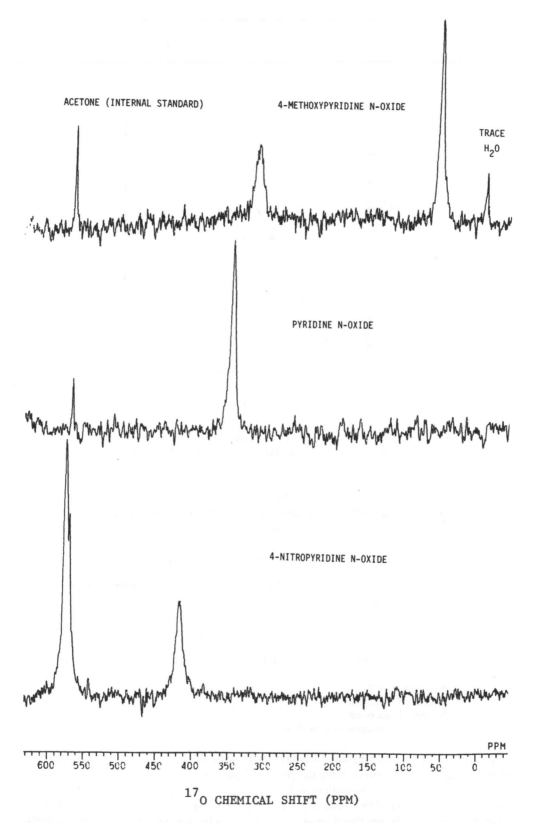

FIGURE 4. ^{17}O NMR spectra of 4-methoxypyridine *N*-oxide (**68**), pyridine *N*-oxide (**72**), and 4-nitropyridine *N*-oxide (**77**).

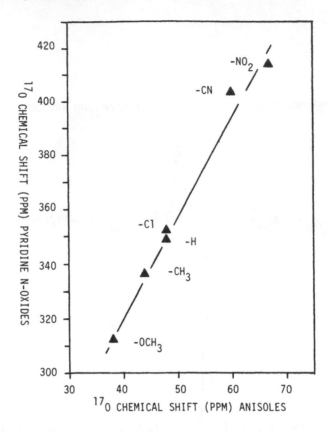

FIGURE 5. Plot of the NO ^{17}O chemical shift of 4-substituted
pyridine *N*-oxides vs. the ^{17}O chemical shift for *p*-substituted anisoles.

TABLE 7
^{17}O Chemical Shift Data Correlations for 4-Substituted Pyridine *N*-Oxides[a]

Correlation	Slope	Intercept	r[b]	n[c]
σ^+ vs. δ_{17O}	63.7 ± 1.9[d]	8.3	0.983	10
$\delta_{17O_{N+-O-}}$ vs. δ_{17O} anisoles	3.7 ± 0.7[d]	173.5	0.992	6

Correlation	ρ_i	$\rho_R{}^o$	f(SD/RMS)	r[b]	n[c]
DSP	67.6 ± 5.1	131.3 ± 8.4	0.16	0.990	9

[a] From Reference 27.
[b] Correlation coefficient.
[c] Number of data points.
[d] 95% confidence limits, error in slope.

delta value for the nitro compound was consistent with the greater double-bond character
for its *N*-oxide group and presumably the greater effect of hydrogen bonding to it. Consistent
with this conclusion is the observation that ^{17}O chemical shifts for the single-bonded oxygen
of benzyl alcohols show little sensitivity to hydrogen bonding effects.[34] Since the ^{17}O chemical shift of the *N*-oxide functional group was reported to be very sensitive to protic solvents,

TABLE 8
[17]O Chemical Shift Data (ppm) for Hindered Heteroaromatic
N-Oxides and Related Unhindered Isomers[a]

Compound no.	Name	Chemical shift (ppm)
72	Pyridine *N*-oxide	349
78	2-Methylpyridine *N*-oxide	350
79	4-Methylpyridine *N*-oxide	336
80	2-Ethylpyridine *N*-oxide	346
81	4-Ethylpyridine *N*-oxide	336
82	2-*n*-Propylpyridine *N*-oxide	342
83	2-*i*-Propylpyridine *N*-oxide	342
84	2-*t*-Butylpyridine *N*-oxide	361
85	4-*t*-Butylpyridine *N*-oxide	338
86	2-Phenylpyridine *N*-oxide	347
87	4-Phenylpyridine *N*-oxide	351
88	2,6-Dimethylpyridine *N*-oxide	350
89	Quinoline *N*-oxide	343
90	8-Hydroxyquinoline *N*-oxide	289
91	4-Methylquinoline *N*-oxide	332
92	2-Methylquinoline *N*-oxide	341
93	6-Methylquinoline *N*-oxide	341
94	8-Methylquinoline *N*-oxide	370
95	Benzo[*f*]quinoline *N*-oxide	344
96	Benzo[*h*]quinoline *N*-oxide	362
97	4-Methylpyrimidine 1-oxide	324
98	4-Methylpyrimidine 3-oxide	335

[a] From References 27 to 29.

[17]O methodology should provide an excellent method for studying the association of *N*-oxide compounds with proton donors or Lewis acids. In view of the sensitivity of the chemical shift of this functional group to hydrogen bonding, it is clearly important to take precautions to exclude protic sources when recording [17]O chemical shifts of *N*-oxide systems in aprotic solvents.

Intramolecular hydrogen bonding has been reported to cause large changes in [17]O chem-

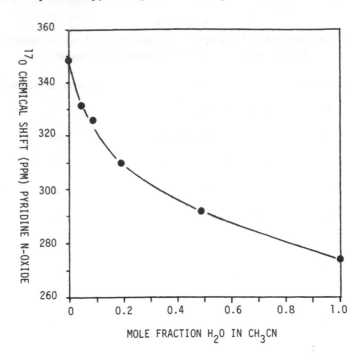

FIGURE 6. Plot of change of pyridine *N*-oxide ¹⁷O chemical shift in acetonitrile on addition of water.

TABLE 9

Comparison of ¹⁷O Chemical Shift Data (ppm) for Pyridine *N*-Oxides in Acetonitrile and in Water[a]

Compound no.	Name	$\delta(N^+-O^-)_{CH_3CN}$	$\delta(N^+-O^-)_{H_2O}$
72	Pyridine *N*-oxide	349	274
68	4-Methoxy pyridine *N*-oxide	312	253
77	4-Nitropyridine *N*-oxide	414	329

[a] From Reference 27.

ical shifts of the carbonyl groups of hydroxynaphthoquinones[35,36] and related compounds (see Chapter 5).[37] One example in which intramolecular hydrogen bonding was considered to be important has been reported for aromatic *N*-oxides.[29] The ¹⁷O NMR signal for the NO function of 8-hydroxyquinoline *N*-oxide appears at 289 ppm substantially shielded from that of quinoline *N*-oxide at 343 ppm (see Figure 7). It is difficult to determine the magnitude of shielding which can be attributed solely to hydrogen bonding since compressional effects of the 8-hydroxy group on the *N*-oxide oxygen would be expected to be deshielding. Thus the delta value of 54 ppm for **89** and **90** may be regarded as the minimum effect due to hydrogen bonding. Proton to oxygen coupling is also observed for the OH signal of **90** (see Figure 7).

D. HYDROXAMIC ACIDS AND OXIMES

The ¹⁷O NMR characteristics of hydroxamic acids are essentially unexplored. In a detailed study of the structure of benzohydroxamic acids, ¹⁷O NMR data on the carbonyl oxygen of four enriched benzohydroxamic acids (**99** to **102**) were recorded.[38] No data were reported for the oxygen bound to nitrogen for these compounds. Table 10 contains the data

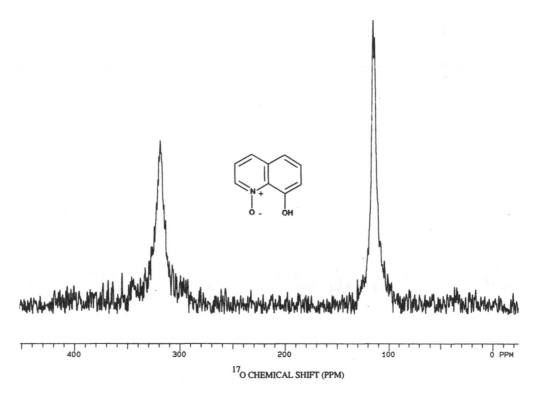

FIGURE 7. ^{17}O NMR spectrum of 8-hydroxyquinoline N-oxide (**90**) which shows ^{17}O-proton coupling for the hydroxy group.

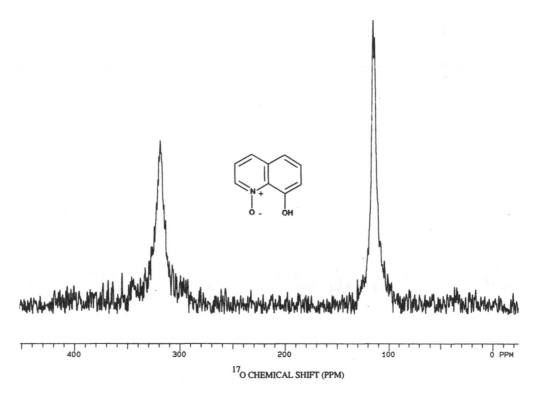

TABLE 10
^{17}O Chemical Shift Data (ppm) for Hydroxamic Acids and Esters[a]

Compound no.	Structure	Chemical Shift (ppm)	
		(C=O)	(NHOH)
99	$C_6H_5CONHOH$	333 (302)[b]	(79)[b]
100	$C_6H_5CON(CH_3)OH$	330	
101	$C_6H_5CONHOCH_3$	341	
102	$C_6H_5CON(CH_3)OCH_3$	347	
103	$CH_3CONHOH$	(311)[b]	(81)[b]

[a]   Data from Reference 38 in dioxane solvent unless otherwise noted.
[b]   Data from 0.5 M acetonitrile solutions by authors.

from the earlier report for **99** to **102** and the data for both types of oxygens for **99** and acetohydroxamic acid **103** obtained at natural abundance in our laboratories. A typical spectrum is shown in Figure 8.

The downfield trend of the carbonyl signal for **99** to **102** corresponds to the trend noted on introduction of methyl groups in simple amides. The carbonyl signal for these hydroxamic acids (**99** to **102**) was shown to be sensitive to both intramolecular and intermolecular hydrogen bonding effects arising from interactions with solvent. The data for the oxygen bound to nitrogen for **99** and **102** show relatively small differences. These signals doubtlessly are also sensitive to hydrogen bonding. Clearly, additional work is needed to characterize the ^{17}O chemical shift properties of this functional group.

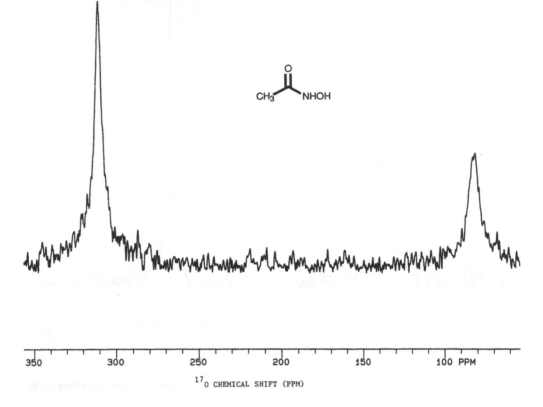

FIGURE 8.   ¹⁷O NMR spectrum of acetohydroxamic acid (**103**).

There are no reports of ¹⁷O NMR studies on oximes in the literature, although the chemical shifts for the *O*-methyl derivative of acetaldoxime ($CH_3CH = N\text{-}OCH_3$) has been reported to be 157 ppm.[39] Table 11 contains the chemical shift values obtained in our laboratories for some representative oximes. Figure 9 shows a typical ¹⁷O NMR spectrum of an oxime. The chemical shift range for the compounds (Table 11) is not large. It remains to be determined if ¹⁷O NMR data can be used to distinguish between *syn* and *anti* forms of the oximes. Since the oxime group can function both as a hydrogen bond donor and acceptor, it will be necessary to carefully work out the influence of hydrogen bonding on chemical shift of the oxygen in this functional group.

E. MISCELLANEOUS
1. Heterocyclic N–O Containing Systems
¹⁷O chemical shift data for substituted isoxazoles were reported (Table 12).[40] The chemical shift variations noted for the methyl-substituted isoxazoles were rationalized in terms of β- and γ-effects analogous to those observed for vinyl ethers.[41] Data for miscellaneous heterocyclic systems containing nitrogen bound to oxygen are shown below.[39,42]

590 PPM   446 PPM

117   118

TABLE 11
^{17}O Chemical Shift Data (ppm) of Oximes[a]

Compound no.	Structure	Chemical shift (ppm)
104	Cyclopentanone oxime	179
105	Cyclohexanone oxime	170
106	Dicyclopropylketone oxime	167
107	2,6-Dimethylbenzaldehyde oxime	188
108	2,6-Dichlorobenzaldehyde oxime	196.5
109	Fluorenone oxime	195

[a] Data from 0.5 M acetonitrile solutions by authors.

2. α,N-Diaryl Nitrones

The ^{17}O chemical shift data for the nitrones **119** to **125** are listed in Table 13.[43] The chemical shift of the parent compound **119**, 337 ppm, was downfield by 28 ppm from that of pyridine N-oxide **72** which indicated that the N–O bond for the nitrones had more double bond character than that of the heterocyclic N-oxides. This result was consistent with expectations since greater delocalization of the N–O bond was expected in the heteroaromatic N-oxides. It is clear from Table 13 that the nitrone chemical shift is quite sensitive to substituents. A range of 55 ppm in chemical shift from the cyano to dimethylamino compound was noted.

The chemical shift of oxygen in the nitrone system was less sensitive to substituent effects than the oxygen for pyridine N-oxide, by a factor of approximately two. The difference was consistent with the fact that the substituents and the N–O function of the nitrones were separated by two more carbon atoms than the substituent and the N–O group of the pyridine N-oxides. The data gave a reasonable correlation when plotted vs. Hammett sigma constants (Figure 10). The slope of the line was consistent with increasing electron density on oxygen resulting in shielding of the ^{17}O chemical shift.

3. Other N–O Containing Organic Systems

Chemical shift data for azoxybenzene and three polyfluoroazoxybenzenes have been reported.[13] The chemical shifts for these systems seem to follow expectations based upon normal effects of substituents, unlike benzyl alcohols and acetates[34] which have the oxygen atom located in the same position relative to the aromatic ring.

456 ppm

O$^-$

C_6H_5—N$\overset{\pm}{\underset{\|}{=}}$N—$C_6H_5$

126

549 ppm

O$^-$

C_6H_5—N$\overset{\pm}{\underset{\|}{=}}$N—$C_6F_5$

127

534 ppm

O$^-$

p-CH$_3$C$_6$F$_4$—N$\overset{\pm}{\underset{\|}{=}}$N—C$_6F_4$-p -CH$_3$

128

578 ppm

O$^-$

p-CF$_3$C$_6$F$_4$—N$\overset{\pm}{\underset{\|}{=}}$N—C$_6$F-p-CF$_3$

129

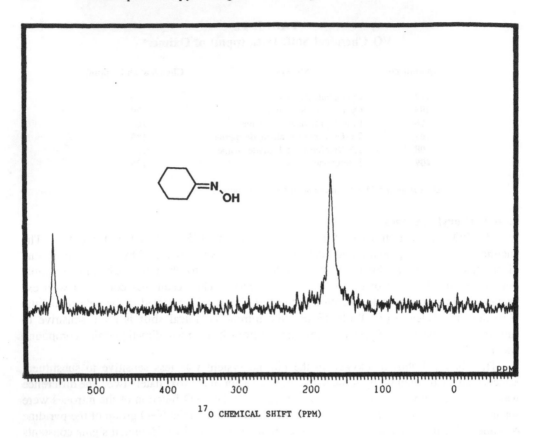

FIGURE 9. ^{17}O NMR spectrum of cyclohexanone oxime (105) [2-butanone internal standard 558 ± 1 ppm].

TABLE 12
^{17}O Chemical Shift Data (ppm) for Isoxazoles and Related Compounds[a]

Compound no.	R_1	R_2	R_3	δ(ppm)
110	H	H	H	350
111	CH_3	H	H	344
112	H	H	CH_3	356
113	CH_3	CH_3	H	338
114	CH_3	H	CH_3	350
115	CH_3	CH_3	CH_3	340
116[b]	1,2-benzisoxazole			306

[a]   Unless otherwise noted from Reference 40.
[b]   Balakrishman, P. and Boykin, D. W., unpublished results.

TABLE 13
[17]O Chemical Shift Data (ppm)
for α, N-Diaryl Nitrones[a]

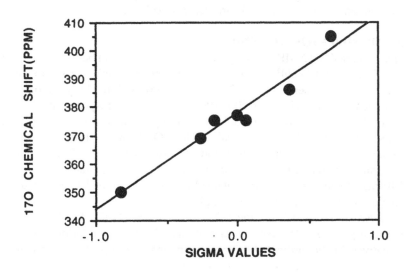

No.	X	δN→O (ppm)
119	H	377
120	Me$_2$N	350
121	MeO	369
122	F	375
123	Me	375
124	Cl	386
125	CN	405

[a] From Reference 43.

FIGURE 10.   Plot of [17]O chemical shifts of α-N-diaryl nitrones vs. Hammett sigma values.

II. OXYGEN BOUND TO OXYGEN

A. INTRODUCTION

^{17}O NMR spectroscopy is rapidly becoming an important method[1-3] for examining mechanistic and structural problems[4-6] in many classes of organic compounds. Interestingly, ^{17}O NMR data on peroxidic material are very limited in number. For example, prior to 1980, only a few isolated spectra on peroxides and hydroperoxides had been published (*vide infra*). Presumably, this was due, in part, to the difficulties inherent in the synthesis and handling of peroxidic compounds. Since high concentrations of peroxy compounds are to be avoided (*Danger!*) and samples often can not be subjected to elevated temperatures, additional difficulties are present that have deterred investigators from readily acquiring ^{17}O NMR data. In addition, broad signals are likely to be obtained under these constraints, which increase the instrument time required to obtain reasonable signal-to-noise ratios and further increase the complications due to sample decomposition.

Some of these problems can be overcome by the use of ^{17}O-enriched compounds, especially since ^{17}O-enriched molecular oxygen is readily available commercially. The newer generation of instruments is much more sensitive, and, thus, natural abundance ^{17}O NMR spectra can be obtained on even moderately stable compounds. Recent ^{17}O NMR studies on several classes of peroxy compounds have shown that ^{17}O NMR data can provide new mechanistic and structural insights into peroxide chemistry (*vide infra*). Cumulative ^{17}O NMR data to date on peroxidic materials are listed in Table 1.

B. BACKGROUND (PRIOR TO 1980)

The earliest reported ^{17}O NMR data on peroxidic compounds can be found in the extensive survey study of Christ and Diehl[7] in 1961. The ^{17}O chemical shifts for hydrogen peroxide (**1**), 174 ppm, *t*-butyl hydroperoxide (**2**), 260 ppm (broad), and di-*t*-butyl peroxide (**3**), 269 ppm were obtained with linewidths of 200, 1000, and 300 Hz, respectively.

Many years later, ^{17}O NMR methodology was applied to the study of O_2F_2 and the subsequent disproof of the "O_3F_2" molecule.[8] The ^{17}O chemical shift value for O_2F_2 (**4**) was found to be 647 ppm. In addition, this study reported the ^{17}O NMR data on ozone (**5**), 1032 ppm (central oxygen) and 1598 (terminal oxygens). Two independent studies showed[9,10] that ^{17}O NMR data could not be readily obtained for the binding of ^{17}O-enriched molecular oxygen with Vaska's-type compounds (*vide infra*).

In 1979, Iwamura showed[11] that the ^{17}O data for O_3(**5**), O_2F_2 (**4**), di-*t*-butyl peroxide (**3**), and hydrogen peroxide (**1**) gave an "approximate" linear correlation with the sum of the electronegatives for dicoordinated oxygen compounds, in apparent agreement with expectations based on the Karplus-Pople equation. The ^{17}O NMR chemical shift value of **3** was reported as 276 ppm, in reasonable agreement with that of the earlier report.

C. RECENT STUDIES

The work of Postel on oxo-peroxo Mo(VI) compounds showed[12] that ^{17}O NMR methodology would be of value for the investigation of oxygen chemistry in transition elements. Synthesis of Mo(VI) compounds with specific ^{17}O-enrichment of the peroxo and/or oxo sites showed that these sites did not exchange oxygen atoms. The ^{17}O NMR spectrum of Mo(VI) compound **6** showed a signal at 487 ppm (1800 Hz) for the peroxo group.

$$O$$
$$(CN)_4 Mo \overset{O}{\underset{O}{\diagdown}} \quad L_2$$

L = N(PPh_3)_2

6

t-Bu
|
O
|
O
\   /
Pt   Pt
/   \
O
|
O
|
t-Bu

7

In addition, the ^{17}O NMR data for the "Pt-O-O-t-But" group[13] of a large molecular weight platinum compound **7** were retaken: 1250 ppm (1400 Hz) and 230 ppm (1000 Hz). The signal at 1250 was assigned to the platinum-bound oxygen. The assignment for the other peroxy oxygen was established by comparison with the data for **3** and a Pd-O-O-t-But compound, in agreement with the earlier work.[13] The authors further pointed out[12] that the lack of observable ^{17}O NMR signals in Vaska's-type complexes and the $PtO_2(L)_2$ and $PdO_2(L)_2$ cases[12] was not due to slow molecular tumbling nor quadrapolar coupling effects. Rather, exchange with molecular oxygen (reversible binding) was suggested as the cause of extreme line-broadening producing "undetectable" signals. However, the recent observation[14] of ^{17}O NMR signals for reversible, Vaska's-type complexes has invalidated the latter suggestion. For example, ^{17}O NMR signals were reported[14] at 325 ppm for Vaska's compound [Ir(CO)Cl(O_2)(PPh_3)_2]; at 350 for Ir(CO)I(O_2)(PPh_3)_2; and at 385 for Pt(O_2)(PPh_3)_2, **(7a)**, with linewidths of 9,000 to 10,000 Hz.

A study by Curci et al.,[15] followed up the work of Postel et al.[12] and Bregeault and Mimourn.[13] The ^{17}O NMR data for $MoO(O_2)_2L$ **(8)** and two $CrO(O_2)_2L$ **(9a** and **b)** complexes were obtained. In addition ^{17}O-enriched t-butyl hydroperoxide **(2)** and di-t-butyl peroxide **(3)** were prepared and the ^{17}O NMR chemical shifts obtained: for **2** 254 ppm (450 Hz) and 221 ppm (400 Hz); and for **3** 281 ppm (600 Hz). The molybdenium compound **8** showed one ^{17}O NMR signal at 458 ppm (1500 Hz) for the peroxo groups in agreement with the results of Postel.[12] The Cr(VI) compounds showed two signals for the peroxo groups: for **9a** 722 ppm (1000 Hz) and 820 (1000); and for **9b** 726 ppm (700 Hz) and 785 ppm (700 Hz). The spectra did not show a temperature dependance other than the expected line-broadening. No significant exchange between the oxo and peroxo position was found for the Cr(VI) cases in agreement with the Mo(VI) results. The ^{17}O NMR data suggested that the Cr(VI) solid-state structures could be obtained in solution. It was suggested[15] that the electron charge distribution at the O—O bond might be linked to the ^{17}O chemical shift.

8   L = HMPA

9a L = HMPA
9b L = Pyr.

TABLE 1
Cumulative ^{17}O NMR Data for Peroxidic Materials

Compound	Structure	δ ppm ($\nu_{1/2}$); solvent	Ref.
1	H_2O_2	174 (200)	7
2	t–But–OOH	260 (1000)	7
		254 (450); 221 (400)	15
3	t–But–O–O–t–But	269 (300)	7
		276	11
		260 (340)	17
		281 (600)	15
		270; $CDCl_2$	18
4	O_2F_2	647	8
5	O_3	1032 central; 1590 terminal	8
6	$(CN)_4MoO(O_2)L_2$	487 (1800)	12
7	"Pt–OO–t–But"	1250 (1400); 230 (1000)	12,13
7a	$Pt(O_2)(PPh_3)_2$	385 (9000)	14
Vaska's	$Ir(CO)Cl(O_2)(PPh_3)_2$	325 (10,000)	14
8	$MoO(O_2)_2HMPA$	458 (1500)	15
9a	$CrO(O_2)_2HMPA$	772 (1000); 820 (1000)	15
9b	$CrO(O_2)_2Pyr$	726 (700); 785 (700)	15
10a	p–MeOArCH(OOH)N=N–Ph	251, 206; CH_3CN	16
10b	PhCH(OOH)N=N–Ph	254, 204; CH_3CN	16
10c	p–Br–ArCH(OOH)N=N–Ph	249, 196, CH_3CN	16
11a	EtOOEt	253 (300)	17
		261; Benzene	18
11b	n-PrOOn-Pr	249 (300)	17
11c	i-PrOOi-Pr	255 (360)	17
12	$(CH_2)_3O_2$	200 (100)	17
13	$(CH_2)_4O_2$	254 (130)	17
14	$(CH_2)_n$ n=1	303 (100)	17
15	14 n = 2	250 (200)	17
16	14 n = 3	259 (200)	17
17	$(CH_2)_n$ n=2	265 (200)	17
18	17 n = 2	277 (227)	17
19	17 n = 3	232 (100)	17
20		310 (346)	17
21		318 (513)	17
22		283 (701)	17

TABLE 1 (continued)
Cumulative ^{17}O NMR Data for Peroxidic Materials

Compound	Structure	δ ppm ($\nu_{1/2}$); solvent	Ref.
23	Isopropylideneadamantane dioxetane	275	17
24	Tetramethyl-1,2-dioxetane	~287; Benzene	18
25	Dimethyldioxirane	302 (113)	19
26	Acetone diperoxide	263 (350)	19
		265; CDCl$_3$	18
27	''-FeO$_2$-''hindered prophyrin complex	1755; 2488	21

Other than the report of t-butyl hydroperoxide,[7,15] only one ^{17}O NMR spectroscopic study[16a] of alkyl hydroperoxides has appeared. The ^{17}O NMR data on ^{17}O-enriched α-azo hydroperoxides showed[16] an interesting solvent dependence. α-Azo hydroperoxides **10a** to **c** were prepared in high yield by autoxidation of the corresponding phenylhydrazones with one equivalent of ^{17}O-enriched molecular oxygen. Compound **10a** was shown to be an effective ^{17}O-labeling reagent. The oxidation of alkenes, sulfides, and phosphines yielded ^{17}O-enriched epoxides, sulfoxides, and phosphine oxides in good to excellent yields.

$$X-Ar-\underset{\underset{**}{OOH}}{CH}-N=N-Ph$$

10a X = p-MeO

10b X = H

10c X = p-Br

The ^{17}O NMR spectra of ^{17}O-enriched α-azo hydroperoxides **10a** to **c** were reported[16] in benzene, acetonitrile, and methanol at 31°. Representative spectra in the three differing solvents for **10a** are shown in Figure 1. The ^{17}O NMR signals for **10a** to **c** were poorly resolved in benzene. In acetonitrile and methanol two broad (~1600 Hz), well-resolved signals were observed for **10a** to **c**. The ^{17}O NMR data for **10a** to **c** in acetonitrile are summarized in Table 1.

The observed solvent dependence of the ^{17}O NMR data for **10a** to **c** was thought to be indicative of hydrogen bonding effects. In benzene, α-azo hydroperoxides are intramolecularly hydrogen-bonded. Disruption of this internal hydrogen bond by acetonitrile or methanol could result in the observed chemical shift changes. The high reactivity of α-azo hydroperoxides in ionic electrophilic oxygen-atom transfer reactions has been ascribed[16] to a mechanism in which intramolecular proton transfer (hydrogen bonding) in the transition state is involved. Thus, the hydrogen-bonding differences observed by ^{17}O NMR spectroscopy should affect the kinetics of these ionic oxidations. A reasonable correlation between the ^{17}O NMR solvent dependence and solvent effects on kinetic data on oxygen-atom transfer reactions of **10a** has been suggested.[16a]

A recent ^{17}O NMR study by Salomon[17] has greatly increased the chemical shift data on dialkyl peroxides. ^{17}O NMR chemical shifts were reported for symmetrical dialkyl peroxides, ROOR (**3, 11a** to **c**); cyclic peroxides (**12** and **13**); two series of bicyclic peroxides (**14** to

p-MeO Ar-CH(O-O-H)-N=N-Ph

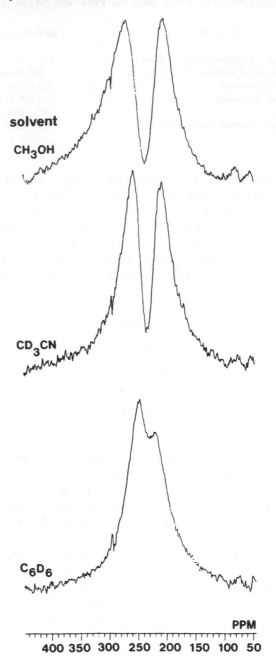

FIGURE 1.   ^{17}O NMR spectra for [17]O-enriched acyclic α-azo hydroperoxide (**10a**) in benzene-d₆, acetonitrile-d₆ and methanol.

16, 17 to **19**); several miscellaneous bicyclic peroxides (**20** to **22**); and one dioxetane (**23**). The ^{17}O NMR data of the dialkyl peroxides were found to be insensitive to changes in solvent polarity. In addition, little or no concentration dependence on the chemical shifts was noted. The [17]O and ^{13}C chemical shifts were correlated; the acyclic compounds showed a different, less-sensitive relationship than the bicyclic peroxides in relation to the ^{13}C data (see Chapter 3 for a discussion of conformational effects[4a]).

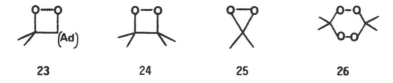

R-O-O-R

3 R=$\underline{t}$-Bu

11a R=Et   **12** n=3   **14** n=1   **17** n=2

11b R=$\underline{n}$-Pr   **13** n=4   **15** n=2   **18** n=3

11c R=$\underline{i}$-Pr   **16** n=3   **19** n=4

Interestingly, the ^{17}O NMR chemical shift[17] for dioxetane (**23**) at 275 ppm is remarkably similar to that of **3** and **12**. The ^{17}O NMR data for tetramethyl-1,2-dioxetane (**24**) in benzene has been found[18] at ~287 ppm (Figure 1). Curci has recently reported[19] the ^{17}O NMR chemical shift of dimethyldioxirane (**25**) at 302 ppm (113 Hz). The [17]O chemical shift for acetone diperoxide (**26**) was found at 263 ppm (350 Hz) similar to that of acyclic dialkyl peroxides. The ^{17}O NMR results were interpreted to be consistent with the cyclic structure of **25** rather than a "carbonyl oxide" structure. In a subsequent paper, Adam[20] reported the calculated value for **25** as 330 ppm in reasonable agreement with the observed value of 302. The dioxetane parent ring system was predicted to show an ^{17}O NMR signal at 268 ppm.

23   **24**   **25**   **26**

Gerothanassis and Momenteau have recently reported[21] the ^{17}O NMR spectrum of a single-face, hindered iron porphyrin-dioxygen complex **27** in solution (toluene). Two broad signals at 1755 and 2488 ppm were observed for the "FeO_2" linkage of this dioxygen-1-methylimidazole-iron (d^6) porphyrin complex. The data were interpreted to rule out sideways triangular bonding in favor of an end-on angular bonding arrangement. The data were explained in terms of bonding models in which the electrons of the FeO_2 moiety were totally paired. These observations clearly indicate that ^{17}O NMR spectroscopy is a valuable, promising method for the study of oxygen carriers in biological systems.

ACKNOWLEDGMENTS

Acknowledgment of partial support of our research is made to the Donors of the Petroleum Research Fund, administered by the American Chemical Society, to NSF (CHE-8506665), and to the Georgia State University Research Fund. A.L.B. was a fellow of the Camille and Henry Dreyfus Foundation, 1981 to 1986. Instrumentation used in our studies was supported, in part, by a NSF Equipment Grant (CHE 8409599).

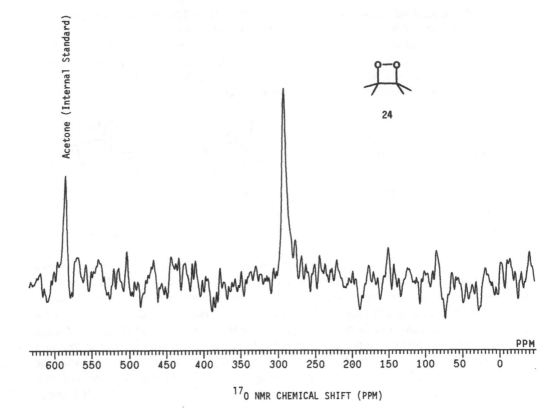

FIGURE 2.   ¹⁷O NMR spectrum (natural abundance) of tetramethyl-1,2-dioxetane (**24**) in benzene at 40° [acetone as internal standard].

REFERENCES

Section I

1. **Kintzinger, J.-P.**, Oxygen NMR characteristic parameters, in *NMR-17. Oxygen-17 and Silicon-29*, Diehl, P., Fluck, E., and Kosfeld, R., Eds., Springer-Verlag, New York, 1981, 1.
2. **Delseth, C. and Kintzinger, J. P.**, Resonance magnetique nucleaire de 17-O. Aldehydes et cetones aliphatiques: additivite des effects de substitution et correlation avec la ¹³C-RMN, *Helv. Chim. Acta,* 59, 466, 1976.
3. **Subramanian, T., Baumstark, A. L., and Boykin, D. W.**, Natural abundance ¹⁷O NMR study of α-substituted methyl acetates, *Spectrochim. Acta,* 45A, 335, 1989.
4. **Boykin, D. W.**, 17-O NMR investigation of alkyl isocyanates: electronic and steric effects, *J. Chem. Res.,* 1987, p. 338.
5. **Levy, G. C., Lichter, R. L., and Nelson, G. L.**, *Carbon-13 Nuclear Magnetic Resonance Spectroscopy,* 2nd ed., John Wiley & Sons, New York, 1980, 140.
6. **Nogaj, B. and Schroeder, G.**, Studies of intramolecular electronic effects in chloroacetophenones $C_6H_5COCH_3$-n-Cl and p-ClC$_6$H$_4$COCH$_{3n}$-Cl$_n$ by 35ClNQR, *Magn. Reson. Chem.,* 25, 565, 1987.
7. **Andersson, L.-O. and Mason, J.**, Oxygen-17 nuclear magnetic resonance. I. Oxygen-nitrogen grouping, *J. Chem. Soc. Dalton Trans.,* 202, 1974.
8. **Lipkowitz, K. B.**, A reassessment of nitrobenzene valence bond structures, *J. Am. Chem. Soc.,* 104, 2647, 1982.
9. **Fraser, R. R., Ragauskas, A. J., and Stothers, J. B.**, Nitrobenzene valence bond structures: evidence in support of 'through-resonance,' *J. Am. Chem. Soc.,* 104, 6475, 1982.

10. **Craik, D. J., Levy, G. C., and Brownlee, R. T. C.,** Substituent effects on 15-N and 17-O chemical shifts in nitrobenzenes: correlations with electron densities, *J. Org. Chem.*, 48, 1601, 1983.

11. **Boykin, D. W., Baumstark, A. L., Balakrishnan, P., Perjessy, A., and Hrnciar, P.,** 17-O NMR studies of 4- and 4'-substituted chalcones and p-substituted beta-nitrostyrenes, *Spectrochim. Acta,* 40A, 887, 1984.

12. **Balakrishnan, P. and Boykin, D. W.,** Relationship of aromatic nitro group torsion angles to 17-O chemical shifts, *J. Org. Chem.*, 50, 3661, 1985.

13. **Furin, G. G., Rezvukhin, A. I., Fedotov, M. A., and Yakobson, G. G.,** 15-N, 17-O, 31-P and 77-Se nuclear magnetic resonance spectra of polyfluoroaromatic compounds, *J. Fluorine Chem.*, 22, 231, 1983.

14. **Balakrishnan, P. and Boykin, D. W.,** Natural Abundance ^{17}O NMR Spectroscopy of Heteroaromatic Nitro Compounds, *J. Heterocycl. Chem.*, 23, 191, 1986.

15. **Lippma, E., Magi, M., Novikov, S. S., Lebedev, O. V., and Epishina, L. V.,** 13-C, 14-N, 15-N, and 17-O NMR Spectra of nitropyrroles and nitroimidazoles, *Org. Magn. Reson.*, 4, 153, 1972.

16. **Boykin, D. W. and Rhodes, R. A.,** 17-O NMR investigation of acetyl and formyl thiophenes, furans and pyrroles, *J. Heterocycl. Chem.*, 25, 643, 1988.

17. **Abramovitch, R. A. and Smith, E. M.,** in *Pyridine and its Derivatives, Supplement* Part II, John S. Wiley & Sons, New York, 1974.

18. **Katritzky, A. R. and Lagowski, J. M.,** *Chemistry of the Heterocyclic N-Oxides,* Academic Press, New York, 1971.

19. **Ochiai, E.,** *Aromatic Amine N-Oxides,* Elsevier, Amsterdam, 1967.

20. **Smith, D. M.,** *Comprehensive Organic Chemistry,* Vol. 4 (part 16.1), Heterocyclic *Compounds,* Sammes, P. G., Ed., Pergamon Press, Elmsford, New York, 1979.

21. **Katritzky, A. R. and Lagowski, J. M.,** N-Oxides and related compounds. XVIII. Proton nuclear magnetic resonance specta of 4-substituted pyridines and pyridine 1-oxides, *J. Chem. Soc.*, p. 43, 1961.

22. **Yavari, I. and Roberts, J. D.,** Nitrogen-15 nuclear magnetic resonance spectroscopy. Pyridine N-oxides and quinoline N-oxides, *Org. Magn. Reson.*, 12, 87, 1979.

23. **DiGiola, A. J., Furst, G. T., Psota, L., and Lichter, R. L.,** Natural abundance nitrogen-15 nuclear magnetic resonance spectroscopy. Nitrogen chemical shifts of alkyl pyridines, picolinuim and lutidinium ions and picoline N-oxides, *J. Phys. Chem.*, 82, 1644, 1978.

24. **Anet, F. A. L. and Yavari, I.,** Carbon-13 nuclear magnetic resonance study of pyridine N-oxide, *J. Org. Chem.*, 41, 3589, 1976.

25. **Herlocker, D. W., Drago, R. S., and Meek, V. I.,** A study of the donor properties of 4-substituted pyridine N-oxides, *Inorg. Chem.*, 5, 2009, 1966.

26. **Kubota, T. and Miyazaki, H.,** The effect of substituents on the half-wave potential of the polarographic reduction of pyridine N-oxide derivatives, *Bull. Chem. Soc. Jpn,* 39, 2057, 1966.

27. **Boykin, D. W., Baumstark, A. L., and Balakrishnan, P.,** 17-O NMR spectroscopy of 4-substituted pyridine N-oxides: substituent and solvent effects, *Magn. Reson. Chem.*, 23, 276, 1985.

28. **Boykin, D. W., Balakrishnan, P., and Baumstark, A. L.,** 17-O NMR spectroscopy of heterocycles. Steric effects for N-oxides, *Magn. Reson. Chem.*, 23, 695, 1985.

29. **Boykin, D. W., Balakrishnan, P., and Baumstark, A. L.,** Natural abundance 17-O NMR spectroscopy of heterocyclic N-oxides and di N-oxides. Structural effects, *J. Heterocycl. Chem.*, 22, 981, 1985.

30. **Sawada, M., Takai, Y., Kimura, S., and Misumi, S.,** A 17-O NMR study. SCSs of 4-substituted pyridine 1-oxides in DMSO: importance of dual enhanced resonance contributions with pi-donor and pi-acceptor substituents, *Tetrahedron Lett.,* 1986, p. 3013.

31. **Katoh, M., Sugawara, T., Kawada, Y., and Iwamura, H.,** 17-O nuclear magnetic resonance studies. V. 17-O shieldings of some substituted anisoles, *Bull. Chem. Soc. Jpn.*, 52, 3475, 1977.

32. **Karplus, M. and Pople, J. A.,** Theory of carbon NMR chemical shifts in conjugated molecules, *J. Chem. Phys.*, 38, 2803, 1963.

33. **Reuben, J.,** Hydrogen-bonding effects on oxygen-17 chemical shifts, *J. Am. Chem. Soc.*, 91, 5725, 1969.

34. **Balakrishnan, P., Baumstark, A. L., and Boykin, D. W.,** 17-O NMR spectroscopy: unusual substituent effects in *para*-substituted benzyl alcohols and acetates, *Tetrahedron Lett.,* p. 169, 1984.

35. **Chandrasekaran, S., Wilson, W. D., and Boykin, D. W.,** 17-O NMR studies on polycyclic quinones and related cyclic ketones: models for anthracycline intercalators, *Org. Magn. Reson,* 22, 757, 1984.

36. **Jaccard, G. and Lauterwein, J.,** Intramolecular hydrogen bonds of the C=O . . . H-O type as studied by oxygen-17 NMR, *Helv. Chim. Acta,* 69, 1469, 1987.

37. **Lapachev, V. V., Mainagashev, I. Y., Stekhova, S. A., and Fedotov, V. P.,** 17-O NMR studies of enol and phenol compounds with intramolecular hydrogen bonds, *J. Chem. Soc. Chem. Commun.,* p. 494, 1985.

38. **Lipczynska-Kochany, E. and Iwamura, H.,** Oxygen-17 nuclear magnetic resonance studies of the structures of benzohydroxamic acids and benzohydroxamate ions in solution, *J. Org. Chem.*, 47, 5277, 1982.

39. **Christ, H. A., Diehl, P., Schneider, H. R., and Dahn, H.,** Chemische Verschiebungen in der Kernmagnetischen Verbindunger, *Helv. Chim. Acta,* 44, 865, 1961.

40. **Chimichi, S., Nesi, R., and DeSio, F.,** 17-O nuclear magnetic resonance study of some five-membered heterocyclic derivatives, *Org. Magn. Reson.,* 22, 55, 1984.

41. **Kalahin, G. A., Kushnarev, D. T., Valeyen, R. B., Trofimov, B. A., and Fedotov, M. A.,** 17-O NMR investigation of p, pi-interactions in α,β-unsaturated and aromatic ethers, *Org. Magn. Reson.,* 18, 1, 1982.

42. **Stefaniak, L., Witanowski, and Roberts, J. D.,** Oxygen-17 and carbon-13 NMR study of *N*-methylsydone and related structures, *Spectrosc. Int. J.,* 2, 178, 1983.

43. **Thenmozhi, M., Sivasubramanian, S., Balakrishnan, P., and Boykin, D. W.,** Natural abundance 17-O NMR spectroscopy of (z)-(substituted benzylidene)phenylamine *N*-oxides (α,*N*-diaryl nitrones), *J. Chem. Res.,* p. 43, 1986.

Section II

1. **Kitzinger, J.-P.,** Oxygen-17 NMR, Application of ^{17}O NMR spectroscopy to structural problems, in *NMR of Newly Accessible Nuclei,* Vol. 2, Laszlo, P., Ed., Academic Press, New York, 1983.

2. **Klemperer, W. G.,** in *The Multinuclear Approach to NMR Spectroscopy,* Lambert, J. B. and Riddel, F. G., Eds., Reidel, Dordrecht, Holland, 1983, 245.

3. **Kintzinger, J.-P.,** Oxygen NMR characteristic parameters, in *NMR-17. Oxygen-17 and Silicon-29,* Diehl, P., Fluck, E., and Kosfeld, R., Eds., Springer-Verlag, New York, 1981.

4. (a) **Boykin, D. W., and Baumstark, A. L.,** Applications of ^{17}O NMR spectroscopy to structural problems in organic chemistry: torsion angle relationships, in *^{17}O NMR Spectroscopy in Organic Chemistry,* Boykin, D. W., Ed., CRC Press, Boca Raton, FL, 1990; (b) **Baumstark, A. L. and Boykin, D. W.,** Applications of ^{17}O NMR spectroscopy to structural problems in rigid, planar organic molecules, in *^{17}O NMR Spectroscopy in Organic Chemistry,* Boykin, D. W., Ed., CRC Press, Boca Raton, FL, 1990.

5. **Boykin, D. W. and Baumstark, A. L.,** 17-O NMR spectroscopy: assessment of steric perturbations of structure of organic compounds, *Tetrahedron,* 45, 3613, 1985.

6. **Woodard, R. W.,** ^{17}O NMR as a mechanistic probe to investigate chemical and bioorganic problems, in *^{17}O NMR Spectroscopy in Organic Chemistry,* Boykin, D. W., Ed., CRC Press, Boca Raton, FL, 1990.

7. **Christ, H. A., Diehl, P., Schneider, H. R., and Dahn, H.,** Chemische Verschiebungen in der Kernmagnetischen Verbindunger, *Helv. Chim. Acta,* 44, 865, 1961.

8. **Salomon, I. J., Keith, J. N., Kacmarck, A. J., and Raney, J. K.,** Additional studies concerning the existence of O_3F_2, *J. Am. Chem. Soc.,* 90, 5408, 1968.

9. **Lapidot, A. and Irving, C. S.,** Oxygen-17 nuclear magnetic resonance spectroscopy and iridium and rhodium molecular oxygen complexes, *J. Chem. Soc. Dalton Trans.,* p. 668, 1972.

10. **Lumpkin, D., Doxon, W. T., and Poser, T.,** Oxygen-17 nuclear quadrupole resonances in molecular oxygen reversibly bonded to iridium carrier, *Inorg. Chem.,* 18, 982, 1979.

11. **Sugawara, T., Kawada, Y., Katoh, M., and Iwamura, H.,** Oxygen-17 nuclear magnetic resonance. III. Oxygen atoms with a coordination number of two, *Bull. Chem. Soc. Jpn.,* 52, 3391, 1979.

12. **Postel, M., Brevard, C., Arzoumanian, H., and Riess, J. G.,** 17-O NMR as a tool for studying oxygenated transition-metal derivatives: first direct 17-O NMR observations of transition-metal-bonded peroxidic oxygen atoms. Evidence for the absence of oxo-peroxo oxygen exchange in molybdenum (VI) compounds, *J. Am. Chem. Soc.,* 105, 4922, 1983.

13. **Bregeault, J. M. and Mimour, H.,** Platinum *tert*-butyl peroxide trifluoroacetate: synthesis, characterization and oxidation of terminal olefins to methyl ketones, *N. J. Chim.,* 5, 287, 1981.

14. **Lee, H. C. and Oldfield, E.,** High-field 17-O-NMR spectroscopic observation of 17-O_2 in Vaska's compound and in $Pt(O_2)(PPh_3)_2$, *J. Magn. Reson.,* 69, 367, 1986.

15. **Curci, R., Frisco, G., Sciacovelli, and Troisi, L.,** 17-O NMR of organic peroxides: chromium (VI) and molybdenum (VI) oxide diperoxides as compared to simple organic peroxides, *J. Molec. Cat.,* 32, 251, 1985.

16. (a) **Baumstark, A. L., Vasquez, P. C., and Balakrishnan, P.,** 17-O-enriched alpha-azohydroperoxides: 17-O NMR spectroscopy, 17-O labeling reagents, *Tetrahedron Lett.,* p. 2051, 1985; (b) **Baumstark, A. L.,** Oxygen-atom transfer chemistry of alpha-azohydroperoxides, *J. Bioorg. Chem.,* 14, 326, 1986.

17. **Zagorski, M. G., Allan, D. S., Salomon, R. G., Clennan, E. L., Heah, P. C., and L'Esperance, R. D.,** Oxygen-17 nuclear magnetic resonance chemical shifts of dialkyl peroxides: large conformational effects, *J. Org. Chem.,* 50, 4485, 1985.

18. (a) **Baumstark, A. L. and McCloskey, C. J.,** unpublished results; (b) Baumstark, A. L. and Beeson, M., unpublished results,

19. **Cassidei, L., Fiorentino, M., Mello, R., Sciacovelli, O., and Curci, R.,** Oxygen-17 and carbon-13 identification of the dimethyldioxirane intermediate arising in the reaction of potassium caroate with acetone, *J. Org. chem.,* 52, 699, 1987.

20. **Adam. W., Chan, Y.-Y., Cremer, D., Gauss, J., Schautzow, D., and Schindler, M.,** Spectral and chemical properties of dimethyldioxirane as determined by experiment and *ab initio* calculations, *J. Org. Chem.*, 52, 2800, 1987.

21. **Gerothanassis, I. P., Momenteau, M.,** 17-O NMR spectroscopy as a tool for studying synthetic oxygen carriers related to biological systems: application to a synthetic single-face hindered iron porphyrin-dioxygen complex in solution, *J. Am. Chem. Soc.*, 109, 6944, 1987.

Chapter 10

OXYGEN-17 NUCLEAR MAGNETIC RESONANCE (NMR) SPECTROSCOPY OF ORGANOSULFUR AND ORGANOPHOSPHORUS COMPOUNDS

Slayton A. Evans, Jr.

TABLE OF CONTENTS

I. INTRODUCTION

Oxygen-17 nuclear magnetic resonance (NMR) spectroscopy is rapidly assuming a prominent role as a pivotal tool for accessing useful information in evaluating the importance of bonding and structural characteristics of organosulfur and organophosphorus compounds.[1,2] The introductory theory attending ^{17}O NMR has been addressed in Chapter 1 and will not be repeated here.

It is, however, important to mention that despite its spin, I of 5/2, its extremely low natural abundance of 0.037%, its quadrupole moment, Q, and its low sensitivity (2.91 × 10^{-2} times that for ^{1}H), NMR spectroscopy involving the [17]O nucleus is rapidly becoming an invaluable probe and an analytical tool for a host of novel applications. The expansion and refinement of ^{17}O NMR techniques in the broader scientific spectrum continue unabated, and future prospects for new and intriguing applications are encouraging. The current furor in ^{17}O NMR spectroscopy is linked, in part, to the development of new pulse sequences[3] and the influx of new NMR probe technology which addresses problems associated with the deleterious effects of acoustic ringing and pulse breakthrough.[4]

II. BACKGROUND

Oxygen-17 nuclear shieldings (σ_{tot}^{O}) in organosulfur and organophosphorus compounds are probably best understood from the results of early LCAO-MO treatments of NMR shifts[5] that stress the additive importance of the local diamagnetic, σ_{d}^{O}, and paramagnetic, σ_{p}^{O}, contributors (Equation 1).

$$\sigma_{tot}^{O} = \sigma_{d}^{O} + \sigma_{p}^{O} \tag{1}$$

The results of several studies indicate that ^{17}O NMR shifts of monocoordinated oxygens (i.e., N→O, S=O, C=O, P=O) are largely controlled by the paramagnetic component (Equation 2).[6]

$$\sigma_{p}^{O} = -e^{2}\hbar^{2}/2m^{2}c^{2} \, (\Delta E^{-1}) \, \langle r^{-3}\rangle_{2pO} \, \Sigma \, Q_{xo} \text{ (where X = S, P)} \tag{2}$$

ΔE is referred to as the "average energy" approximation and expressed as $\Sigma \, \Delta E^{-1}$ where ΔE is the excitation energy difference characterizing the electronic ground state and the various excited states of increasing energy. Generally, the first excited state (lowest energy) serves as a reasonable approximation of ΔE (i.e., $\Delta E_1 = \Delta E$) since it is unlikely that contributions from higher energy electronic states are significant. The expectation value of the inverse cube of the mean radius of the atomic 2p orbital of oxygen is $\langle r^{-3} \rangle_{2p}$. Effects which lead to expansion of the 2p orbital radius on oxygen tend to diminish the significance of the σ_p^0 component. Q_{x_0} describes the elements of charge density on oxygen and the extent of multiple bond contributions involving the sulfur or phosphorus heteroatoms. Thus, a reduced multiple bond contribution in the S=O or P=O fragment results in a dimunition in the σ_p^0 term as well. All of the variables in Equation 2 may not be mutually exclusive, and unfortunately, not enough information is available to utilize the extensive predictive potential of this equation. Nevertheless, current trends in ^{17}O NMR shifts and their correlation with other physical properties and refined quantum mechanical calculations provide the basis for useful discussions on oxygen-heteroatom bonding, complexation potentials, applications in structural and conformational analyses, etc.

This review focuses on ^{17}O NMR parameters of oxygens **monocoordinated** to sulfur and phosphorus. A number of useful ^{17}O NMR chemical shift trends have emerged and provided exciting research opportunities for direct analysis of oxygen heteroatoms involved in static and dynamic chemical processes.

III. THE SULFINYL (–S=O) GROUP

A. OXYGEN-17 NMR SHIFT RATIONALE FOR THE SULFINYL GROUP

The ^{17}O NMR chemical shifts of aliphatic, acyclic sulfinyl compounds occur between δ 25 and -25 ppm relative to external and naturally abundant $H_2{}^{17}O$. This is particularly striking when ^{17}O NMR shift comparisons are made between dialkyl ketones (δ 530 to 580 ppm)[7] and sulfoxides because of the relative similarities in the Allred "average" electronegativities of sulfur (2.58) and carbon (2.55).[8] If the difference in the range of ^{17}O NMR shifts for ketones and alkyl sulfoxides is controlled largely by the magnitude of the ΔE term (Equation 2), the observed low-energy UV transitions involving the nonbonding electrons on oxygen (e.g., n→π*) in both sulfoxides and ketones would predict that sulfoxides would be more shielded than the analogous ketones. Furthermore, assuming that the dipolar description of the sulfinyl linkage (i.e., S^+–O^-) is of major importance,[9] multiple bond contributions should be less than that in the carbonyl analog, and the increase in electron density at the sulfinyl oxygen should initiate an expansion of the oxygen 2p orbital radius, affording a subsequent reduction in the magnitude of the $\langle r^{-3} \rangle_{2p}$ term when compared to ketones. The cumulative result of these effects translates to a reduction in the paramagnetic shielding contribution in sulfoxides and thus an *upfield* ^{17}O shielding effect compared to the carbonyl analogs.

B. SUBSTITUENT-INDUCED CHEMICAL SHIFT (SCS) EFFECTS

In symmetrically substituted, acyclic sulfoxides, the ^{17}O nuclei experience upfield shifts when a hydrogen atom which is attached to the β carbon (i.e., Cβ) is replaced by methyl or methylene groups (i.e., γ-Me effect).

$$C\delta\text{–}C\gamma\text{–}C\beta\text{–}SO\text{–}C\beta\text{–}C\gamma\text{–}C\delta$$

The ^{17}O NMR chemical shifts for dimethyl sulfoxide (**1**), diethyl sulfoxide (**2**), and diisopropyl sulfoxide (**3**) reflect additive γ-Me effects of -18 ppm for sulfoxide **2** and -34 ppm for sulfoxide **3** (or γ-Me effect = -8 to -9 ppm/methyl). By contrast, the ^{17}O NMR shift for di-*tert*-butyl sulfoxide (**4**) is -2 ppm and only 14 ppm to higher field than the ^{17}O shift for dimethyl sulfoxide[10] (Table 1).

In this latter case, it is reasoned that the steric repulsive interactions between the large

TABLE 1
Oxygen-17 NMR Chemical Shifts of Sulfinyl Groups in Acyclic Organosulfur Compounds

Entry	Compound	$\delta^{17}O$ (ppm)	$W_{1/2}$ (Hz)	Solvent	Temp.,°C	Ref.
1	$CH_3S(O)CH_3$	12 ± 1	—	Neat	—	a
		19 ± 1	—	Neat	100	a
		13	120	Neat	—	b
		20	—	Neat	—	c
		15.7	—	Neat	25	d
	$CD_3S(O)CD_3$	14.9	—	Neat	25	d
2	EtS(O)Et	-6 ± 1	—	CHCl₃	25	a
3	i-PrS(O)-i-Pr	-20 ± 1	—	Neat	25	a
4	t-BuS(O)-t-Bu	-2 ± 1	—	CH_2Cl_2	25	a
5	n-PrS(O)n-Pr	-6 ± 1	—	CH_2Cl_2	25	a
6	n-BuS(O)-n-Bu	-7 ± 1	—	MeCN	25	a
7	$(CH_2=CHCH_2CH_2)_2SO$	-9 ± 1	—	MeCN	25	a
8	MeS(O)SMe	73 ± 2	—	Neat	Amb	e
		74 ± 2	—	CH_2Cl_2	Amb	e
9	EtS(O)SEt	64 ± 2	—	CH_2Cl_2	Amb	e
10	i-PrS(O)S-i-Pr	57 ± 2	—	CH_2Cl_2	Amb	e
11	MeS(O)NMe₂	78.8	—	Neat	—	f
12	MeOS(O)OMe	176	35	Neat	—	b
13	Cl-S(O)-Cl	291	45	Neat	—	b
		292 ± 2	—	Neat	27	g
14	MeS(O)Cl	188	—	Neat	Amb	e
15	EtS(O)Cl	180	—	Neat	Amb	e
16	MeS(O)OMe	135	—	Neat	Amb	e
17	*cis*-17	$2.1 \pm 2-3$	100—200	Toluene	100	h
18	*trans*-17	$2.4 \pm 2-3$	100—200	Toluene	100	h
19	*cis*-18	$5.1 \pm 2-3$	100—200	Toluene	100	h
20	*trans*-18	$-5.0 \pm 2-3$	100—200	Toluene	100	h

Note: a: Dyer, J. C., Harris, D. L., and Evans, S. A., Jr., *J. Org. Chem.*, 47, 3660, 1982; b:Christ, H. A., Deihl, P., Schneider, H. R., and Dahn, H., *Helv. Chim. Acta.*, 44, 865, 1961; c: Block, E., Bassi, A. A., Lambert, J. B., Wharry, S. M., Andersen, K. K., Dittmer, D. L., Patwardhan, B. H., and Smith, D. J. H., *J. Org. Chem.*, 45, 4807, 1980; d: Aime, S., Santucci, E., and Fruttero, R., *Magn. Reson. Chem.*, 24, 919, 1986; e: (1) Bass, S. W., Ph.D. dissertation, University of North Carolina, Chapel Hill, 1980; (2) Evans, S. A., Jr., *Magnetic Resonance. Introduction. Advanced Topics and Applications to Fossil Energy*, Petrakis, L. and Fraissard, J.-P., Eds., Advanced NATO Institute, Reidel, 1984, 757; f: Hakkinen, A.-M. and Ruostesuo, P., *Magn. Reson. Chem.*, 23, 424, 1985; g: Figgis, B. N., Kidd, R. G., and Nyholm, R. S., *Proc. R. Soc. London*, A269, 469, 1962; h: Manoharan, M. and Eliel, E. L., *Magn. Reson. Chem.*, 23, 225, 1985.

tert-butyl groups is relieved through expansion of the C-S-C bond angle, and as this angle increases from 90° to 120°, the π_{SO} orbital achieves enhanced stability.[11] This implies better overlap between the $2p_O - 3d_S$ orbitals and suggests an increase in both the Q_{SO} and $\langle r^{-3}\rangle_{2pO}$ terms which combine to enhance the deshielding of the ¹⁷O nucleus.

Methyl or methylene groups farther removed than the γ carbon (i.e., Cδ) in the acyclic sulfoxides have essentially no effect on the ¹⁷O NMR shifts of sulfinyl oxygens. This is clearly seen when ¹⁷O NMR shift comparisons are made between diethyl sulfoxide (δ -6 ppm), di-*n*-propyl sulfoxide (**5**; δ − 6), di-*n*-butyl sulfoxide (**6**; δ − 7), and di-3-butenyl sulfoxide (**7**; δ − 9). Within experimental error these ¹⁷O NMR shifts are effectively identical (Table 1).

Thiolsulfinates, RSS(O)R, are easily prepared by the oxidation of the corresponding disulfides, and they exhibit ¹⁷O NMR shifts approximately 60 to 70 ppm to lower field than their sulfoxide counterparts. For example, $CH_3S(O)SCH_3$ (δ 73), $CH_3CH_2S(O)SCH_2CH_3$ (δ

64 ppm), and $(CH_3)_2CHS(O)SCH(CH_3)_2$ (δ 57 ppm) exhibit ^{17}O NMR shifts which are also responsive to γ methyl shielding interactions.[12] In the corresponding sulfoxides, the γ-Me effect contributes -8 to -9 ppm to the sulfinyl oxygen ^{17}O chemical shift. Within the thiolsulfinates, the average γ methyl value is -8 ppm and it is clear that in these acyclic organosulfur compounds only the *single* γ methyl contributes significantly to the overall ^{17}O NMR shift. The δ methyl [i.e., $-S(O)SCH_{n-1}(CH_3)_n$] is apparently not positioned in a sterically congested array so as to transmit a noticeable δ compression effect. What factors control the ^{17}O NMR shifts in these systems are a matter of considerable speculation, but it seems certain that S-S multiple bonding, extended conjugative interactions between the sulfenyl sulfur and the sulfinyl group (S-SO), and polarization effects between sulfenyl sulfur and the sulfinyl oxygen may have unique importance.

C. SOLVENT EFFECTS

It is well documented that the position of the ^{17}O NMR resonance of carbonyl compounds is sensitive to solvent effects and especially protic solvents that exert a pronounced shielding effect through intermolecular hydrogen bonding. For example, Reuben[13] and DeJeu[14] have demonstrated that the ^{17}O NMR absorption of acetone is shielded by 52 to 57 ppm in aqueous media. The ^{17}O NMR shifts of sulfoxide **1** in several solvents have been reported,[10] and while its ^{17}O NMR absorption is less sensitive than the oxygen of acetone, it is shifted upfield by 10 to 11 ppm in protic media (i.e., MeOH and EtOH).

Interestingly, the ^{17}O NMR shift for *anhydrous*, neat sulfoxide **1** is δ 19 ppm; however, the ^{17}O NMR resonance is shifted to higher field (more shielded) in slightly polar solvent as well as protic solvents (e.g., $CHCl_3$ and H_2O, CH_3OH). It seems likely that *homogeneous* dimethyl sulfoxide should exist as a dimer[15a] or form "association polymers."[15b,c] This would suggest that the sulfinyl oxygen in neat sulfoxide **1** may actually resemble the divalent S-O-S fragment in the absence of solvent and more of the dipolar form (i.e., S^+-O^-) in polar media (Table 1).

It has also been reported[16] that substitution of deuterium for hydrogen in sulfoxide **1** results in a shielding effect on the sulfinyl ^{17}O nucleus, and this isotope effect is calculated as -0.13 ppm/deuterium [$CH_3S(O)CH_3$:δ 15.7 ppm; $CD_3S(O)CD_3$: δ 14.9 ppm]. This observation emphasizes the sensitivity of the sulfinyl ^{17}O NMR shift to an increase in the fractional mass of the γ atoms.

An interesting and quite revealing report [17] describes the solvent-dependent ^{17}O NMR shifts of N,N-dimethylmethanesulfinamide (**8**) and, to a lesser extent, N,N-dimethylmethanesulfonamide (**9**). For sulfinamide **8**, the ^{17}O NMR shifts range from δ 78.8 (neat) to 69.7 ppm as a 1:1 mixture of **8**: $(CF_3)_2CHOH$. From Figure 1, it is clear that the ^{17}O NMR shifts for **8** are roughly linear in relation to the solvent Kosower Z values. It was also reported that the solvent-induced changes in the ^{17}O NMR shifts for sulfonamide **9** are smaller than those for sulfinamide **8**. This observation is consistent with the lower hydrogen bond-forming potential of sulfonamides.[18]

$$\underset{\textbf{8}}{\overset{\displaystyle O \atop \displaystyle \|}{CH_3\text{-}S\text{-}NMe_2}} \qquad\qquad \underset{\textbf{9}}{\overset{\displaystyle O \atop \displaystyle \|}{\underset{\displaystyle O}{CH_3\text{-}S\text{-}NMe_2}}}$$

The Kamlet-Dickinson-Taft approach[19] was used to evaluate the effect of solvent-solute interactions on ^{17}O NMR shifts. This method employs (1) a polarity-polarizability parameter, π^*, (2) a hydrogen bond-donor parameter, α, and (3) a hydrogen bond-acceptor parameter, β as described in Equation 3.

$$\delta = \delta_o + s\,\pi^* + a\,\alpha + b\,\beta \tag{3}$$

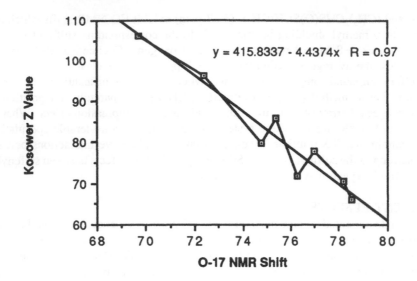

FIGURE 1..   The solvent-dependent ^{17}O NMR chemical shifts of sulfinamide **8**.[17]

For sulfinamide **8**, neglecting bβ since a proton donating group is absent in **8**, followed by implementation of a multiple linear regression analysis of the data affords Equation 4.

$$(^{17}O)\delta \ = \ 78.8 \ - \ 0.526 \ \pi^* \ - \ 4.35 \ \alpha \qquad (4)$$

The essence of Equation 4 nicely demonstrates that the ^{17}O NMR shift of sulfinamide **8** is controlled largely by the hydrogen bond-donating ability of the solvent, α.[17]

D. ACYCLIC AROMATIC SULFOXIDES

From Table 2, it is evident that the ^{17}O NMR shifts for diphenyl sulfoxide (**10**; δ 2), phenyl methyl sulfoxide (**11**; δ − 1), and *p*-methoxyphenyl methyl sulfoxide (**12**; δ − 1) vary only slightly suggesting that the shielding contribution of the phenyl group to the ^{17}O resonance must be controlled by nonbonding steric interactions rather than significant resonance contributions. The ^{17}O NMR shifts of several alkyl phenyl sulfoxides have been reported,[20] and their ^{17}O NMR shifts are essentially invariant in comparison with the long-chain alkyl homologues. For example, R-SO-C_6H_5 and R = Et, *n*-Pr, *n*-Bu, and *i*-Bu all exhibit ^{17}O NMR shifts between δ − 3 to − 5 ppm highlighting a trend which is consonant with the suggestion that beyond Cγ, in conformationally flexible substrates, the carbon chain has minimal impact on the ^{17}O NMR shifts.

i-Propylphenyl sulfoxide (**13**) exhibits a ^{17}O NMR resonance at δ − 10 ppm and *tert*-butylphenyl sulfoxide (**14**) occurs at δ − 4 ppm in keeping with expectations of the SCS effects arising from (1) an increase in the γ-Me steric interactions (γ-Me shieldings) in sulfoxide **13** and (2) C-S-C angle widening effect which deshields the ^{17}O NMR nucleus in sulfoxide **14**.

TABLE 2
Oxygen-17 NMR Chemical Shifts of Sulfinyl Groups in Acyclic Aryl Organosulfur Compounds

Entry	Compound	$\delta^{17}O$ (ppm)	$W_{1/2}$ (Hz)	Solvent	Temp.,°C	Ref.
1	Ph_2SO	2 ± 1	—	CH_2Cl_2	25	a
2	PhS(O)Me	-1	—	CH_2Cl_2	25	a
		7	210	$(CH_2Br)_2$	70	b
3	p-MeOC$_6$H$_4$S(O)Me	-1	—	CH_2Cl_2	25	a
4	PhS(O)Et	-4	320	$(CH_2Br)_2$	70	b
5	PhS(O)-n-Pr	-4	320	$(CH_2Br)_2$	70	b
6	PhS(O)-n-Bu	-5	270	$(CH_2Br)_2$	70	b
7	PhS(O)-i-Bu	-3	310	$(CH_2Br)_2$	70	b
8	PhS(O)-i-Pr	-10	210	$(CH_2Br)_2$	70	b
9	PhS(O)-t-Bu	-4	330	$(CH_2Br)_2$	70	b
10	cyclohexyl–S(=O)–Ph	-11	270	$(CH_2Br)_2$	70	b
11	(t-Bu)cyclohexyl–S(=O)–Ph	-4	170	$(CH_2Br)_2$	70	b
12	(t-Bu)cyclohexyl–S(=O)–Ph	-4	200	$(CH_2Br)_2$	70	b
13	Ph–S(=O)–CH(Me)Ph	3	150—200	$CDCl_3$	60	c
14	Ph–S(=O)–CH(Me)Ph	7	150—200	$CDCl_3$	60	c
15	Ph–S(=O)–CH(Et)Ph	3	150—200	—	142-3	c
16	Ph–S(=O)–CH(Et)Ph	19	150—200	—	90-3	c
17	PhS(O)NMe$_2$	65	—	Neat	—	d
18	PhS(O)Cl	221	150	Neat	—	e

Note: a: Dyer, J. C., Harris, D. L., and Evans, S. A., Jr., *J. Org. Chem.*, 47, 3660, 1982; b: Duddeck, H., Korek, V., Rosenbaum, D., and Drabowicz, J., *Magn. Reson. Chem.*, 24, 792, 1986; c: Kobayashi, K., Gugawara, T., and Iwamura, H., *J. Chem. Soc. Chem. Commun.*, p. 479, 1981; d: Hakkinen, A.-M. and Ruostesuo, P., *Magn. Reson. Chem.*, 23, 424, 1985; e: Christ, H. A., Diehl, P., Schneider, H. R., and Dahn, H., *Helv. Chim. Acta*, 44, 865, 1961.

13   **14**

Interestingly, both *cis-* and *trans-*1-phenylsulfinyl-4-*tert*-butylcyclohexane (**15**) exhibit their ^{17}O NMR shifts at δ − 4 ppm.[20] This is particularly surprising since the isosteric analog, sulfoxide (**13**; δ − 10 ppm) and phenylcyclohexyl sulfoxide

cis-15   **trans-15**   **16**

(**16**; δ − 11 ppm) have essentially identical ^{17}O NMR shifts. Furthermore, the conformational free energy ($− \Delta G°$) of the C_6H_5SO group in the cyclohexyl system is 1.9 kcal/mol favoring the equatorial conformation.[21] Barring any unusual distortional effects caused by the C-4-*tert*-butyl group in *trans-***15**, cyclohexyl sulfoxide **16** and *trans-***15** are expected to exhibit nearly identical ^{17}O NMR chemical shifts.

While the lack of a substantial ^{17}O NMR shift difference between *cis-* and *trans-***15** is interesting, the source of this phenomenon is not solely exceptional. It is noteworthy that *cis-* and *trans-*1-methylsulfinyl-4-*tert*-butylcyclohexanes (**17**) display a similar ^{17}O NMR shift trend affording nearly identical ^{17}O NMR shifts as well [e.g., *trans-***17** (equatorial sulfoxide), δ 2.4 ppm; *cis-***17** (axial sulfoxide), δ2.1 ppm].[22] It has been previously demonstrated[23] that sulfoxide *cis-***17** favors a C-S bond rotamer where the sulfinyl sulfur's sp^3 lone pair electrons are directed inside of the ring (nearer the 3,5-synaxial hydrogens) to minimize repulsive steric interactions between the axial $S(O)CH_3$ group and the 3,5-synaxial hydrogens. Presumably, for the same reason, the equatorial sulfinyl group prefers a C-S rotamer where the sulfur sp^3 lone pair is proximal to the 2,6-synaxial hydrogens of the ring. In this manner, the two sulfinyl oxygen atoms (equatorial and axial) experience similar steric and electronic environments which give rise to similar ^{17}O NMR shifts.[22]

cis-17   **trans-17**

While this rationale seems reasonable, it obviously does not apply to the ^{17}O NMR chemical shift difference between the *cis-* and *trans-*2-isopropyl-5-methylsulfinyl-1,3-dioxanes (**18**).[22] The ^{17}O NMR absorption for the sulfinyl oxygen in *cis-***18** occurs at δ 5.1 ppm and is slightly more deshielded compared to the sulfinyl oxygen in *cis-***17** (δ 2.1 ppm). It seems apparent that the electrostatic repulsion between the sulfinyl and dioxanyl ring oxygens may be responsible for this slight deshielding effect. However, the ^{17}O NMR shift comparison, $\Delta\delta(^{17}O)$ = 7.4 ppm, between *trans-***17** (δ 2.4 ppm) and *trans-***18** (δ − 5.0 ppm) strongly imply that the equatorial sulfinyl oxygen in *trans-***18** prefers the C-S rotamer which orients its sulfinyl oxygen in a position to take advantage of the electrostatic attraction caused by the polarized C-O bonds of the dioxanyl ring (cf. **19**). It seems reasonable that such an attraction would stimulate an expansion of the sulfinyl oxygen's 2p orbitals and cause preferential shielding of the ^{17}O nucleus in *trans*-18.

trans-18   **cis-18**

19

E. ACYCLIC DIASTEREOMERIC SULFOXIDES

In, perhaps, the first effort to differentiate diastereomeric [^{16}O, ^{17}O] sulfonyl oxygens in an acyclic sulfone by ^{17}O NMR spectroscopy,[24] chiral ($\pm$)-phenyl 1-phenylethyl sulfide (**20**) was oxidized with $C_6H_5ICl_2/H_2^{17}O$/pyridine to the corresponding diastereomeric sulfoxides **21** and **22** whose configurational assignments had been previously determined.[25] The individual ^{17}O NMR shifts for sulfoxides **21** and **22** were δ 3 and 7 ppm, respectively (Table 1). While the ^{17}O NMR shift difference ($\Delta\delta_{SO} = 4$ ppm) is small, it is significant and has implications regarding the importance of preferred rotamers based on SCS effects. Assuming that both sulfoxides prefer conformations where the two phenyl groups are oriented gauche to each other rather than antiperiplanar,[26] the high field ^{17}O NMR shift attending diastereomer **21** (δ 3 ppm) results from a γ-Me effect. In sulfoxide **21**, the sulfinyl oxygen is gauche to the methyl group in both of the preferred rotamers, whereas only one of the two predominant rotamers experience the γ-Me effect in sulfoxide **22**. Thus, the more significant γ-Me effect is in **21** resulting from a weighted average of the two "contributing" rotamers.[24]

20   **21**   **22**

F. SUBSTITUTED THIOLANE 1-OXIDES

Certainly, the possibility for facile conformational mobility (pseudorotation)[27] within the thiolane S-oxide ring makes half-chair and envelope conformations equally probable.[28] This facile, dynamic conformational activity implies that distinct γ-CH_3 or γ-CH_2 interactions involving sulfinyl oxygens may not be acute enough to provoke a large ^{17}O NMR shift response. This is clearly evident in the ^{17}O NMR shift comparison of *cis*- and *trans*-2-methylthiolane 1-oxides (**23**).[28] The ^{17}O NMR shift for *cis*-**23** compared to thiolane 1-oxide (δ 16 ppm) dramatizes the shielding contribution arising from the proximity of a γ-methyl group and a gauche sulfinyl oxygen (Table 3). By contrast, the ^{17}O NMR shift of *trans*-**23** (δ 17 ppm) signifies that the C-2 methyl is γ anti and distant from the sulfinyl oxygen. As expected, the C-3 or δ methyl effects are also minimal. Considering the conformational flexibility of the thiolane ring, it is not too surprising that the orientation of the methyl group in *cis*- and *trans*-3-methylthiolane 1-oxides or 3,3-dimethylthiolane 1-oxide have no effect on the ^{17}O NMR shifts of the sulfinyl group (δ 15 ppm).[28] It is, however, surprising that

TABLE 3
Oxygen-17 NMR Chemical Shifts of Sulfinyl Groups in Cyclic Organosulfur Compounds

Entry	Compound	δ¹⁷O (ppm)	Solvent	Temp.,°C	Ref.
1		−71	CDCl₃	—	a
		−70	MeC(O)Me	—	a
2		61	CDCl₃	—	a
		68	MeC(O)Me	—	a
		66	CH₂Cl₂	Amb	b
3		15 ± 1	CDCl₃	—	a
		12 ± 1	MeC(O)Me	—	a
		16	Neat	Amb	b
		11 ± 1	—	—	c
4		−7 ± 1	CHCl₃	—	c
5		17 ± 1	CHCl₃	—	c
6		15 ± 1	CHCl₃	—	c
7		15 ± 1	CHCl₃	—	c
8		15 ± 1	CHCl₃	—	c
9		15 ± 1	CHCl₃	—	c
10		9 ± 1	CHCl₃	—	c

TABLE 3 (continued)
Oxygen-17 NMR Chemical Shifts of Sulfinyl Groups in Cyclic Organosulfur Compounds

Entry	Compound	$\delta^{17}O$ (ppm)	Solvent	Temp.°C	Ref.
11		−3	CDCl$_3$	Amb	a
		−2	MeC(O)Me	Amb	a
		1	MeCN	Amb	b
		−4 ± 1	CHCl$_3$	35	d
12		5.6	CH$_2$Cl$_2$	Amb	b
13		−11.4	CH$_2$Cl$_2$	Amb	b
14		2 ± 1	CHCl$_3$	35	d
15		−35 ± 1	CHCl$_3$	35	d
16		−14 ± 1	CHCl$_3$	35	d
17		13 ± 1	CHCl$_3$	35	d
18		−20 ± 1	CHCl$_3$	35	d
19		5 ± 1	CHCl$_3$	35	d
20		14 ± 1	CHCl$_3$	35	d
21		7 ± 1	CHCl$_3$	35	d
22		11 ± 1	CHCl$_3$	35	d

TABLE 3 (continued)
Oxygen-17 NMR Chemical Shifts of Sulfinyl Groups in Cyclic Organosulfur Compounds

Entry	Compound	δ^{17}O (ppm)	Solvent	Temp.°C	Ref.
23		13	CD₃CN	70	e
24		31	CD₃CN	70	e
25		2 ± 1	CHCl₃	35	d
26		− 14 ± 1	CHCl₃	35	d

Note: a: Block, E., Bassi, A. A., Lambert, J. B., Wharry, S. M., Andersen, K. K., Dittmer, D. L., Patwardhan, B. H., and Smith, D. J. H., *J. Org. Chem.*, 45, 4807, 1980; b: Dyer, J. C., Harris, D. L., and Evans, S. A., Jr., *J. Org. Chem.*, 47, 3660, 1982; c: Barbarella, G., Rossini, S., Bongini, A., and Tugnoli, V., *Tetrahedron*, 41, 4691, 1985; d: Barbarella, G., Dembech, P., and Tugnoli, V., *Org. Magn. Reson.*, 22, 402, 1984; e: Quin, L. D., Szewczyk, J., Linehan, K., and Harris, D. L., *Magn. Reson. Chem.*, 25, 271, 1987.

2,2-dimethylthiolane 1-oxide (**24**; δ 15 ppm) with a γ gauche methyl has the same ^{17}O NMR chemical shift as the C-3 methyl derivatives.[28]

cis-23   **trans-23**   **24**

G. SUBSTITUTED THIANE 1-OXIDES

Evidence demonstrating the importance of γ gauche methyl or methylene shielding effects on sulfinyl oxygens in a six-membered ring is reflected in a comparison of the ^{17}O NMR shifts for the diastereomeric sulfoxides of *trans*-1-thiadecalin.[29] The ^{17}O NMR shift for **25β** (δ − 11.4 ppm) with the axial sulfinyl oxygen is substantially more shielded [Δδ (SO) = 17.0 ppm] than that for **25α** (δ 5.6 ppm) with the equatorial sulfinyl oxygen.[10]

25β   **25α**

2-,3-, and 4-Methyl substituted thiane 1-oxides give rise to *cis* and *trans* diastereomeric sulfoxides where the axial sulfinyl oxygens are shielded in excess of 20 ppm compared to the equatorial S-O's.[30] (Table 3). The shielding effect due to the presence of the γ-methyl group (i.e., at C-2) compared to the δ-methyl group is clearly evident in the ^{17}O NMR shift comparisons between *trans*-2-methylthiane 1-oxide (**trans-26**) vs. *cis*-3-methylthiane 1-oxide (**cis-27**) and *cis*-2-methylthiane 1-oxide (**cis-26**) vs. *trans*-3-methylthiane 1-oxide (**trans-27**).

trans-26   **cis-26**   **trans-27**   **cis-27**

Assuming that the diequatorial conformations are preferred in both **trans-26** and **cis-27**, it is obvious that the γ gauche Me (eq)/S=O (eq) array is shielding (Δδ = 11 ppm). The apparent influence of the γ Me in **cis-25** [Me(eq)/S=O(ax)] is dramatic and translates to a 21 ppm shielding of the sulfinyl oxygen compared to the conformationally homogeneous sulfinyl group in **trans-27**. The ^{17}O NMR shielding difference (Δδ = 37 ppm) between *cis*-2-methylthiane 1-oxide (**26**) and its diastereomer, *trans*-2-methylthiane 1-oxide (**26**) might result from three γ CH_2/CH_3 interactions in **cis-26** involving the axial sulfinyl oxygen compared to the single γ-CH_3 interaction in **trans-26**.[30]

The apparent influence of the single γ gauche Me/SO interaction in **cis-26** is, in fact, quite dramatic exerting a γ Me/S=O of approximately 21 ppm when compared to **trans-27**! The following suggestion might be useful in rationalizing this phenomenon. The ''energy minimized'' axial sulfinyl oxygen in diastereomer **cis-26** is expected to minimize the interactions between the 1,3-synaxial hydrogens and protrude away from the interior of the thianyl ring. This movement forces the axial S=O towards the γ equatorial C-2 methyl and serves to enhance the magnitude of the net shielding influence.

Finally, it is noteworthy that the conformational chair interchange between the axial and equatorial sulfinyl groups of thiane 1-oxide afford a time-averaged ^{17}O NMR shift of δ −4 ppm. Comparison of this shift with those from the axial (δ −14) and equatorial (δ 7 ppm) sulfinyl oxygens of the *cis*-3,5-dimethylthianyl substructure predicts an approximate 1:1 ratio of axial to equatorial conformer, which is in accord with the value obtained independently from ^1H NMR studies at −75°C in $CDCl_3$ solvent.[30]

H. SUBSTITUTED 2-OXO-1,3,2-DIOXATHIANES

Mattinen et al.[31] have shown that selected methyl substituted 2-oxo-1,3,2-dioxathianes can (1) dictate a strong preference for chair conformations with an axial sulfinyl group, (2) secure the equatorial array for the S=O group particularly for C-4,6 disubstitution, and (3) tolerate a mixture of S=O (ax) and S=O (eq) *chair conformers* through rapid conformational interchange. Thus, using methyl substitution to enforce conformational homogeneity, the ^{17}O NMR chemical shifts of axial and equatorial S=O groups in sulfites can be examined[31] (Table 4).

Several interesting features have emerged from this current study.[31] First, from the ^{17}O NMR shifts for sulfites **28** to **31**, it is clear that the axial sulfinyl oxygens are consistently more *deshielded* than the equatorial ones (Cf. **28** vs. **29**).[32] Secondly, assuming that the chair conformation is preserved, the equatorial S=O group in **32** responds to the perturbation of the axial methyl group at C-4. This seems clear from a comparison of the ^{17}O NMR shifts for **29** (δ 175.9 ppm) and **32** (δ 183.2 ppm). Finally, the ^{17}O NMR shift comparison between **28** (δ 182.3 ppm) and **33** (δ 196.4 ppm) shows the large deshielding δ-Me effect of approximately 14 ppm for the 1,3-synaxial SO and CH_3 groups.

TABLE 4
Oxygen-17 NMR Chemical Shifts of Sulfinyl Groups in 1,3,2-Dioxathiolane, Thiane, and Thiepane 2-Oxides

Entry	Compound	δ^{17}O (ppm)	W$_{1/2}$ (Hz)	Solvent	Temp.,°C	Ref.
1		213.2	≤270	CDCl$_3$	33	a
2		224.4;226.2	≤270	CDCl$_3$	33	a
3		217.8;218.7	≤270	CDCl$_3$	33	a
4	cis, trans	214.3 (cis,trans)	≤270	CDCl$_3$	33	a
5	trans, trans	217.8 (trans,trans)	≤270	CDCl$_3$	33	a
6		222.1	≤270	CDCl$_3$	33	a
7		179.8 180.6	≤270 —	CDCl$_3$ CDCl$_3$	33 35	a,b c
8		180.2 180.8	≤270 —	CDCl$_3$ CDCl$_3$	33 35	a c
9		180.0 181.8	≤270 —	CDCl$_3$ CDCl$_3$	33 35	a,b c
10		181.1 176.9	≤270 — .	CDCl$_3$ CDCl$_3$	33 35	a c
11		179.8 180.3	≤270 —	CDCl$_3$ CDCl$_3$	33 35	a c

TABLE 4 (continued)
Oxygen-17 NMR Chemical Shifts of Sulfinyl Groups in 1,3,2-Dioxathiolane, Thiane, and Thiepane 2-Oxides

Entry	Compound	$\delta^{17}O$ (ppm)	$W_{1/2}$ (Hz)	Solvent	Temp.,°C	Ref.
12		200.9	≤270	CDCl₃	33	a
		194.1	—	CDCl₃	35	c
13		174.1	≤270	CDCl₃	33	a,b
		177.9	—	CDCl₃	35	c
14		181.8	≤270	CDCl₃	33	a,b
15		179.0	≤270	CDCl₃	33	a
16		210.5	≤270	CDCl₃	33	a,b
		182.3	—	CDCl₃	35	c
17		179.4	≤270	CDCl₃	33	a,b
		195.1	—	CDCl₃	35	c
18		176.8	≤270	CDCl₃	33	a
		175.9	—	CDCl₃	35	c
19		195.7	≤270	CDCl₃	33	a,b
		196.4	—	CDCl₃	35	c
20		180.0	≤270	CDCl₃	33	a,b
		183.2	—	CDCl₃	35	c
21		196.1	≤270	CDCl₃	33	a

TABLE 4 (continued)
Oxygen-17 NMR Chemical Shifts of Sulfinyl Groups in 1,3,2-Dioxathiolane, Thiane,
and Thiepane 2-Oxides

Entry	Compound	δ^{17}O (ppm)	W$_{1/2}$ (Hz)	Solvent	Temp.,°C	Ref.
22		207.8	≤270	CDCl$_3$	33	a
23		192.3	≤270	CDCl$_3$	33	a
24		179.8	≤270	CDCl$_3$	33	a
25		177.1	≤270	CDCl$_3$	33	a
26		180.5	≤270	CDCl$_3$	33	a
27		197.8	≤270	CDCl$_3$	33	a
28		183.6	≤270	CDCl$_3$	33	a
29		178.2	≤270	CDCl$_3$	33	a
30		180.0	≤270	CDCl$_3$	33	a

TABLE 4 (continued)
Oxygen-17 NMR Chemical Shifts of Sulfinyl Groups in 1,3,2-Dioxathiolane, Thiane, and Thiepane 2-Oxides

Entry	Compound	δ¹⁷O (ppm)	W₁/₂ (Hz)	Solvent	Temp.,°C	Ref.
31		182.0	—	CDCl₃	35	c
32		179.9	—	CDCl₃	35	c
33		181.1	≤270	CDCl₃	33	a
		181.8	≤300	CDCl₃	30	d
34		180.0	≤300	CDCl₃	30	a
35		179.4	≤300	CDCl₃	30	d
36		184.3	≤300	CDCl₃	30	d
37		181.7	≤300	CDCl₃	30	d
38		181.2	≤300	CDCl₃	30	d
39		182.5	≤300	CDCl₃	30	d
40		194.0	≤300	CDCl₃	30	d

TABLE 4 (continued)
Oxygen-17 NMR Chemical Shifts of Sulfinyl Groups in 1,3,2-Dioxathiolane, Thiane, and Thiepane 2-Oxides

Entry	Compound	$\delta^{17}O$ (ppm)	$W_{1/2}$ (Hz)	Solvent	Temp.,°C	Ref.
41	(trans,trans)	176.2 (*trans,trans*)	≤300	CDCl$_3$	30	d
42	(cis,trans)	180.2 (*cis,trans*)	≤300	CDCl$_3$	30	d
43	(cis,cis)	179.8 (*cis,cis*)	≤300	CDCl$_3$	30	d
44	(trans,trans)	179.2 (*trans,trans*)	≤300	CDCl$_3$	30	d
45	(cis,trans)	181.0 (*cis,trans*)	≤300	CDCl$_3$	30	d
46	(cis,cis)	183.9 (*cis,cis*)	≤300	CDCl$_3$	30	d
47		180.4	≤300	CDCl$_3$	30	d

TABLE 4 (continued)

TABLE 4 (continued)
Oxygen-17 NMR Chemical Shifts of Sulfinyl Groups in 1,3,2-Dioxathiolane, Thiane, and Thiepane 2-Oxides

Entry	Compound	$\delta^{17}O$ (ppm)	$W_{1/2}$ (Hz)	Solvent	Temp.,°C	Ref.
48	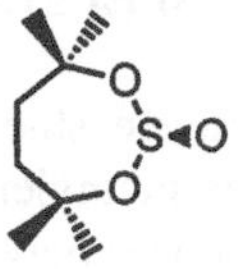	195.1	≤300	CDCl₃	30	d

Note: a: Hellier, D. G., and Liddy, H. G., *Magn. Reson. Chem.*, 26, 671, 1988; b: Hellier, D. G., *Magn. Reson. Chem.*, 24, 163, 1986; c: Mattinen, J. and Pihlaja, K., *Magn. Reson. Chem.*, 25, 569, 1987; d: Hellier, D. G. and Liddy, H. G., *Magn. Reson. Chem.*, 27, 431, 1989.

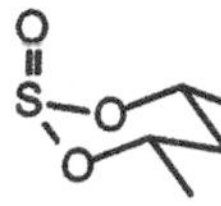

28 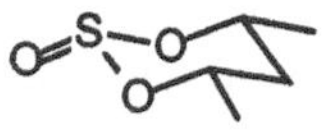 29 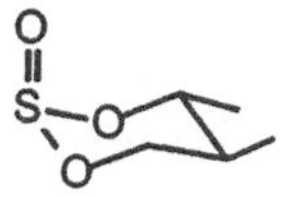 30

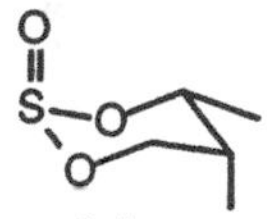

31 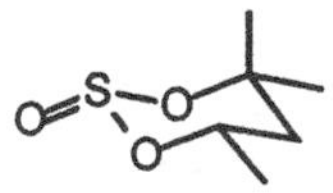 32 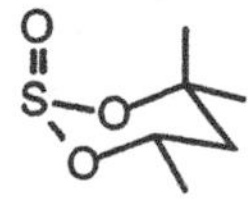 33

I. SUBSITUTED 1,3,2,-DIOXATHIEPANE 2-OXIDES

The ^{17}O NMR shifts (approximately δ 180 ppm) for a series of substituted 1,3,2-dioxathiepane 2-oxides (cf., **34**) have been reported,[33] and their ^{17}O NMR shifts are comparable to the average value reported for six-membered ring sulfites. However, it is notable that substantial methyl substitution on the ring provokes a sizeable deshielding effect on the sulfinyl oxygen, apparently arising from a δ-Me compression effect (Table 4).

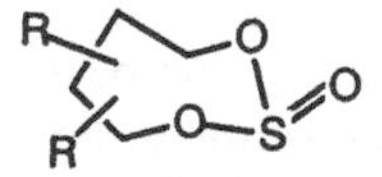

34

IV. THE SULFONYL (–SO₂–) GROUP

A. OXYGEN-17 NMR SHIFT RATIONALE FOR THE SULFONYL GROUP

The ^{17}O NMR shifts of acyclic sulfones appear between δ 120 and 190 ppm relative to external, naturally abundant $H_2^{17}O$.[10] The rather large downfield shifts exhibited by the aliphatic, acyclic sulfones (compared to the sulfoxides) appear consistent with the expectations from other physical data which indicate that the sulfonyl S=O bonds possess more multiple bond character than the sulfinyl bond.[34] The increase in the double bond character of the S=O bonds in sulfones undoubtedly arises from the influence of the "contracted" 3d orbitals of the sulfonyl sulfur which encourages a more efficient overlap with the smaller oxygen 2p orbitals.[35]

Accordingly, the lower field shifts of the –SO₂– oxygens are in agreement with expectations based on the apparent increase in the contributions of the S_{3d}–O_{2p} π bond order (i.e., greater Q_{SO} contribution) and $\langle r^{-3} \rangle_{2pO}$ terms caused by the inductive effect of the sulfonyl sulfur. Possible deshielding effects arising from geminal oxygen-oxygen lone pair interactions[36] may also be important. Dialkyl sulfones are transparent in the near UV[37] and one might reasonably conclude that the range of sulfonyl ^{17}O NMR shifts is not critically associated with contributions from ΔE when compared to the ^{17}O NMR shifts of the sulfoxides.

B. SUBSTITUENT-INDUCED CHEMICAL SHIFT (SCS) EFFECTS

1. Alkyl Substituents

In acyclic sulfones, the sulfonyl oxygen nuclei experience shielding effects when a hydrogen atom attached to $C\beta$ is replaced by a methyl or methylene group (i.e., γ-Me effect).[10] Comparisons of the ^{17}O NMR chemical shift differences between dimethyl sulfone (**35**), diethyl sulfone (**36**), and diisopropyl sulfone (**37**) reveal γ-Me effects of -24 ppm for sulfone **36** and -44 ppm for sulfone **37** [γ-Me/Me $= -11 - (-12)$ ppm].

$$C\delta - C\gamma - C\beta - SO_2 - C\beta - C\gamma - C\delta$$

This γ-Me effect at the sulfonyl^{17}O nuclei is qualitatively analogous to that described for a similar γ shift effect involving the sulfinyl group.[10]

35 **36** **37** **38** **39**

However, the ^{17}O NMR shift of di-*tert*-butyl sulfone (**38**; δ 122) compared with that of sulfone **35** (δ 164 ppm) translates into a γ-Me effect/Me of *only* -7 ppm (Table 5).

The γ-Me ^{17}O NMR shift effects in the acyclic sulfones are slightly larger than those calculated for the sulfoxides, which may result from an increase in γMe/SO interactions caused by the steric congestion about the two sulfonyl oxygens. The diminution in the γ-Me effect with increased C-S-C bond angles has already been addressed in connection with a series of aliphatic sulfoxides.[11] Presumably, a similar alteration in the stability of the sulfonyl πSO orbitals impact the Q_{SO} and $\langle r^{-3} \rangle_{2pO}$ contributions from Equation 2 and translate to enhanced deshielding.

As probably expected, alkyl homologation beyond the $C\gamma$ position has essentially no effect on the ^{17}O NMR shifts of acyclic sulfones. For example, diethyl sulfone, di-*n*-propyl sulfone, and di-*n*-butylsulfone exhibit ^{17}O NMR shifts in the relatively narrow range of δ 140-148 ppm, implying that a methylene or methyl group farther removed than $C\gamma$ in the acyclic systems have either no influence or exhibit a slight deshielding effect (i.e., δ-effect) on the ^{17}O chemical shifts of the sulfonyl oxygens.[10]

The ^{17}O NMR chemical shifts of diethyl sulfone and divinyl sulfone (**39**) are virtually identical (δ 140 and 141 ppm, respectively). The obvious conclusion is that if there is a substantial 2p-3dπ interaction between the C=C and SO_2 groups in divinyl sulfone **39**, the ^{17}O NMR shift is not sensitive enough probe for its detection.[10]

2. Thiolsulfonates (RSO_2SR)

The ^{17}O NMR shifts (Table 5) for simple thiolsulfonates are 35 to 60 ppm to lower field than the corresponding dialkyl sulfones (*vide supra*), and it is clear that the sulfenyl (–S–) sulfur plays a unique role in controlling the ^{17}O NMR shifts of the sulfonyl moiety (cf. the ^{17}O NMR shifts for the thiolsulfinates).[12] The trend in the ^{17}O NMR SCS effects mirror that observed for the thiolsulfinates. For example, $CH_3SO_2SCH_3$ (δ 199), $CH_3CH_2SO_2SCH_2CH_3$ (δ 190), and $(CH_3)_2CHSO_2SCH(CH_3)_2$ (δ 183) afford an average γ methyl shielding effect of -8 ppm.[12]

TABLE 5
Oxygen-17 NMR Chemical Shifts of Sulfonyl Groups in Organosulfur Compounds

Entry	Compound	$\delta^{17}O$ (ppm)	$W_{1/2}$ (Hz)	Solvent	Temp., °C	Ref.
1	SO_2	513 ± 5	—	Neat	27	a
2	SO_3	188 ± 1	—	Stabilized liq.	27	a
3	$MeSO_2Me$	163	—	$CHCl_3$	Amb	b
		164	—	$CDCl_3$	Amb	c
		165	—	$MeC(O)Me$	Amb	c
4	$EtSO_2Et$	140	—	CH_2Cl_2	Amb	b
5	$i\text{-}Pr\text{-}SO_2\text{-}i\text{-}Pr$	120	—	Neat	Amb	b
6	$t\text{-}Bu\text{-}SO_2\text{-}t\text{-}Bu$	122	—	CH_2Cl_2	Amb	b
7	$n\text{-}Pr\text{-}SO_2\text{-}n\text{-}Pr$	147	—	MeCN	Amb	b
8	$n\text{-}Bu\text{-}SO_2\text{-}n\text{-}Bu$	148	—	MeCN	Amb	b
		145	—	$CDCl_3$, $MeC(O)Me$	Amb	c
9	$(CH_2{=}CH)_2SO_2$	141	—	Neat	Amb	b
10	$(ClCH_2)_2SO_2$	150 ± 1	—	$CDCl_3$	Amb	c
11	$MeSO_2SMe$	199	—	Neat	Amb	d
12	$EtSO_2SEt$	190	—	Neat	Amb	d
13	$i\text{-}Pr\text{-}SO_2\text{-}S\text{-}i\text{-}Pr$	183	—	Neat	Amb	d
14	$PhSO_2CH_3$	157	100	$(CH_2Br)_2$	70	e
15	$PhSO_2Et$	145	100	$(CH_2Br)_2$	70	e
16	$PhSO_2\text{-}n\text{-}Pr$	145	—	$CHCl_3$	Amb	b
		148	200	$(CH_2Br)_2$	70	e
17	$PhSO_2\text{-}n\text{-}Bu$	147	190	$(CH_2Br)_2$	70	e
18	$PhSO_2\text{-}i\text{-}Bu$	151	180	$(CH_2Br)_2$	70	e
19	$PhSO_2\text{-}i\text{-}Pr$	136	160	$(CH_2Br)_2$	70	e
20	$PhSO_2\text{-}t\text{-}Bu$	131	130	$(CH_2Br)_2$	70	e
21	$PhSO_2\text{-}n\text{-}Hex$	145	—	$CHCl_3$	Amb	b
22	Ph_2SO_2	138	—	$CHCl_3$	Amb	b
		137	—	$MeC(O)Me$	Amb	c
		139	—	$CDCl_3$	Amb	c
		140	137	$CHCl_3$	28-30	f
23	$(p\text{-}MeC_6H_4)_2SO_2$	141	159	$CHCl_3$	28-30	f
24	$(p\text{-}MeC(O)OC_6H_4)_2SO_2$	142	190	$CHCl_3$	28-30	f
25	$(p\text{-}BrC_6H_4)_2SO_2$	142	188	$CHCl_3$	28-30	f
26	$p\text{-}MeC_6H_4SO_2Ph$	141	152	$CHCl_3$	28-30	f
27	$PhSO_2CH_2Ph$	152	193	$CHCl_3$	28-30	f
28	$(PhCH_2)_2SO_2$	152	164	$CHCl_3$	28-30	f
29	$PhCH_2SO_2\text{-}n\text{-}Bu$	149	152	$CHCl_3$	28-30	f
		147	190	$(CH_2Br)_2$	70	e
30	$PhSO_2CH_2SO_2Ph$	158	227	$CHCl_3$	28-30	f
31	$PhSO_2CH_2CH_2SO_2Ph$	147	181	$CHCl_3$	28-30	f

32		137 ± 1	150—200	C_6H_6	Amb	g
		134 ± 1	150—200	$CHCl_3$	Amb	g
		132 ± 1	150—200	MeOH	Amb	g
		127 ± 1	150—200	CF_3COOH	Amb	g

● $= {}^{17}O$

33		141 ± 1	150—200	C_6H_6	Amb	g
		139 ± 1	150—200	$CHCl_3$	Amb	g
		137 ± 1	150—200	MeOH	Amb	g
		133 ± 1	150—200	CF_3COOH	Amb	g

TABLE 5 (continued)
Oxygen-17 NMR Chemical Shifts of Sulfonyl Groups in Organosulfur Compounds

Entry	Compound	δ¹⁷O (ppm)	W₁/₂ (Hz)	Solvent	Temp., °C	Ref.
34		136 ± 1	150—200	C_6H_6	Amb	g
		134 ± 1	150—200	$CHCl_3$	Amb	g
		132 ± 1	150—200	MeOH	Amb	g
		124 ± 1	150—200	CF_3COOH	Amb	g
35		142 ± 1	150—200	C_6H_6	Amb	g
		140 ± 1	150—200	$CHCl_3$	Amb	g
		139 ± 1	150—200	MeOH	Amb	g
		134 ± 1	150—200	CF_3COOH	Amb	g
36	$PhSO_2CH(Me)Et$	141	100—200	Toluene	100	h
37	$PhSO_2CH(Ph)CH_2Ph$	140,145	100—200	Toluene	100	h
38	$PhSO_2CH(Me)Ph$	140,143	100—200	Toluene	100	h
39		149	100—200	$CHCl_3$	Amb	f
40		163	100—200	$CHCl_3$	Amb	f
41	1:1 Mixture	159	184	$CHCl_3$	Amb	f
42		138	186	$CHCl_3$	Amb	f
		145	339	$CHCl_3$	Amb	f
43	$PhSO_2SPh$	152	103	$CHCl_3$	Amb	f

Note: a: (1) Wasylishen, R. E., MacDonald, J. B., and Friedrich, J. O., *Can. J. Chem.*, 62, 1181, 1984; (2)Figgis, B. N., Kidd, R. G., and Nyholm, R. S., *Proc. R. Soc. London*, A269, 469, 1962; b: Dyer, J. C., Harris, D. L., Evans, S. A., Jr., *J. Org. Chem.*, 47, 3660, 1982; c: Block, E., Bassi, A. A., Lambert, J. B., Wharry, S. M., Andersen, K. K., Dittmer, D. L., Patwardhan, B. H., and Smith, D. J. H., *J. Org. Chem.*, 45, 4807, 1980; d: Evans, S. A., Jr., *Magnetic Resonance Introduction. Advanced Topics and Applications to Fossil Energy*, Petrakis, L. and Fraissard, J. P., Eds., Advanced NATO Institute, Reidel, 1984, 757; e: Duddeck, H., Korek, V., Rosenbaum, D., and Drabowicz, J., *Magn. Reson. Chem.*, 24, 792, 1986; f: Kelly, J. W. and Evans, S. A., Jr., *Magn. Reson. Chem.*, 25, 305, 1987; g: Kobayashi, K., Sugawara, T., and Iwamura, H., *J. Chem. Soc. Chem. Commun.*, p.479, 1981; h: Manoharan, M. and Eliel, E. L., *Magn. Reson. Chem.*, 23, 225, 1985.

3. Aromatic Sulfones

Substitution of a single *n*-alkyl group of a di-*n*-alkyl sulfone with a phenyl group results in shielding of the sulfonyl oxygens but does not necessarily reflect e⁻ density contributions via aryl $2p \rightarrow 2p - 3d\pi$ interactions between the C_6H_5 and SO_2 groups.[38] For example, comparison of the ^{17}O NMR shifts of di-*n*-propyl sulfone (δ 147) and *n*-propyl phenyl sulfone (**40**; δ 145 ppm) argues against a substantial conjugation involving the phenyl $2p\pi$ electrons and the orbitals of the S=O bond in the sulfonyl group.[38]

This conclusion may initially seem at odds with the ^{17}O NMR shift for the SO_2 group in diphenyl sulfone (**41**; δ 140 ppm), which is shielded compared to the oxygens in *n*-propyl phenyl sulfone (**40**). Kelly and Evans[38] suggested that an important source of shielding in the diphenyl sulfones arises from γ-steric interactions between the sulfonyl oxygens and the *ortho*-CH hydrogens of the phenyl ring (i.e., γ-ortho-CH effect) (cf. **42**). In support of this view, mildly electron-withdrawing or π electron-donating *p*-substituents do not influence the ^{17}O NMR shifts of the sulfonyl oxygens within experimental error.

An orbital interactive view coupled with conformational and structural input favorable for electron delocalization involving the sulfonyl group seems instructive. The results of early MO calculations[39] suggest that the nearly planar structure for thioxanthene 10,10-dioxide (**43**) should represent a favorable orientation where the sulfonyl group is in the "orbital interactive conformation" **A**. This conformation provides for convenient S 3d orbital (viz. $3d_{xy}$) and oxygen $2p_x$ orbital overlap with the carbon $2p\pi$ orbitals of the aromatic ring. Thus, conformation **A** encourages π electron delocalization over the contiguous carbon, sulfur, and oxygen atoms, appropriately visualized as dipolar structure **I**.

Coplanarity of the two phenyl rings in diphenyl sulfone **41** (cf. **B**) is severely discouraged because of the repulsive transannular steric interactions between the ortho positions. This means that sulfone **41** should exhibit characteristics of the "orbital interactive conformations" **A** and **C**. In conformer **C**, only overlap between either C and S (by $3d_{x2-y2}$) or S and O (by $3d_{xy}$ or $3d_{xz}$) orbitals is possible. Thus, sulfones having contributions from interactive conformer **C** will exhibit π conjugative interactions involving carbon and sulfur with the negative charge delocalized on sulfur (cf. **III**). The net effect is that diphenyl sulfone should be less likely to experience electron delocalization from the aryl ring to the sulfonyl oxygens compared with the S,S-dioxide **43**. Hence, charge density arguments do not support the high-field ^{17}O NMR shift observed for diphenyl sulfone (δ 140) compared with the S,S-dioxide **43** (δ 149 ppm). The most reasonable conclusion is that π conjugative interactions leading to excess charge on oxygen is minimal in acyclic diaryl sulfones. Thus, the mainstay of the ^{17}O NMR shielding contribution to the sulfonyl oxygens from an aryl group probably arises from the γ ortho CH/SO₂ steric interaction (i.e., γ-ortho CH effect).[38]

If one, then two methylene groups (CH₂) separate the phenyl ring from the sulfonyl group, the sulfonyl oxygens are no longer positioned to experience the shielding arising from the γ ortho CH effect and they are systematically deshielded [cf. diphenyl sulfone (δ 140 ppm), benzyl phenyl sulfone (**44**; δ 145 ppm), and dibenzyl sulfone (**45**; δ 152 ppm)]. These observations stress the absence of the γ steric shielding contribution of the *o*-CH fragment in sulfones **44** and **45**.

A   B   C

I   II   III

44   45

Interestingly, the presence of the benzyl group influences the sulfonyl ^{17}O NMR chemical shift in a manner symbolic of a long-chain alkyl group. This point is highlighted by comparisons of the ^{17}O NMR shifts of benzyl *n*-butyl sulfone (δ 149 ppm) and di-*n*-butyl sulfone (δ 148 ppm).

4. The Effects of Heteroatoms and Functional Groups on the ^{17}O NMR Shift of the Sulfonyl Group

Electron-withdrawing functional groups and heteroatoms which are attached to the sulfonyl group have a pronounced effect on the sulfonyl ^{17}O NMR chemical shifts (Table 6). For example, Barbarella et al.[40] have demonstrated that ^{17}O NMR shifts for organosulfur compounds possessing the X–SO$_2$–Y fragment experience a deshielding effect as the electronegativity of both ligands (i.e., X, Y) increase, although a linear relationship between ^{17}O NMR shifts and the sum of the group electronegativities ($\chi_x + \chi_y$) could not be established. Nevertheless, some insightful trends emerge when the ^{17}O NMR shift data for the –SO$_2$– groups in (1) MeSO$_2$Y, (2) MeOSO$_2$Y, and (3) ClSO$_2$Y are compared with χ_Y.

For MeSO$_2$Y, an increase in χ_y is accompanied by a deshielding of the sulfonyl ^{17}O nuclei, apparently caused by the σ-inductive electron withdrawal potential of the Y substituent. By contrast, the sulfonyl nuclei in organosulfur compounds, MeOSO$_2$Y and ClSO$_2$Y, are consistently shielded with increasing electronegativity of Y. While the reason for this latter phenomenon is still unclear, the ^{17}O NMR shifts for the ClSO$_2$Y compounds are generally deshielded by about 60 ppm relative to the MeOSO$_2$Y series despite the similarities in group electronegativities for Cl and MeO.[41] Barbarella et al.[40] rationalize this latter trend by suggesting that an interaction between the oxygen's 2p lone pairs and the low-lying σ^*_{S-Cl} antibonding orbital decreases the charge density on oxygen and translates to a net deshielding effect.

Other examples demonstrating the effect of electron-withdrawing groups on the ^{17}O NMR shifts of proximal sulfonyl oxygens have been reported. For example, the chemical shift comparison between thioxanthene 10,10-dioxide (**43**; δ 149) and thioxanthone 10,10-dioxide (**46**; δ 163) is a clear example of this deshielding effect. Both of these molecules prefer the shallow boat conformation with S,S-dioxide **46** favoring the more planar conformation.[42] It is clear that the carbonyl group deshields the sulfonyl oxygens, presumably through an inductive effect. Finally, comparison of the ^{17}O NMR shifts between

TABLE 6
Oxygen-17 NMR Chemical Shifts of Sulfonyl Groups Adjacent to Electronegative Heteroatoms

Entry	Compound	$\delta^{17}O$ (ppm)	$W_{1/2}$ (Hz)	Solvent	Temp., °C	Ref.
1	Cl-SO$_2$-Cl	296	100—300	CHCl$_3$	30	a
		304	—	Neat	27	b
		298	45	Neat	—	c
2	F-SO$_2$-F	148	—	Neat	−100	d
3	Me-SO$_2$-F	186	100—300	CHCl$_3$	30	a
4	Me-SO$_2$-Cl	238	100—300	CHCl$_3$	30	a,e
5	MeOSO$_2$-F	148	100—300	CHCl$_3$	30	a
6	MeSO$_2$OMe	170	100—300	CHCl$_3$	30	a
		169	—	CH$_2$Cl$_2$	Amb	e
7	MeOSO$_2$OMe	140	100—300	CHCl$_3$	30	a
		150	70	—	—	c
8	MeOSO$_2$Cl	219	100—300	CHCl$_3$	30	a
9	Me$_2$NSO$_2$Cl	208	100—300	CHCl$_3$	30	a
10	MeSO$_2$NMe$_2$	156	—	CHCl$_3$	—	f
		157	—	DMSO	—	f
		160	—	CH$_2$Cl$_2$	Amb	e
11	CF$_3$SO$_2$Cl	211	100—300	CHCl$_3$	30	a
12	EtSO$_2$Cl	228	100—300	CHCl$_3$	30	a
13	EtOSO$_2$Cl	220	100—300	CHCl$_3$	30	a
14	CF$_3$SO$_2$OMe	147	100—300	CHCl$_3$	30	a
15	Et$_2$NSO$_2$OMe	143	100—300	CHCl$_3$	30	a
16	EtSO$_2$OMe	159	100—300	CHCl$_3$	30	a
17	PhSO$_2$NH$_2$	159	—	Acetone	—	f
		158	—	MeCN	—	e
18	PhSO$_2$NMe$_2$	139	—	CHCl$_3$	—	f
		140	—	Acetone	—	f
19	p-MeC$_6$H$_4$SO$_2$NH$_2$	160	135	CHCl$_3$	—	g
20	FSO$_2$OSO$_2$F	167	—	Neat	25	d
21	FSO$_2$OOF	152	—	Neat	−10	d
22	PhSO$_2$Cl	226	—	MeCN	—	e
23	p-BrC$_6$H$_4$SO$_2$Cl	223	—	MeCN	—	e
24	p-MeC$_6$H$_4$SO$_2$Cl	224	—	MeCN	—	e
25	p-NO$_2$C$_6$H$_4$SO$_2$Cl	220	—	MeCN	—	e
26	p-MeC$_6$H$_4$SO$_2$OMe	154	—	CH$_2$Cl$_2$	—	e

Note: a: Barbarella, G., Chatgilialoglu, C., Rossini, S., and Tugnoli, V., *J. Magn. Reson.*, 70, 204, 1986; b: Figgis, B. N., Kidd, R. G., and Nyholm, R. S., *Proc. R. Soc. London Ser. A.*, 269, 469, 1962; c: Christ, H. A., Diehl, P., Schneider, H. R., and Dahn, H., *Helv. Chim. Acta.*, 44, 865, 1961; d: Solomon, I. G., Kacmarek, A. J., and Raney, J., *Inorg. Chem.*, 7, 1221, 1968; e: Bass, S. W., Ph.D. dissertation, University of North Carolina, Chapel Hill, 1980; f: Hakkinen, A.-M. and Ruostesuo, P., *Magn. Reson. Chem.*, 23, 424, 1985; g: Kelly, J. W. and Evans, S. A., Jr., *Magn. Reson. Chem.*, 25, 305, 1987.

bis(benzenesulfonyl)methane (**47**; δ 158 ppm) and bis(benzenesulfonyl)ethane (**48**; δ 147 ppm) indicates that the close proximity of the two sulfonyl groups causes substantial deshielding as well.

46   **47**   **48**

The ¹⁷O NMR chemical shifts of sulfonamides [RSO$_2$NR'$_2$] also occur in the same general region as alkyl and aryl sulfones.[43] (Table 6) SCS effects are also evident and predictable within the sulfonamides. For example, comparison of the ¹⁷O NMR shifts for benzenesulfonamide (**49**; δ 159 ppm) and *N,N*-dimethylbenzenesulfonamide (**50**; δ 139 ppm) appears indicative of a substantial γ steric shift effect resulting from interactions between the *N*-methyl groups and the sulfonyl oxygens. The existence of the γ-ortho-CH effect (cf. **51**) seems apparent when comparing the ¹⁷O NMR shifts for *N,N*-dimethylmethanesulfon-amide (**9**; δ 156 ppm) and sulfonamide **50** (δ 139 ppm)[38,43]

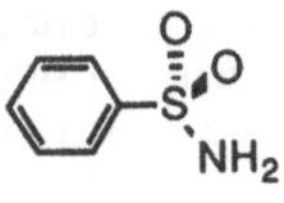

49 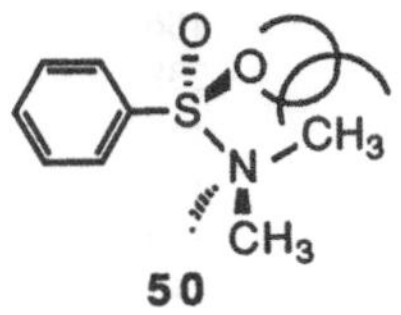   **50**

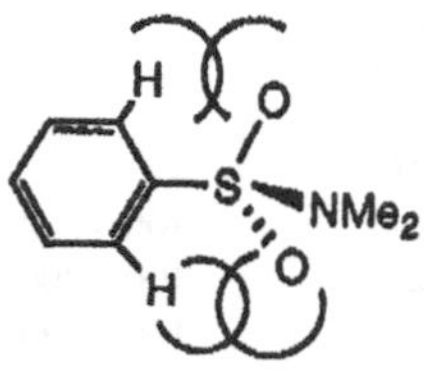

51

C. THE ¹⁷O NMR SHIFTS OF THE SULFONYL GROUP IN CONFORMATION-ALLY RIGID MOLECULES. THE CYCLOHEXYL AND 1,3-DIOXANYL RING SYSTEMS

The ¹⁷O NMR spectra for *cis*- and *trans*-4-*tert*-butylcyclohexylthiomethyl 1,1-dioxides (**52**)[22] exhibit unique ¹⁷O NMR shift differences (Table 7). The low field shift of ***cis*-52** (δ 162.8 ppm) compared to ***trans*-52** (δ 153.0) is rationalized by the suggestion that one of the sulfonyl oxygens points inside the ring and experiences a δ-compression effect from C-3,5 methylene hydrogens resulting in a downfield shift. The ¹⁷O NMR shift trend is similar for *cis*- and *trans*-1-phenylsulfonyl-4-*tert*-butylcyclohexanes (***cis*-53**:δ 149 and ***trans*-53**:δ 138 ppm).[20]

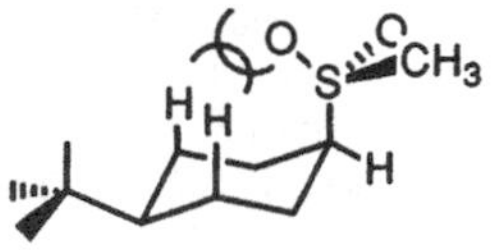

cis-52 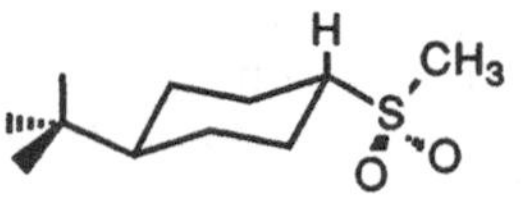   **trans-52**

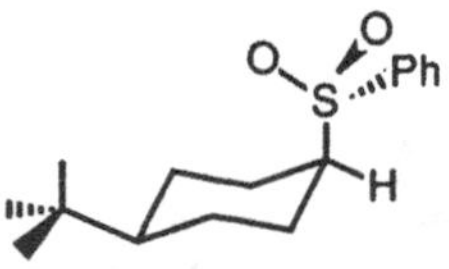

cis-53 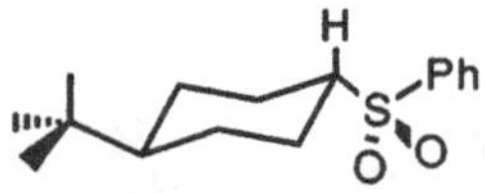   **trans-53**

TABLE 7
Oxygen-17 NMR Chemical Shifts of Sulfonyl Substituents in Cyclic Organosulfur Compounds

Entry	Compound	$\delta^{17}O$ (ppm)	$W_{1/2}$ (Hz)	Solvent	Temp., °C	Ref.
1		138	200	$(CH_2Br)_2$	70	a
2		138	200	$(CH_2Br)_2$	70	a
3		149	330	$(CH_2Br)_2$	70	a
4		153.0	100—200	Toluene	100	b
5		162.8	100—200	Toluene	100	b
6		161.5	100—200	Toluene	100	b
7		159.8	100—200	Toluene	100	b
8		138.0	100—200	Toluene	100	c
9		139.5	100—200	Toluene	100	c

Note: a: Duddeck, H., Korek, V., Rosenbaum, and Drabowicz, J., *Magn. Reson. Chem.*, 24, 792, 1986; b: Manoharan, M. and Eliel, E. L., *Magn. Reson. Chem.*, 23, 225, 1985; c: Juaristi, E., Martinez, R., Mendez, R., Toscano, R. A., Soriano-Garcia, M., Eliel, E. L., Petsom, A., and Glass, R. S., *J. Org. Chem.*, 52, 3806, 1987.

However, this trend is not repeated in comparing the ^{17}O NMR shifts for conformational homogeneous *cis-* and *trans-2-iso-*propyl(5-methylsulfonyl)-1,3-dioxanes (54).[22] Here, $\Delta\delta$ SO_2 (*cis-trans*) = − 1.7 ppm indicating that the sulfonyl oxygens of *cis-54* are more shielded than those in *trans-54*. The fact that the ^{17}O NMR shift for *cis-54* is δ159.8 ppm and only slightly more shielded than *trans-54* (δ 161.5 ppm) seems entirely reasonable, since the methyl group of MeSO₂- in *cis*-**54** resides "inside" the cavity of the dioxanyl ring with the sulfonyl oxygens removed from the repulsive electrostatic interactions of the 1,3-dioxanyl oxygens. In *cis*-**54**, the long range (W) coupling constant, $^4J = 1.14 \pm 0.02$ Hz, between the sulfonyl methyl and the C-5 hydrogen of the dioxanyl ring strongly supports this view.[44] In this orientation, the sulfonyl oxygens in *cis*-**54** are γ-gauche to the C-4,6 methylenes. In much the same manner, the equatorial −SO₂Me group in *trans*-**54** can also experience similar γ-gauche interactions with the C-4,6 methylenes which may account for the similarities in their ¹⁷O NMR shifts.

cis-54   **trans-54**

(^4J = 1.14 Hz)   (^4J = 0.39 Hz)

Both the cis and trans diastereomers of 2-*t*-butyl(5-*t*-butylsulfonyl)-1,3-dioxane (**55**) exhibit interesting ¹⁷O NMR shifts for the -SO₂- group.[45] For example, the ¹⁷O NMR resonance for *cis*-**55** with the axial *tert*-butylsulfonyl group occurs at δ 139.5 ppm while the ¹⁷O resonance for *trans*-**55** is slightly more shielded, δ 138.0 ppm. While there is no adequate explanation for this small ¹⁷O NMR shift difference [$\Delta\delta_{cis-trans} = 1.5$ ppm], it is, nevertheless, quite surprising considering (1) the anticipated rotameric preference about the -C-SO₂-*t*-Bu bond and (2) the importance of electrostatic repulsions coupled with the anticipated deshielding effect on the "participating" oxygens.

cis-55   **trans-55**

For *cis*-**55**, the X-ray diffraction data is enlightening but adds a bit of mystery to factors most likely to control the ¹⁷O NMR shifts of the sulfonyl group. The results from the X-ray study indicate that the sulfonyl *tert*-butyl group is removed from the interior of the ring (i.e., away from the dioxanyl oxygens).[45] The average torsional angle O-S-C-C of 8.25 ± 2.35° reflecting a nearly eclipsed array for these bonds is also surprising. Presumably, this preferred rotamer stems from the "driving force" attending the relaxation of the severe steric interactions between the CH₃'s and the dioxanyl oxygens despite the "anticipated" destablizing effect arising from the repulsive electrostatic interactions between the proximal electronegative oxygens.

Finally, the −SO₂− ¹⁷O NMR shifts in *cis*- and *trans*-**54** are approximately 20 ppm more deshielded than the analogous 5-*tert*-butylsulfonyl dioxanes, *cis*- and *trans*-**55**. This shielding effect in the diastereomeric dioxanes **55** arises from the increased number of γ-gauche interactions between the sulfonyl oxygens and the *tert*-butyl methyls.

The ¹⁷O NMR shift differences are useful in determining conformational free energies (−ΔG°'s) in systems undergoing dynamic conformational interchange. The conformational free energy (−ΔG°) for the −SO₂Ph group in the cyclohexyl system is 2.5 kcal/mol[21] favoring the equatorial conformation. The corresponding ¹⁷O NMR shift for phenyl cyclohexyl sulfone (δ 138 ppm) when compared to those of *cis*- and *trans*-4-*tert*-butyl-1-phenylsulfonylcyclohexanes (^{17}O NMR shifts at δ 149 and 138 ppm, respectively),[20] is in accord with the exclusive preference of the $-SO_2Ph$ group for the equatorial conformation (Table 7).

D. DIASTEREOTOPIC SULFONYL OXYGENS IN ACYCLIC SULFONES

A novel, unequivocal ^{17}O NMR differentiation of diastereotopic sulfonyl oxygens was performed by Kobayashi et al.[24] Using ^{17}O NMR, coupled with specific ^{17}O labeling techniques, the diastereotopicity of 1-phenylethyl sulfone **56** was observed and [^{16}O, ^{17}O] configurational assignments were established (Table 5). At 10.782 MHz, the ^{17}O NMR spectrum of sulfone **56** (= **57**) showed a broadened singlet which was devoid of sufficient resolution to allow for a reasonable determination of the ^{17}O NMR shift difference.

However, oxidation of phenyl 1-phenylethyl sulfide (**20**) with iodobenzene dichloride and $H_2{}^{17}O$ (10 atom%) in pyridine[46] affords the diastereomeric phenyl 1-phenylethyl[^{17}O] sulfoxides, **21** and **22**, of known configuration.[25]

Subsequent *m*-chloroperoxybenzoic acid (*m*CPBA) oxidation of sulfoxides **21** and **22** affords two diastereomeric [^{16}O, ^{17}O] sulfones, **56** and **57**, with individual shifts at δ 137 and 141 ppm (C_6H_6), respectively. Since diastereomers **56** and **57** differ only in the location of ^{16}O and ^{17}O attached to sulfonyl sulfur, their rotameric populations are identical and the ^{17}O NMR chemical shift difference (Δδ = 4 ppm) is a measure of the $-SO_2-$ ^{17}O NMR shift anisochronism. It is also noteworthy that the increasing upfield ^{17}O NMR shifts of sulfones **56** and **57** with increasing solvent polarity are in accord with the expectations attending intermolecular hydrogen bonding by protic solvents (*vide supra*). For sulfones **56** and **57**: δ 134 and 139 ppm ($CHCl_3$); δ 132 and 137 ppm (MeOH); δ 127 and 133 ppm (CF_3COOH), respectively.

Application of a similar oxidation protocol to phenyl 1-phenylpropyl sulfide using $H_2{}^{17}O/$ $C_6H_5ICl_2$ followed by the appropriate ^{17}O NMR assignments for the diastereomeric phenyl 1-phenylpropyl sulfoxides, and subsequent *m*CPBA oxidation afforded the diastereomeric [^{16}O, ^{17}O] sulfones **58** and **59**[24] (Table 5). As expected, their individual and solvent-induced ^{17}O NMR shifts were similar to those observed for the diastereomeric phenyl 1-phenylethyl [^{16}O, ^{17}O] sulfones. Thus, for **58** and **59**: δ 136 and 142 ppm (C_6H_6); δ 134 and 140 ppm ($CHCl_3$); δ 132 and 139 ppm (MeOH); and δ 124 and 134 ppm (CF_3COOH), respectively.

$$\underset{\textbf{58}}{\text{O}\overset{^{17}\text{O}}{\underset{\text{Ph}}{\underset{|}{\overset{||}{S}}}}\text{---}\underset{\text{Ph}}{\overset{H}{\underset{|}{C}}}\text{---CH}_2\text{CH}_3} \quad + \quad \underset{\textbf{59}}{{}^{17}\text{O}\overset{\text{O}}{\underset{\text{Ph}}{\underset{|}{\overset{||}{S}}}}\text{---}\underset{\text{Ph}}{\overset{H}{\underset{|}{C}}}\text{---CH}_2\text{CH}_3}$$

The magnitude of the ¹⁷O NMR sulfonyl anisochronism in similar acyclic sulfonyl oxygens has been assessed employing higher NMR field strengths.[22] For example, phenyl 1-phenylethyl sulfone (**56** = **57**) exhibits ¹⁷O NMR shifts at δ 140.1 and 143.3 ppm. Larger substituents attached to the α carbon enhance the magnitude of the anisochronous −SO₂− oxygens slightly. For example, phenyl 1,2-diphenylethyl sulfone (**60**) exhibits ¹⁷O NMR shifts at δ 140.3 and 145.4 ppm. Groups of similar steric size (i.e., methyl vs. ethyl) exert a minimum conformational effect on the rotameric preferences and, thus, the impact communicated to the sulfonyl ¹⁷O shift nonequivalence is minor. For example, in phenyl 2-*sec*-butyl sulfone (**61**) the diastereomeric sulfonyl oxygens are isochronous (δ 141 ppm) (Table 5).

56   **60**   **61**

E. DIASTEREOTOPIC SULFONYL GROUPS IN CYCLIC SULFONES
1. Four-Membered Ring Sulfonyl Compounds

Block et al.[47] described the effect of ring size on the ¹⁷O NMR shifts of 3- to 8-membered ring sulfones and highlighted thietane 1,1-dioxide (**62**) as possessing the most deshielded sulfonyl oxygens (δ 182 ppm) resulting from the "four-membered ring effect". Incidentally, the sulfonyl oxygens in thiirane S,S-dioxide (**63**) appeared at a considerably higher field, δ 111 ppm. Several other four-membered ring organosulfur compounds exhibited sizeable ¹⁷O chemical shift nonequivalence of the sulfonyl group (Table 8). For example, ¹⁷O NMR shifts of δ 210 and 243 ppm in CHCl₃ for thiolsulfonate **64** demonstrate the influence of an α sulfur atom combined with a specific γ steric shift effect between one of the sulfonyl oxygens and an α-Et group. In 3-thiophenylthietane 1,1-dioxide (**65**), the sulfonyl oxygens are diastereotopic with ¹⁷O NMR shifts at δ 182 and 198 ppm (acetone). Comparison of these shifts with those in 3-hydroxythietane 1,1-dioxide (**66**) (e.g., δ 184 and 187 ppm; acetone) imply that the more deshielded sulfonyl oxygens in both of these compounds may be due to nonbonding electrostatic interactions (O/S in sulfone **65** and O/O in sulfone **66**) between the proximal heteroatoms. The absence of ¹⁷O NMR shift nonequivalence in 1,3-dithiane 1,1-dioxide (**67**) and 3,3-dimethylthietane 1,1-dioxide (**68**) point to rapid conformer interconversion on the ¹⁷O NMR time scale.

62   **63**   **64**   **65**   **66**   **67**   **68**

2. Substituted Thiolane 1,1-Dioxides

In some ring systems, the position of a stereogenic center as well as constraints on the conformational flexibility of the ring system have an impact on the magnitude of the sulfonyl oxygen ^{17}O shift nonequivalence.[28] For example, 2-methylthiolane 1,1-dioxide (**69**) exhibits two distinctive ^{17}O NMR shifts at δ 150 and 158 ppm while 3-methylthiolane 1,1-dioxide (**70**) shows sulfonyl oxygens which are isochronous (δ 168 ppm). In fact, the similarities in ^{17}O NMR shifts for thiolane 1,1-dioxide (δ 164 ppm), thiolane 1,1-dioxide **70** (δ 168 ppm), 3,3-dimethylthiolane 1,1-dioxide (**71**; δ 171 ppm), and *trans*-2-thiahydrindan-2,2-dioxide (**72**; δ 169 ppm) point to a fairly flexible thiolane ring (i.e., half-chair→envelope) such that steric interactions between the sulfonyl oxygens and the ring substituents can be minimized if not entirely avoided (Table 8) and ultimately having essentially no effect on the ^{17}O NMR shifts of the sulfonyl oxygens.

The ^{17}O NMR spectrum of 3,4-epoxythiolane 1,1-dioxide (**73**) requires special comment.[48] The ^{17}O NMR shift of the *ethereal* oxygen is deshielded (δ 15.9 ppm) when compared with the ^{17}O NMR shift for cyclopentene oxide (δ −8 ppm).[49] This 24 ppm deshielding effect may arise from a combination of inductive and field effects between the sulfonyl group and the epoxide oxygen. The sulfonyl oxygens are distinctive with ^{17}O NMR shifts at δ 179.3 and 188.3 ppm. It seems likely that the sulfonyl oxygen with the δ 188.3 ppm resonance is proximal to the ethereal oxygen, and its shift results from O/O electrostatic repulsions which probably translates into a contraction of the 2p orbitals on each oxygen and ultimately a deshielding effect.[48]

An ^{17}O NMR shift comparison between thiolane 1,1-dioxide (δ 168.5 ppm) and 2,5-dihydrothiophene 1,1-dioxide (**74**; δ 169.9 ppm) shows that the presence of a β,γ-olefinic group in a thiolanyl S,S-dioxide ring makes no meaningful contribution to the sulfonyl ^{17}O NMR shift. However, 4,5-dihydrothiophene 1,1-dioxide (**75**) exhibits an ^{17}O NMR shift at δ 176.8 ppm and points to an unusual deshielding[48a] (but unexplained!)[48b] contribution to the sulfonyl ^{17}O NMR shift.

3-Hydroxy- and 3-alkoxythiolane 1,1-dioxides [R = OH, OMe, OEt, O-*n*-Bu, and O-*i*-Pr] have intrinsic diastereotopic sulfonyl oxygens and exhibit ^{17}O NMR differences 1.3 to 2.1 ppm[48] (Table 8). Interestingly, the level of ^{17}O NMR anisochronism within the sulfonyl oxygens can be enhanced through competitive binding with the paramagnetic ion, Eu^{3+}.

TABLE 8
Oxygen-17 NMR Chemical Shifts of Sulfonyl Oxygens in Cyclic Organosulfur Compounds

Entry	Compound	$\delta^{17}O$ (ppm)	$W_{1/2}$ (Hz)	Solvent	Temp., °C	Ref.
1		111	—	CDCl$_3$	—	a
2		183	—	CH$_2$Cl$_2$	—	b
		182	—	CHCl$_3$	—	a
		176	—	D$_2$O	—	a
3		164	—	MeCN	—	b
		165	—	CHCl$_3$	—	a
		164	—	MeC(O)Me	—	a
		168	28	Toluene	80	c
4		150	—	CHCl$_3$	—	a
		154	—	MeC(O)Me	—	a
5		134	—	CHCl$_3$	—	a
		137	—	MeC(O)Me	—	a
6		181	—	CDCl$_3$	—	a
		180	—	MeC(O)Me	—	a
7		210,243,	—	CHCl$_3$	—	a
8		182,198	—	MeC(O)Me	—	a
9		184,187	—	MeC(O)Me	—	a
10		150,158	—	CHCl$_3$	—	d
11		168	—	CHCl$_3$	—	d

TABLE 8 (continued)
Oxygen-17 NMR Chemical Shifts of Sulfonyl Oxygens in Cyclic Organosulfur Compounds

Entry	Compound	$\delta^{17}O$ (ppm)	$W_{1/2}$ (Hz)	Solvent	Temp., °C	Ref.
12		145	—	CHCl$_3$	—	d
13		171	—	CHCl$_3$	—	d
14		169	—	CHCl$_3$	—	d
15		179.3; 188.3	70; 68	DMF	100	c
16		165.3	100	CHCl$_3$	50	c
17		177	—	Toluene	80	c
18		171.1	145	DMF	80	c
19		170.9; 172.5 168.1; 170.2	38; 29 156	Toluene CHCl$_3$	100 50	c c
20		170.8; 172.3 168.2; 169.6	50; 38 159	Toluene CHCl$_3$	100 50	c c
21		171.0; 172.3	163	Toluene	100	c

TABLE 8 (continued)
Oxygen-17 NMR Chemical Shifts of Sulfonyl Oxygens in Cyclic Organosulfur Compounds

Entry	Compound	$\delta^{17}O$ (ppm)	$W_{1/2}$ (Hz)	Solvent	Temp., °C	Ref.
22		171.4; 172.8	128	Toluene	100	c
23		180.0	83	DMF	100	c
24		179.7; 181.7	—; 135	DMF	100	c
25		180.9; 182.6	—; 129	DMF	100	c
26		153.1 ± 0.5	—	CDCl$_3$	—	e
27		167.5 ± 0.5	—	CDCl$_3$	—	e
28		161.0 ± 0.5	—	CDCl$_3$	—	e
29		161.0 ± 0.5	—	CDCl$_3$	—	e
30		155.9 ± 0.5	≤270	CDCl$_3$	33	f
31		159.1 ± 0.5	≤270	CDCl$_3$	33	f

TABLE 8 (continued)
Oxygen-17 NMR Chemical Shifts of Sulfonyl Oxygens in Cyclic Organosulfur Compounds

Entry	Compound	$\delta^{17}O$ (ppm)	$W_{1/2}$ (Hz)	Solvent	Temp., °C	Ref.
32		149.0±0.5	≤270	CDCl$_3$	33	f
33		163.0±0.5	≤270	CDCl$_3$	33	f
34		160.7; 165.9	≤270	CDCl$_3$	33	f
35		151.7; 167.9	≤270	CDCl$_3$	33	f
36		171.8	≤270	CDCl$_3$	33	f
37		146.5	≤270	CDCl$_3$	33	f
38		147.4	≤270	CDCl$_3$	33	f
39		147.0	≤270	CDCl$_3$	33	f
40		147.0	≤270	CDCl$_3$	33	f

TABLE 8 (continued)
Oxygen-17 NMR Chemical Shifts of Sulfonyl Oxygens in Cyclic Organosulfur Compounds

Entry	Compound	$\delta^{17}O$ (ppm)	$W_{1/2}$ (Hz)	Solvent	Temp., °C	Ref.
41		146.1	≤270	CDCl₃	33	f
42		159.8	≤270	CDCl₃	33	f
43		159.6	≤270	CDCl₃	33	f
44		163.1	≤270	CDCl₃	33	f

Note: a: Block, E., Bassi, A. A., Lambert, J. B., Wharry, S. M., Andersen, K. K., Dittmer, D. L., Patwardhan, B. H., and Smith, D. J., H., *J. Org. Chem.*, 45, 4807, 1980; b: Dyer, J. C., Harris, D. L., and Evans, S. A., Jr., *J. Org. Chem.*, 46, 3660, 1982; c: Sammakia, T. H., Harris, D. L., and Evans, S. A., Jr., *Org. Magn. Reson.*, 22, 747, 1984; d: Barbarella, G., Rossini, S., Bongini, A., and Tugnoli, V., *Tetrahedrom*, 41, 4691, 1985; e: Lowe, G. and Salamone, S. J., *J. Chem. Soc. Chem. Commun.*, p. 1392, 1983; f: Hellier, D. G. and Liddy, H. G., *Magn. Reson. Chem.*, 26, 671, 1988.

Oxygen-17 NMR shifts are influenced by lanthanide shift reagents and respond to (1) the complex formation shift, (2) the contact shift, and (3) the pseudocontact shift interactions.[50] The overall effect, assuming that the contact shift dominates the observed ¹⁷O NMR shift increments, with increasing Eu^{3+} concentration results in progressive upfield shifts of the coordinated ¹⁷O nuclei. In this way, *internal* accidental ¹⁷O NMR chemical shift coincidence may be resolved.

Application of this Eu^{+3}-induced ¹⁷O NMR shift technique to 3-*iso*-propoxythiolane 1,1-dioxide (**76**) provided evidence for a tentative ¹⁷O NMR sulfonyl oxygen-shift assignments as well as insight into some conformational factors. At 100°C in toluene solvent, the ¹⁷O NMR shifts for the sulfonyl oxygens in **76** differ by 1.3 ppm (δ 172.3 and 171.0 ppm). However, in the presence of 7.8×10^{-2}M Eu(fod)₃, the ¹⁷O NMR shifts are δ 156.6 and 132.7 ppm with $\Delta\delta$ SO₂ = 23.9 ppm. It is suggested that the more shielded sulfonyl oxygen responds to a stronger coordination with Eu^{3+} and is, therefore, more sterically accessible[48] (Scheme 1).

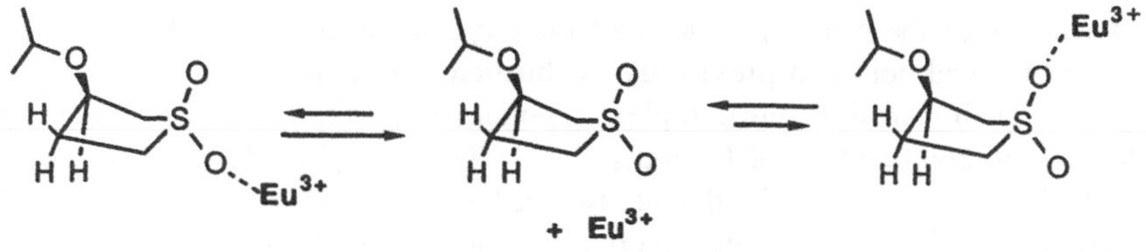

SCHEME 1. Equilibria between sulfone **76** and Eu^{3+}.

A similar ^{17}O NMR/lanthanide shift reagent probe has been applied to resolve a 1:1 component mixture containing the enantiotopic sulfonyl oxygens of **77** and the diastereotopic sulfonyl oxygens of **78** where both compounds exhibit the same ^{17}O NMR resonance at δ 159 ppm.[38] This is a case of *internal* and *external* accidental ^{17}O NMR shift coincidence. When a solution containing **77** and **78** is admixed with sufficient Eu(fod)$_3$, three ^{17}O NMR resonances are easily identified. Sulfone **77** exhibits a single resonance at δ 135 ppm for the enantiotopic oxygens and the diastereotopic sulfonyl oxygens in **78** are observed separately at δ 142 and 152 ppm with a given quantity of Eu(fod)$_3$.[38] It is also assumed that the high field ^{17}O NMR resonance for **78** (δ 142 vs. 152 ppm) is sterically more accessible to Eu^{3+} and presumably anti to the C-2 methyl group.

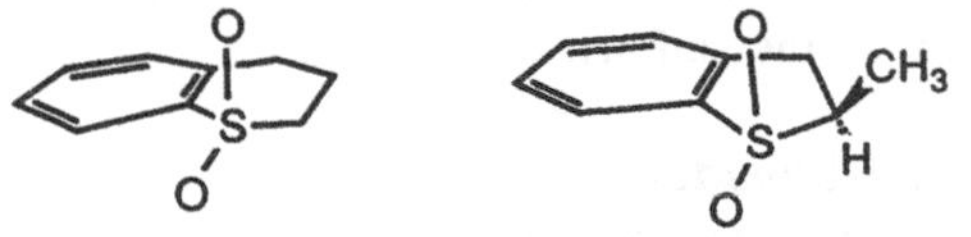

77   **78**

3. 2,2-Dioxa-1,3,2-Dioxathiolanes

Lowe and Salamone[51] have employed a clever use of ^{17}O isotopic labeling combined with lanthanide-induced ^{17}O NMR shifts to describe the stereorelationship of the ^{17}O-labeled nucleus within a 2,2-dioxa-1,3,2-dioxathiolane (Table 8). For example, an 88:12 diastereomeric mixture of *trans*- and *cis*-2-oxo-*cis*-4,5-diphenyl-1,3,2-dioxathiolanes (**79**) was oxidized with Ru^{17}O$_4$ to afford an 88:12 mixture of 2,2[^{16}O,^{17}O]dioxa-*cis*-4,5-diphenyl-1,3,2-dioxathiolanes (**80** and **81**). From the ratio of the ^{17}O NMR resonance intensities, it was established that the more intense (88%) δ 153.1 ppm resonance was derived from *trans* 2-oxo-*cis*-4,5-diphenyl-1,3,2-dioxathiolane (*trans*-**79**) and the minor (12%) δ 167.5 ppm resonance was consistent with *cis*-sulfite **79**. *trans*-Sulfite **79** had been previously identified as the major isomer of the mixture from the influence of the sulfinyl diamagnetic anisotropic effect on the ^{1}H NMR chemical shift of the ring protons.[52]

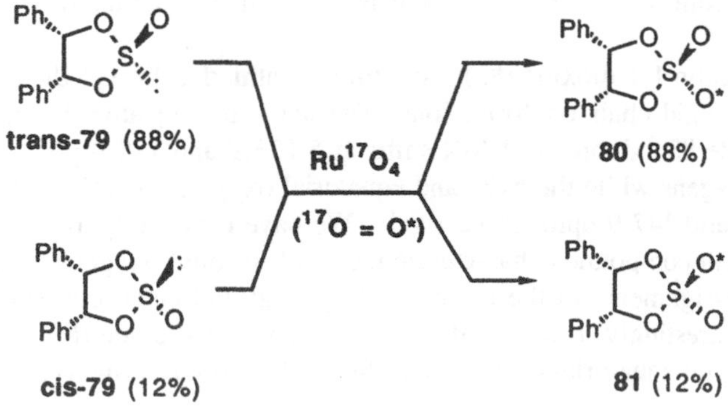

trans-79 (88%)   Ru^{17}O$_4$ **80** (88%)

(^{17}O = O*)

cis-79 (12%)   **81** (12%)

However, since the stereospecificity of the oxygen transfer from RuO_4 to the sulfinyl sulfur in sulfites had not been previously established, an approach to addressing this issue was envisioned using the dynamic complexing potential and ^{17}O NMR perturbing capabilities of $Eu(fod)_3$. The coordination of $Eu(fod)_3$ to the ^{17}O sulfonyl nuclei is expected to initiate upfield ^{17}O NMR shifts (i.e., lanthanide-induced shifts, LIS), and presumably, $Eu(fod)_3$ will preferentially ligate to the least sterically hindered of the diastereotopic ^{17}O-labeled oxygens of the diastereomeric mixture containing **80** and **81**. In this way, the larger ^{17}O NMR shift difference ($\Delta\delta$) will reflect the stronger binding to the least sterically hindered sulfonyl oxygen. With a given quantity of $Eu(fod)_3$, the LIS is 4.0 ppm for the minor isomer (i.e., **81**) resonating at δ 153.1 ppm, and LIS equals 0.5 ppm for the major isomer (i.e., **80**) with the shift at δ 167.5 ppm. These results also confirm that the oxidation of the sulfinyl moiety with $Ru^{17}O_4$ to the sulfonyl groups is stereospecific occurring with retention of stereochemistry.[51]

Recently, Hellier and Liddy[53] reported the ^{17}O NMR shifts for several sulfites (i.e., 2-oxo-1,3,2-dioxathiolanes) and sulfates (i.e., 2,2-dioxa-1,3,2-dioxathiolanes) and concluded that the average ^{17}O NMR shift difference $[\Delta\delta(=SO_2-SO)]$ is 58 ppm for the ethylene derivatives and 31 ppm for the trimethylene compounds, where the 2-oxo-1,3,2-dioxathiolane and 2-oxo-1,3,2-dioxathianes are the most deshielded in each case (Table 8).

These trends are in contrast to those observed for acyclic as well as five- and six-membered ring sulfoxides and sulfones. In the latter cases, the sulfinyl oxygens are considerably more shielded than the sulfonyl oxygens.

4. Substituted Thiane 1,1-Dioxides

Barbarella et al.[30] have demonstrated that placement of methyl substituents on the thiane 1,1-dioxide ring creates preferences for certain conformational isomers such that high levels of ^{17}O NMR shift nonequivalence are realized (Table 9). For example, the rapid-interconverting thiane 1,1-dioxide (**82**) exhibits a time-averaged ^{17}O NMR shift at δ 144 ppm.

82

However, 3-methyl-, 4-methyl-, or 3,5-dimethyl substitution on the thianyl ring favors the more stable equatorial conformation. Consequently, methyl substitution discourages any conformer interchange that would place a methyl group in the axial array and, as a result, the diastereotopic sulfonyl oxygens reside in essentially "conformationally homogeneous" environments.[30] ^{17}O NMR shift differences ($\Delta\delta SO_2$) for the axial and equatorial sulfonyl oxygens ranged from 8 to 11 ppm where it is assumed that the axial oxygen resonates at higher field.

trans-Thiadecalin 1,1-dioxide (**83**)[29] and *trans*-oxathiadecalin 1,1-dioxide (**84**)[48] exist in conformationally rigid chair conformations. The axial and equatorial sulfonyl oxygens in thiadecalin dioxide **83** exhibit ^{17}O NMR shifts at δ 123.9 and 138.9 ppm for the respective diastereotopic oxygens while the axial and equatorial oxygens in sulfone **84** have ^{17}O NMR shifts at δ 131.5 and 147.9 ppm, respectively. We have reasonably assumed that high field resonances for both compounds characterize the axial sulfonyl oxygens by analogy with our ^{17}O NMR shift assignments to the corresponding axial and equatorial sulfinyl compounds (cf. **25α,β**).[10] Interestingly, *trans*-oxathiadecalin-2-ene 1,1-dioxide (**85**) also has a sulfonyl group with axial and equatorial oxygens ($\Delta\delta$ SO_2 = 20.4 ppm); however, assignment of the

TABLE 9
Oxygen-17 NMR Chemical Shifts of Sulfonyl Oxygens in Thiane 1,1-Dioxides

Entry	Compound	$\delta^{17}O$ (ppm)	$W_{1/2}$ (Hz)	Solvent	Temp., °C	Ref.
1		146	—	MeCN	—	a
		142	—	CDCl$_3$	—	b
		149	—	MeC(O)Me	—	b
		144 ± 1	—	CHCl$_3$	35	c
2		123.9; 138.9	—	CDCl$_3$	28—30	d
3		131.5; 147.9	—	CDCl$_3$	28—30	d
4		139.4; 159.8	—	CDCl$_3$	28—30	d
5		139; 150 ± 1	—	CHCl$_3$	35	c
6		138; 146 ± 1	—	CHCl$_3$	35	c
7		139; 149 ± 1	—	CHCl$_3$	35	c

TABLE 9 (continued)
Oxygen-17 NMR Chemical Shifts of Sulfonyl Oxygens in Thiane 1,1-Dioxides

Entry	Compound	δ¹⁷O (ppm)	W₁/₂ (Hz)	Solvent	Temp., °C	Ref.
8		154 ± 1	—	CHCl₃	35	c
9		151 ± 1	—	CHCl₃	35	c
10		142 ± 1	—	CHCl₃	35	c
11		143 143	— —	CDCl₃ MeC(O)Me	— —	b b

Note: a: Dyer, J. C., Harris, D. L., and Evans, S. A., Jr., *J. Org. Chem.*, 47, 3660, 1982; b: Block, E., Bassi, A. A., Lambert, J. B., Wharry, S. M., Andersen, K. K., Dittmer, D. L., Patwarden, B. H., and Smith, D. J. H., *J. Org. Chem.*, 45, 4807, 1980; c: Barbarella, G., Dembech, P., and Tugnoli, V., *Org. Magn. Reson.*, 22, 402, 1984; d: Sammakia, T. H., Harris, D. L., and Evans, S. A., Jr., *Org. Magn. Reson.*, 22, 747, 1984.

axial or equatorial sulfonyl oxygen requires knowledge concerning the effect of α,β-olefinic groups on sulfonyl oxygen chemical shifts. At present, not enough is known to make a definitive ¹⁷O NMR chemical shift assignment.[48]

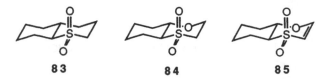

As anticipated, *trans*-3,6-dimethyl-, 3,3-dimethyl-, and 4,4-dimethylthiane 1,1-dioxide afford only one ¹⁷O NMR resonance.[30] The interesting feature is that the time-averaged ¹⁷O NMR shift for thiane-1,1-dioxide (**82**) and 4,4-dimethylthiane 1,1-dioxide (**86**) are essentially the same (δ 144 vs. 142 ppm, respectively); however, the time-averaged ¹⁷O NMR shifts for *trans*-3,6-dimethyl- and 3,3-dimethylthiane 1,1-dioxides are both deshielded (δ 154 and 151 ppm, respectively) compared to dioxide **82**. If one assumes that the ¹⁷O NMR shift for the equatorial sulfonyl oxygen is relatively unchanged in these sulfones, then the axial sulfonyl oxygen must experience a deshielding δ-effect or compression effect induced by the syn axial methyl group.

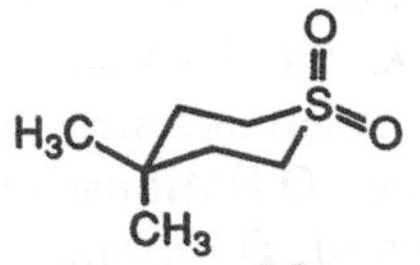

86

V. THE PHOSPHORYL GROUP (–P=O)

A. OXYGEN-17 NMR SHIFT RATIONALE FOR ORGANOPHOSPHORUS COMPOUNDS CONTAINING THE PHOSPHORYL GROUP

The importance of ^{17}O NMR as a structural probe in biomolecules containing oxidized phosphorus atoms is well documented.[54] For substantive details and an overview, the reader is referred to the excellent discussion on the ^{17}O NMR of biophosphates presented by Tsai and Bruzik.[55] This section of the present review will focus on the ^{17}O NMR chemical shifts and ^{31}P–^{17}O couplings involving the phosphoryl group (i.e., –P=O) in phosphine oxides and phosphorus esters.

It is assumed that σ_P^O (Equation 2) also dominates the ^{17}O nuclear shieldings of the phosphoryl group such that the ^{17}O NMR shift trends observed for sulfoxides and sulfones (*vide supra*) would be largely manifested here as well. In addition, the one-bond ^{17}O–^{31}P coupling constants ($^1J_{PO}$) in the phosphoryl bonds seem to correlate with the π_{p-o} bond order, although convincing theoretical calculations point to a strong contribution to the coupling by the Fermi contact mechanism.[56] In the main, $^1J_{PO} > 150$ Hz for P=O and $^1J_{PO} < 90$ Hz for P–O bonds in the esters of phosphoric acid.[55]

In some instances, $^1J_{PO}$ for P–O single bonds may or may not be observable, and this is directly related to the differences in the ^{17}O nuclear quadrupolar coupling constants (i.e., $Q = e^2Qq/h$) for P–O and P=O. For example, Cheng and Brown[57] measured the ^{17}O quadrupolar coupling constants in triphenylphosphate [$(PhO)_3PO$] and reported that Q = 3.825 MHz for P=^{17}O ($\eta = 0.098$) and Q = 9.176 MHz for the P–^{17}O ($\eta = 0.644$) at 77 K. The width-at-half-height ($W_{1/2}$) of the ^{17}O resonance is inversely related to the quadrupolar relaxation time, T_q, but directly related to Q^2 (Equation 5).[58] Thus, large Q's can be the primary source of significant $W_{1/2}$'s if other parameters (i.e., η, τ_r) make minor contributions.

$$W_{1/2} \cong \frac{1}{\pi T_q} \cong \frac{12 \pi}{125} (1 + \frac{\eta 2}{3}) (\frac{e^2qQ}{h})^2 \tau_r \qquad (5)$$

B. THE SUBSTITUENT-INDUCED CHEMICAL SHIFT (SCS) EFFECTS

From a perusal of the ^{17}O NMR data shown below, it is apparent that an increase in the negatively charged oxygens around phosphorus leads to a systematic deshielding of the ^{17}O resonances. For example, the ^{17}O NMR shifts of various ions of phosphoric acid (e.g., H_3PO_4) typify this trend.[58]

$$ClO_4^- \ P(OH)_4^+ \quad < \quad K^+ \ OPO(OH)_2^- \quad <$$
$$\delta\ 77.3 \qquad\qquad\qquad \delta\ 83.6$$
$$(^1J_{PO} = 83.0\ Hz) \quad (^1J_{PO} = 87.9\text{–}88.7\ Hz)$$

$$2\ K^+ \ OPO_2OH^{2-} \quad < \quad 3\ K^+ \ OPO_3^{3-}$$
$$\delta\ 98.1 \qquad\qquad\quad \delta\ 114.2$$
$$(^1J_{PO} = 95.0\ Hz) \quad (^1J_{PO} = 96.6\ Hz)$$

The two immediate trends require comment. First, the upfield shift of the P=^{17}O resonance responding to protonation has been interpreted in terms of negative charge on the phosphoryl oxygens. In essence, the correlation between ^{17}O NMR shifts and charges on oxygen indicate that as the charge per oxygen decreases, the ^{17}O NMR resonances move upfield.[59] Secondly, the magnitudes of $^1J_{PO}$ range from 83.0 to 96.6 Hz (see above) and imply that the extent of π_{PO} bonding diminishes from right to left as reflected in the decreasing coupling constants.

A similar trend is observed in the ^{17}O NMR parameters for the phosphoryl group (–P=O) in the methyl esters of phosphoric acid. For example,

$$(MeO)_3PO \quad < \quad (MeO)_2PO_2^- \quad < \quad MeOPO_3^{2-}$$
$$\delta\ 74 \qquad\qquad \delta\ 85.2 \qquad\qquad \delta\ 98.5$$
$$(^1J_{PO} = 153.8\ Hz)^{58} \quad (^1J_{PO} = 112\ Hz)^{60} \quad (^1J_{PO} = 98\ Hz)^{60}$$

While a more precise explanation of these ^{17}O NMR shift trends must await more profound advances in the theory and predictability of heteroatom NMR shifts, it does seem reasonable to suggest that electrostatic repulsions between the negatively charged oxygens may have an impact on the $\langle r^{-3} \rangle_{2pO}$ term in Equation 2. This qualitative notion implies that electrostatic repulsions between the negatively charged phosphoryl oxygens would promote a contraction of the oxygen 2p orbitals and a reduction in the σ_p^0 term.

The similarity in ^{17}O NMR shifts between the two series of compounds described above is particularly interesting. From these ^{17}O NMR data, it appears that the replacement H with Me has essentially no effect on the phosphoryl oxygen ^{17}O shift. This is somewhat surprising as it indicates an absence of a γ-steric shift (^{17}O shielding) effect on the phosphoryl oxygens of the methyl phosphoesters (*vide infra*).

By contrast, trimethylphosphine oxide (**87**; δ 66.3 ppm; $^1J_{PO}$ = 150.9 Hz)[56,61] exhibits an ^{17}O NMR shift which is deshielded by 23 ppm compared to tributylphosphine oxide (**88**; δ 43 ppm; $^1J_{PO}$ = 152 Hz).[61] Contrary to the previous discussion, this ^{17}O NMR shift difference implies a γ steric shift (shielding) effect of about 8 ppm/methylene. Comparison of the ^{17}O NMR shifts for Ph$_3$PO (**89**; δ 47.7 ppm; $^1J_{PO}$ = 153.9 Hz)[58,61] and Me$_3$PO (δ 66.3 ppm) provides additional support for the γ-steric shift effect.

87   **88**   **89**

While the total π-electronic charge transfer in the P=O bond for (PhO)$_3$P=O is 0.55 e$^-$ and 0.47 e$^-$ in phosphine oxide **89** based on calculations derived from ^{17}O nuclear quadrupolar resonance (NQR) studies,[57] the invariant $^1J_{PO}$ values for Me$_3$PO and Ph$_3$PO argue against any "extended" 2p-3dπ conjugative interactions between the aryl ring and the phosphoryl phosphorus atom (Table 10).

The ^{17}O NMR shifts for a series of 1-adamantyl phosphoryl derivatives have been reported and provide insights into a number of substituent effects.[62] For example, the strong inductive influence of highly electronegative chlorine and bromine atoms attached to the phosphoryl group is evident in the ^{17}O NMR shift comparisons between **90** (δ 167 ppm), **91** (δ 144 ppm), and **92** (δ 70 ppm). In addition, the smaller $^1J_{PO}$ = 145 Hz for **92** compared to $^1J_{PO}$ = 190-191 Hz for **90** and **91** suggests that there is substantial O$_{2p}$-P$_{2d}\pi$ overlap.[62]

TABLE 10
Oxygen-17 NMR Spectral Parameters of the Phosphoryl Oxygen

Entry	Compound	$\delta^{17}O$ (ppm)	$W_{1/2}$ (Hz)	$^{1}J_{PO}$ (Hz)	Solvent	Temp., °C	Ref.
1	$Me_3P{=}O$	66.3	—	150.9	CD_3CN	70	a
				120 ± 15	—	—	b
2	$Bu_3P{=}O$	43	—	152	CD_3CN	70	a
3	$Ph_3P{=}O$	43.3	—	160	$CDCl_3$	—	a
		47.7	—	153.9	CD_3CN	70	c
4	$Cl_3P{=}O$	216	—	205 ± 3	Neat	—	b,d
		208	—	199	HPO_2F_2	—	e
		202	—	188	CH_3SO_3H	—	e
		187	—	177	HNO_3	—	e
		177	—	173	HSO_3Cl	—	e
5	$Me_2NP(O)Cl_2$	149.8 ± 1.5	—	189 ± 5	—	—	f
6	$(Me_2N)_2P(O)Cl$	125.1		169	—	—	f
7	$(Me_2N)_3P{=}O$	73.4	—	145	—	—	f
8	$MeOP(O)Cl_2$	167.4	—	187	—	—	f
9	$(MeO)_2P(O)Cl$	123.8	—	173	—	—	f
10	$(MeO)_3P{=}O$	68	—	165	—	—	f
11	$FCl_2P{=}O$	177.3	—	201	—	—	f
12	$ClF_2P{=}O$	126.6	—	194	—	—	f
13	$F_3P{=}O$	66.0	—	188[g] (184 ± 3)[b]	—	—	f
14	$BrCl_2P{=}O$	232.1	—	205	—	—	f
15	$ClBr_2P{=}O$	245.0	—	199	CCl_4	—	f
16	$Br_3P{=}O$	259	—	195	CCl_4	—	f
17	(structure: Me,Ph-substituted C bonded to P($=$O)(OEt)(OEt) with S–C($=$O)Me)	85	—	145	—	—	h
18	(structure: phenyl-P($=$O)Cl$_2$)	155	70	180	$CDCl_3$	28	i
19	(structure: adamantyl-P($=$O)Br$_2$)	167	80	190	$CDCl_3$	28	i
20	(structure: adamantyl-P($=$O)Cl$_2$)	144	70	191	$CDCl_3$	28	i

TABLE 10 (continued)
Oxygen-17 NMR Spectral Parameters of the Phosphoryl Oxygen

Entry	Compound	$\delta^{17}O$ (ppm)	$W_{1/2}$ (Hz)	$^1J_{PO}$ (Hz)	Solvent	Temp., °C	Ref.
21		113	110	182	$CDCl_3$	28	i
22		114	140	155	$CDCl_3$	28	i
23		120	150	165	$CDCl_3$	28	i
24		90	150	138	$CDCl_3$	28	i
25		92	130	156	$CDCl_3$	28	i
26		70	200	145	$CDCl_3$	28	i
27		72	200	147 ± 3	$CDCl_3$	28	i

TABLE 10 (continued)
Oxygen-17 NMR Spectral Parameters of the Phosphoryl Oxygen

Entry	Compound	$\delta^{17}O$ (ppm)	$W_{1/2}$ (Hz)	$^1J_{PO}$ (Hz)	Solvent	Temp., °C	Ref.
28		72	200	135	CDCl$_3$	28	i
29		72	250	—	CDCl$_3$	28	i
30		88	200	125	CDCl$_3$	28	i
31		66	200	—	CDCl$_3$	28	i
32		145	140	179	CDCl$_3$	28	i
33		167	160	190 ± 10	CDCl$_3$	28	i

Note: a: Quin, L. D., Szewczyk, J., Linehan, K., and Harris, D. L., *Magn. Reson. Chem.*, 25, 271, 1987; b: Gray, G. A. and Albright, T. A., *J. Am. Chem. Soc.*, 99, 3243, 1977; c: Sammons, R. D., Frey, P. A., Bruzik, K., and Tsai, M.-D., *J. Am. Chem. Soc.*, 105, 5455, 1983; d: Christ, H. A., Diehl, P., Schnieder, H. R., and Dahn, H., *Helv. Chim. Acta*, 44, 865, 1961; e: Hibbert, R. C. and Logan, N., *J. Chem. Soc. Dalton Trans.*, p. 865, 1985; f: Grossman, G., Gruner, M., and Seifert, G., *Z. Chem.*, 16, 362, 1976; g: Christ, H. A. and Diehl, P., *Helv. Phys. Acta*, 36, 170, 1963; h: Creary, X. and Inocencio, P. A., *J. Am. Chem. Soc.*, 108, 5979, 1980; i: Duddeck, H. and Hanna, A. G., *Magn. Reson. Chem.*, 23, 533, 1985.

| 90 | 91 | 92 | 94 | 95 | 96 |

Interestingly, the ^{17}O NMR shift of dichlorophenylphosphine oxide (93; δ 155 ppm; $^1J_{PO}$ = 190 Hz) compared to phosphine oxide 91 (δ 144 ppm) provides clear evidence of a γ-steric shielding effect arising from the γ-gauche interactions between the P=O and the bridge CH_2's in 91. On the other hand, the relatively constant ^{17}O NMR shifts for the phosphoryl oxygens in 92, 94, 95, and 96 (δ 70 to 72 ppm) indicate that alkyl substitution or methylene extension beyond the γ position have no noticeable effect on the P–O ^{17}O NMR shift[62] (Table 10).

93

The ^{17}O NMR shifts and $^1J_{PO}$ values for $Me_2NP(O)Cl_2$ (δ 149.8; $^1J_{PO}$ = 189 Hz), $(Me_2N)_2P(O)Cl$ (δ 125.1 ppm; $^1J_{PO}$ = 169 Hz), and $(Me_2N)_3PO$ (δ 73.4 ppm; $^1J_{PO}$ = 145 Hz)[63] emphasize the importance of p-dπ backbonding between the donor atom nitrogen and the acceptor phosphoryl phosphorus. The increasing upfield ^{17}O NMR shifts may imply a higher charge density on oxygen and an expansion of the 2p orbitals[64] coupled with a diminished P–O multiple bond order as reflected in the $^1J_{PO}$ values.

C. SOLVENT EFFECTS

The ^{17}O NMR shift of trichlorophosphine oxide (Cl_3PO) has been examined in the presence of a variety of protic acids,[65] and it is clear that both the ^{17}O NMR shifts and $^1J_{PO}$ values decrease with increasing acidity of the solvent [(neat: δ 215 ppm; $^1J_{PO}$ = 205 Hz), (HPO_2F_2: δ 208 ppm; $^1J_{PO}$ = 199 Hz), ($MeSO_3H$: δ 202 ppm; $^1J_{PO}$ = 188 Hz), (HNO_3: δ 187 ppm; $^1J_{PO}$ = 177 Hz), and (HSO_3Cl: δ 177 ppm; $^1J_{PO}$ = 173 Hz)]. A reasonable rationale for the ^{17}O NMR shift changes assumes that the n→π* transition, a dominant contributor to the ΔE term in Equation 2, increases with increasing protonation giving rise to an overall smaller σ_P^0 term. Finally, the diminishing $^1J_{PO}$ values which are aligned with increasing effectiveness of protonation result from a gradual decrease in the P=O bond order.

D. DIASTEREOTOPIC PHOSPHORYL OXYGENS

Perhaps the first utilization of ^{17}O NMR spectroscopy to establish the configurational differences of phosphoryl oxygens in a conformationally rigid diester was reported for the diastereomers of cyclic 2'-deoxyadenosine 3',5'-[^{17}O,^{18}O]monophosphate (97 and 98)[66] (Table 11). In diastereomer 97 the axial P=^{17}O occurs at δ 92.8 ppm ($^1J_{PO}$ = 130 Hz) and the equatorial P=^{17}O in diastereomer 98 resonates at δ 91.2 ppm ($^1J_{PO}$ = 102 Hz).

| 97 | 98 |

TABLE 11
Oxygen-17 NMR Spectral Parameters of the Phosphoryl Oxygen in Six-Membered Rings

Entry	Compound	$\delta^{17}O$ (ppm)	$W_{1/2}$ (Hz)	$^1J_{PO}$ (Hz)	Solvent	Temp., °C	Ref.
1	(structure: cyclic phosphate with ^{17}O and ^{18}O, Ade)	92.8 ± 0.2	50	130 ± 7	H_2O	95	a
2	(structure: cyclic phosphate with ^{18}O and ^{17}O, Ade)	91.2 ± 0.2	82	102 ± 7	H_2O	95	a
3	(structure: six-membered ring phosphate, OCH_3)	77.7	—	180	Toluene-d_8	95	b
		81	—	162	Toluene	100	c
4	(structure: six-membered ring phosphate, CH_3O)	85.6	—	174	Toluene-d_8	95	b
		88	—	—	Toluene	100	c
5	(structure: six-membered ring phosphate, $N(CH_3)_2$)	70.4	—	182	Toluene-d_8	31	b
6	(structure: six-membered ring phosphate, $(CH_3)_2N$)	88.4	—	177	Toluene-d_8	95	b
7	(structure: six-membered ring phosphate, H)	105.6	—	180	Toluene-d_8	95	b
8	(structure: six-membered ring phosphate, H)	125.9	—	173	Toluene-d_8	95	b

TABLE 11 (continued)
Oxygen-17 NMR Spectral Parameters of the Phosphoryl Oxygen in Six-Membered Rings

Entry	Compound	$\delta^{17}O$ (ppm)	$W_{1/2}$ (Hz)	$^1J_{PO}$ (Hz)	Solvent	Temp., °C	Ref.
9		112.2	—	184	Toluene-d$_8$	95	b
10		120.8	—	175	Toluene-d$_8$	95	b
11		82	—	156	Toluene	100	c
12		83	—	164	Toluene	100	c
13		88	—	161	Toluene	100	c
14		94	—	161	Toluene	100	c
15		89	—	138	Toluene	100	c
16		90	—	156	Toluene	100	c

TABLE 11 (continued)
Oxygen-17 NMR Spectral Parameters of the Phosphoryl Oxygen in Six-Membered Rings

Entry	Compound	$\delta^{17}O$ (ppm)	$W_{1/2}$ (Hz)	$^{1}J_{PO}$ (Hz)	Solvent	Temp., °C	Ref.
17		87.7	46	160	CD_3CN	80	d
	T= Thymidine						
18		78.2	73	156	CD_3CN	80	d
19		86.2	55	156	CD_3CN	80	d
20		96.0	61	160	CD_3CN	80	d
21		119	157	167	CD_3CN	80	d
22		116	352	—	CD_3CN	80	d

Note: a: Coderre, J. A., Mehdi, S., Demou, P. C., Weber, R., Traficante, D. D., and Gerlt, J. A., *J. Am. Chem. Soc.*, 103, 1870, 1981; b: Bock, P. L., Mosbo, J. A., and Redmon, J. L., *Org. Magn. Reson.*, 21, 491, 1983; c: Eliel, E. L., Chandrasekaran, S., Carpenter, L. E., and Verkade, J. G., *J. Am. Chem. Soc.*, 108, 6651, 1986; d: Sopchik, A. E., Cairns, S. M., and Bentrude, W. G., *Tetrahedron Lett.*, 30, 1221, 1989.

E. CYCLIC PHOSPHATES

The ¹⁷O NMR shifts of a series of diastereomeric 2-R-2-oxo-1,3,2-dioxaphosphorinanes have been reported[67] and two important conclusions have surfaced. First, the equatorial phosphoryl oxygen is more shielded than the axial P= O and the latter exhibits larger ¹J$_{PO}$ values as well. Second, the magnitudes of the axial and equatorial ¹J$_{PO}$ values are consonant with Stec's rule for spin ½ nuclei[68] (Table 11). Eliel et al.[69] have also demonstrated that similarily substituted 2-R-2-oxo-1,3,2-dioxaphosphorinanes exhibit comparable ¹⁷O NMR parameters (Table 11). A specific example which focuses on the importance of a δ compression effect is detailed in the ¹⁷O NMR shift comparison between phosphate esters **99** (δ 88 ppm) and **100** (δ 94 ppm). The 6 ppm deshielding of P=O in **100** appears to result from the interaction between the axial C-4 methyl and the axial P=O.

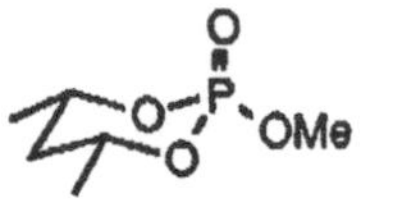

99 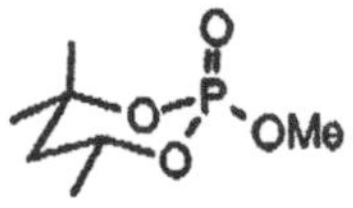   **100**

An interesting attempt to obtain information on the relative basicities of the equatorial vs. the axial P=O functionality was based on the magnitudes of their phenol-induced shifts as determined from IR studies. For example, Δν = 345 cm⁻¹ for **101** and Δν = 315 cm⁻¹ for **102**[70] suggest that the P=O in **101** is more basic; consequently, the equatorial P=O in **102** exhibits more 2p-3dπ character.

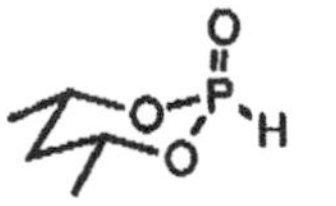

101 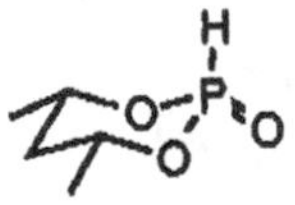   **102**

Finally, Sopchik, et al.[71] have recently described similar ¹⁷O NMR shift parameters for a series of thymidine cyclic 3′,5′ phosphate esters (Table 11).

F. CYCLIC PHOSPHINE OXIDES

The ¹⁷O NMR shifts of the phosphoryl oxygens in cyclic phosphine oxides respond to a variety of different effects including C-P-C angle variations and ring substitution (Table 12). Perhaps, the most striking example of a highly shielded phosphoryl oxygen derived from a phosphine, is that found in 9-(2,4,6-tri-*tert*-butylphenyl)-9-phosphabicyclo[6.1.0]nona-2,4,6-triene 9-oxide (**103**)[72] [δ(¹⁷O) 18.0 ppm, ¹J$_{PO}$ = 166 Hz]. It is noteworthy that thiirane S-oxide (δ −71 ppm)[47] also exhibits the most shielded sulfinyl oxygen of the cyclic sulfoxides.

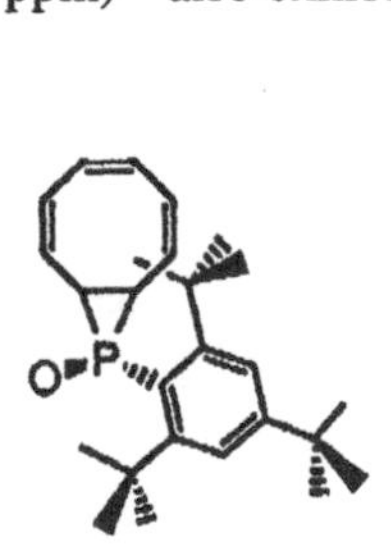

103

While the phosphetane oxides (four-membered ring) exhibit ¹⁷O NMR shifts which occur at lower field (deshielded) than the three-membered ring phosphine oxides, the P=O group

TABLE 12
Oxygen-17 NMR Spectral Parameters of the Phosphoryl Oxygen in Three-, Four-,
and Five-Membered Rings

Entry	Compound	$\delta^{17}O$(ppm)	$W_{1/2}$ (Hz)	$^1J_{PO}$ (Hz)	Solvent	Temp., °C	Ref.
1		18.0	—	166	CD_3CN	50	a
2		74.0	—	175	CD_3CN	70	b
3		58.4	—	203	CD_3CN	70	b
4		93.1	—	179	CD_3CN	70	b
5		66.0	—	164	CD_3CN	70	b
6		53.1	—	167	CD_3CN	70	b

TABLE 12 (continued)
Oxygen-17 NMR Spectral Parameters of the Phosphoryl Oxygen in Three-, Four-,
and Five-Membered Rings

Entry	Compound	$\delta^{17}O$(ppm)	$W_{1/2}$ (Hz)	$^1J_{PO}$ (Hz)	Solvent	Temp., °C	Ref.
7		39.6	—	164	CD_3CN	70	b
8		59.2	—	182	CD_3CN	70	b
9		46.0	—	164	CD_3CN	70	b
10		53	—	170	CD_3CN	70	b
11		a: 108.3	—	137	CD_3CN	70	b
		b: 57.3	—	133	CD_3CN	70	b
12		a: 52.7	—	127	CD_3CN	70	b
		b: 56.5	—	122	CD_3CN	70	b

TABLE 12 (continued)
Oxygen-17 NMR Spectral Parameters of the Phosphoryl Oxygen in Three-, Four-,
and Five-Membered Rings

Entry	Compound	δ¹⁷O(ppm)	$W_{1/2}$ (Hz)	$^1J_{PO}$ (Hz)	Solvent	Temp., °C	Ref.
13		a: 100.8	—	163	CD₃CN	70	b
		b: 71	—	160	CD₃CN	70	b
14		a: 42.1	—	175	CD₃CN	70	b
		b: 71.5	—	163	CD₃CN	70	b
15		116.4	—	183	CD₃CN	70	b
16		80.9	—	181	CD₃CN	70	b

Note: a: Quin, L. D., Yao, E.-Y., and Szewczyk, J., *Tetrahedron Lett.*, 28, 1077, 1987; b: Quin, L. D., Szewczyk, J., Lineham, K., and Harris, D. L., *Magn. Reson. Chem.*, 25, 271, 1987.

responds to γ-methyl shielding effects as described in the ¹⁷O NMR shift comparisons between phosphine oxide **104** (δ 74.0 ppm; $^1J_{PO}$ = 175 Hz) and phosphetane oxide **105** (δ 58.4 ppm; $^1J_{PO}$ = 203 Hz)[61] (Table 12). Here, the γ gauche methyl interactions combine to effect substantial shielding of the phosphoryl oxygen. Bicyclophosphetane oxide **106** exhibits ¹⁷O NMR parameters reflecting substantial deshielding of the P=¹⁷O (δ 93.1 ppm; $^1J_{PO}$ = 179 Hz).[61]

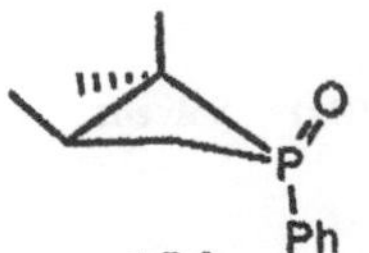

104 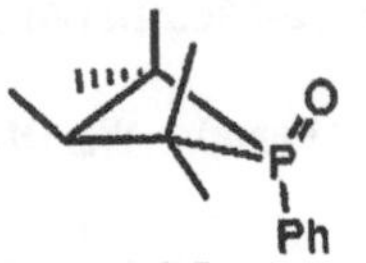   **105** 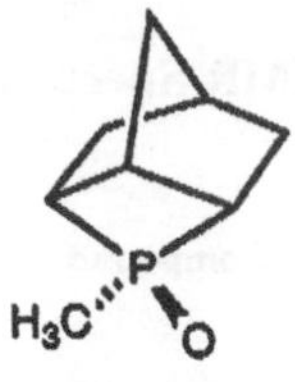   **106**

The ¹⁷O NMR shifts for 3-phospholene oxides (**107** to **110**) exhibit some interesting trends. For example, the P=O's of the phospholene oxides appear to be more shielded than the phosphetane oxides, with the smaller C-P-C angle,[73] and replacement of a *P*-methyl by a *P*-phenyl group in the 3-phospholenes causes a substantial shielding of the P=¹⁷O (Δδ= 13 to 14 ppm). The γ-gauche relationship between a methyl group and the P=¹⁷O in *cis*-phospholene oxides **107** and **108** is responsible for a 6 to 7 ppm shielding of the phosphoryl oxygen when compared to the respective *trans* diastereomers **109** and **110**.[61]

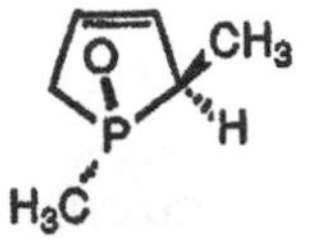

107 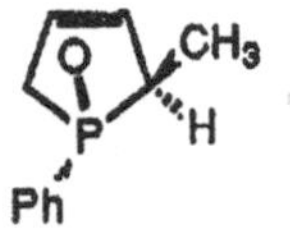   **108** 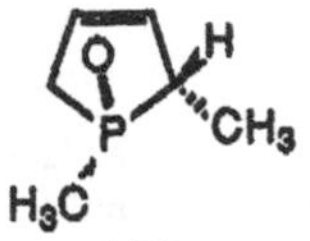   **109** 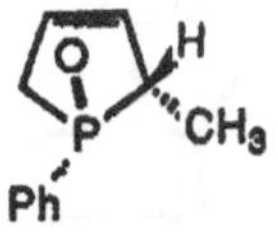   **110**

Bicyclic 3-phospholene oxides with highly contracted C-P-C bond angles exhibit ¹⁷O NMR shifts which are significantly deshielded when compared to other 3-phospholene oxides (Table 12). The influence of this C-P-C angular dependent ¹⁷O NMR shift difference is clearly demonstrated in the ¹⁷O NMR shift comparisons between the two annulated five-membered rings in **111** and **112** as well as **113** and **114**.[61]

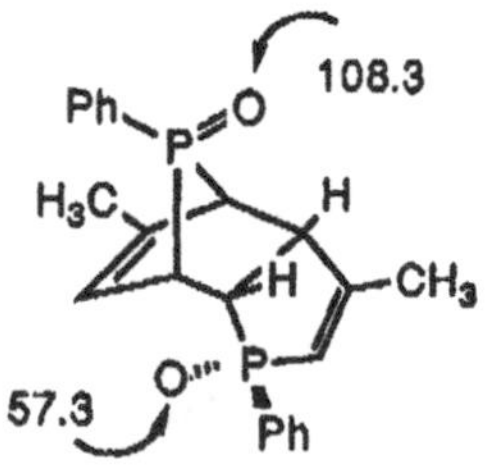

111 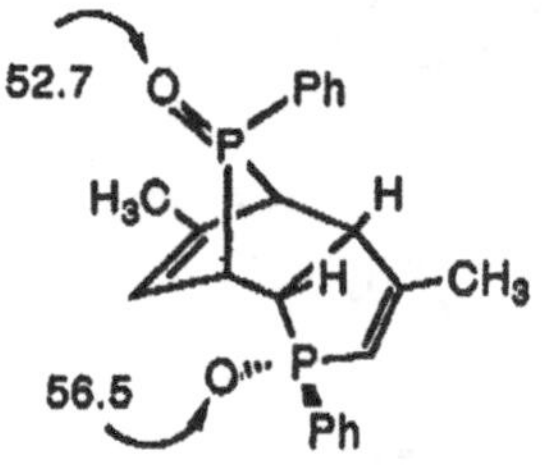   **112**

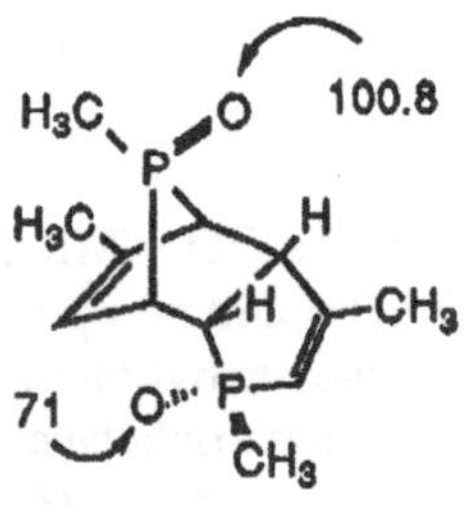

113 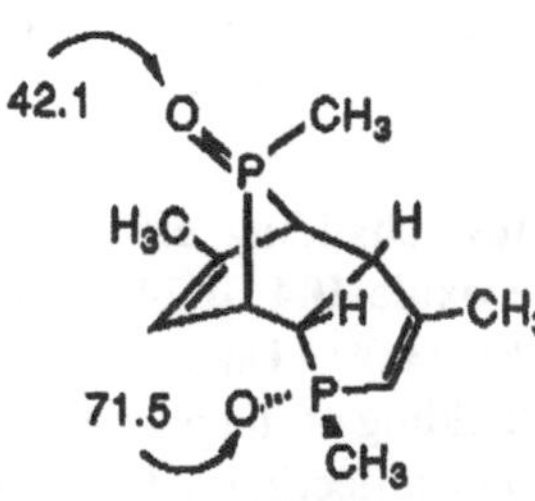   **114**

VI. CONCLUSIONS

The ^{17}O NMR shift trends within the various structural series of organosulfur and organophosphorus compounds offer a reasonable level of confidence in making fairly general stereochemical and conformational assignments. The useful contributions to the ^{17}O NMR shift arising from multiple bonding effects, electronegativity, variable bond angles, and delocalization effects are largely empirical; however, it is anticipated that as more experimental ^{17}O NMR data become available, they will make significant contributions to the refinement of modern NMR theories.

Finally, it is apparent that *solution* ^{17}O NMR spectroscopy is rapidly securing a valuable stronghold in the arsenal of analytical tools for use in stereochemical and conformational structure analyses, the investigation of dynamic exchange processes, the evaluation of reaction mechanism using enriched ^{17}O-labeled substrates, and in providing new insights into the complexities of oxygen-heteroatom bonding arrays. These advances coupled with new opportunities for ^{17}O NMR in the solid-state provide unique potential for correlating structural effects in both media, as well as new clues and insight into a host of fundamental processes.

ACKNOWLEDGMENTS

Acknowledgment is made to the National Science Foundation for support of this research and to the valuable contributions of my co-workers, whose names are cited in the appropriate references.

REFERENCES

1. (a) **Christ, H. A., Diehl, P. D., Schneider, H. R., and Dahn, H.**, Chemische Verschiebungen in der Kernmagnetischen Resonanz von ^{17}O in Organischen Verbindungen, *Helv. Chim. Acta*, 49, 865, 1961; (b) **Christ, H. A., and Diehl, P.**, Lösungsmitteleinflüsse und Spinkopplungen in der Kernmagnetischen Resonanz von ^{17}O, *Helv. Phys. Acta*, 36, 170, 1963; (c) **Kintzinger, J.-P.**, *NMR;17. Oxygen-17 and Silicon-29*, Diehl, P., Fluck, E., and Kosfeld, R., Eds., Springer-Verlag, New York, 1981; (d) **Kintzinger, J.-P.**, in *NMR of Newly Accessible Nuclei*, Vol. 2, Laszlo, P., Ed., Academic Press, New York, 1983, 79.

2. (a) **Rodger, C., Sheppard, N., McFarlane, C., and McFarlane, W.**, *NMR and the Periodic Table*, Harris, R. K. and Mann, B. E., Eds., Academic Press, New York, 1978; (b) **Dyer, J. C.**, Ph.D., Dissertation, University of North Carolina, Chapel Hill, 1980.

3. (a) **Gerothanassis, I. P. and Lauterwein, J.**, An evaluation of various pulse sequences for the suppression of acoustic ringing in oxygen-17 NMR, *J. Magn. Reson.*, 66, 32, 1986; (b) **Goc, R., and Fiat, D**, Simple pulse sequence for suppression of the acoustic ringing in FT NMR, *J. Magn. Reson.*, 70, 295, 1986; (c) **Lowe, E. D., Pollard, K. O. B., Skilling, J., Staunton, J., and Sutkowski, A. C.**, *J. Magn. Reson.*, 72, 493, 1987; (d) **Brevard, C.**, in *NMR of Newly Accessible Nuclei*, Vol. 1, Laszlo, P., Ed., Academic Press, New York, 1983, 3.

4. (a) **Canet, D., Goulon-Ginet, C., and Marchal, J. P.**, Accurate determination of parameters for 17-O in natural abundance studies, *J. Magn. Reson.*, 22, 537, 1976; (b) **Canet, D., Goulon-Ginet, C., and Marchal, J. P.**,, *J. Magn. Reson.*, 25, 397, 1977.

5. **Karplus M. and Pople, J. A.**, Theory of carbon NMR chemical shifts in conjugated molecules, *J. Chem. Phys.*, 38, 2803, 1963.

6. (a) **Grinter, R. and Mason, J.**, Nitrogen-14 and -15 nuclear magnetic resonance spectroscopy. III. The diamagnetic correction and the absolute scale of nitrogen chemical shift, *J. Chem. Soc. A*, p. 2196, 1970; (b) **Andersson, L.-O. and Mason, J.**, Oxygen-17 nuclear magnetic resonance. I. Oxygen-nitrogen groupings, *J. Chem. Soc. Dalton Trans.*, p. 202, 1974; (c) **Andersson, L.-O., Mason, J., and Bronswijk, W.**, Nitrogen nuclear magnetic resonance. I. The nitroso (nitrosyl) group, *J. Chem. Soc. A.*, p. 296, 1970; (d) **Kidd, R. G.**, Relationship between oxygen-17 nuclear magnetic resonance chemical shifts and π-bonding to oxygen in the dichromate ion, *Can. J. Chem.*, 45, 605, 1967.

7. (a) **Klemperer, W. G.**, Oxygen-17 NMR spectroscopy for the solution of chemical problems, *Angew. Chem. Int. Ed. Engl.*, 17, 246, 1978; (b) **Klemperer, W. G.**, in *The Multinuclear Approach to NMR Spectroscopy*, Proc. NATO Adv. Study Inst. on The Multinuclear Approach to NMR Spectroscopy, Lambert, J. B. and Riddell, F. G., Eds., D. Reidel, Dordrecht, Holland, 1983, 245.

8. **Allred, A. L.**, Electronegativity values from thermochemical data, *J. Inorg. Nucl. Chem.*, 17, 215, 1961.

9. **Szmant, H. H.**, *Sulphur in Organic and Inorganic Chemistry*, Vol. 1, Seening, A., Ed., Marcel-Dekker, New York, 1972, 107.

10. **Dyer, J. C., Harris, D. L., and Evans, S. A., Jr.**, Oxygen-17 nuclear magnetic resonance spectroscopy of sulfoxides and sulfones. Alkyl substituent induced chemical shift effects, *J. Org. Chem.*, 47, 3660, 1982.

11. **Bock, H. and Solouki, B.**, Photoelectron spectra and molecular properties. 9. Sulfoxide bond, *Angew. Chem. Int. Ed. Engl.*, 11, 436, 1972.

12. **Evans, S. A., Jr.**, Oxygen-17 and sulfur-33 nuclear magnetic resonance spectroscopy of organosulfur compounds, in *Magnetic Resonance. Introduction. Advanced Topics and Applications to Fossil Energy*, Petrakis, L. and Fraissard, J. P., Eds., Advanced NATO Institute, Reidel, Dordrecht, Holland, 1984, 757.

13. **Reuben, J.**, Hydrogen-bonding effects on oxygen-17 chemical shifts, *J. Am. Chem. Soc.*, 91, 5725, 1960.

14. **De Jeu, W. H.**, Calculation of overlap integrals between slater-type orbitals with nearly equal exponents, *Mol. Phys.*, 18, 31, 1970.

15. (a) **Mislow, K., Green, M. M., Laur, P., and Chrisholm, D. R.**, Optical rotatory dispersion and absolute configuration of dialkyl sulfoxides, *J. Am. Chem. Soc.*, 87, 665, 1965; (b) **Schlafer, H. L. and Schaffernicht, W.**, *Angew. Chem.*, 72, 618, 1960; (c) **Lindberg, J. J., Kenttamaa, J., and Nissema, A.**, Structures of dimethyl sulfoxide-water mixtures in light of thermodynamic and dielectric behavior, *Suom. Kemistil. B.*, 34, 98 and 156, 1961.

16. **Aime, S., Santucci, E., and Fruttero, R.**, Deuterium isotope effects on oxygen-17 chemical shifts, *Magn. Reson. Chem.*, 24, 919, 1986.

17. **Hakkinen, A.-M, Ruostesuo, P., and Kurkisuo, S.**, ^{17}O, ^{15}N, and ^{13}C NMR chemical shifts of *N,N*-dimethylmethanesulphinamide in various solvents, *Magn. Reson. Chem.*, 23, 311, 1985.

18. **Ruostesuo, P. and Karjalainen, J.**, Studies on sulfur-oxygen electron donors. III. Complex formation of diphenyl sulphone and *N,N* dimethylbenzenesulfonamide with phenols, *Spectrochim. Acta*, Part A, 37, 535, 1981.

19. **Kamlet, M. J., Dickinson, C., and Taft, R. W.**, Linear solvation energy relationship. II. An analysis of nitrogen-15 solvent shifts in amides, *J. Chem. Soc. Perkin Trans. 2*, p. 353, 1981.

20. **Duddeck, H., Korek, U., Rosenbaum, D., and Drabowicz, J.**, ^{1}H, ^{13}C, ^{17}O, and ^{33}S NMR investigation of some cyclohexyl phenyl sulphides, sulphoxides, and sulphones, *Magn. Reson. Chem.*, 24, 792, 1986.

21. **Eliel, E. L., Della, E. W., and Rogic, M.**, The stereochemistry and reduction of the 1-oxa-4-thia-8-*t*-butylspiro[4,5]decanes and [5,5]undecanes, *J. Org. Chem.*, 30, 855, 1965.

22. **Manoharan, M. and Eliel, E. L.**, ^{17}O NMR spectra of tertiary alcohols, ethers, sulfoxides, and sulfones in the cyclohexyl and 5 substituted 1,3-dioxanel series and related compounds, *Magn. Reson. Chem.*, 23, 225, 1985.

23. **Eliel, E. L. and Kandasamy, D.**, Conformational analysis 32. Conformational energies of methyl sulfide, methyl sulfoxide and methyl sulfone groups, *J. Org. Chem.*, 41, 3899, 1976.

24. **Kobayashi, K., Gugawara, T., and Iwamura, H.**, Diastereomeric differentiation of sulphone oxygens by ^{17}O N.M.R. spectroscopy, *J. Chem. Soc. Chem. Commun.*, p. 479, 1981.

25. **Modena, G., Quintily, U., and Scorrano, G.**, A novel route to racemization of sulfoxides, *J. Am. Chem. Soc.*, 94, 202, 1972.

26. **Kodama, Y., Zushi, S., Nishihata, K., and Nishio, M.**, A general preference for gauche alkyl-phenyl interactions. The use of lanthanide shift reagents in determining the preferred conformations of some alkyl 1-phenylethyl sulphoxides, *J. Chem. Soc. Perkin Trans. 2*, p. 1306, 1980.

27. **Romers, C., Altona, C., Buys, H. R., and Havinga, E.**, Geometry and conformational properties of some five- and six-membered heterocyclic compounds containing oxygen or sulfur, *Top. Stereochem.*, 4, 39, 1969.

28. **Barbarella, G., Rossini, S., Bongini, A., and Tugnoli, V.**, Force field and multinuclear NMR study of the conformational properties of thiolane-1-oxide and its mono and dimethyl derivatives, *Tetrahedron*, 41, 4691, 1985.

29. **Rooney, R. P. and Evans, S. A., Jr.**, Carbon-13 nuclear magnetic resonance spectra of *trans*-1-thiadecalin, *trans*-1,4-dithiadecalin, *trans*-1,4-oxathiadecalin, and the corresponding sulfoxides and sulfones, *J. Org. Chem.*, 45, 180, 1980.

30. **Barbarella, G., Dembech, P., and Tugnoli, V.**, ^{13}C and ^{17}O chemical shifts and conformational analysis of mono- and di-methyl substituted thiane 1-oxide and thiane 1,1-dioxide, *Org. Magn. Reson.*, 22, 402, 1984.

31. **Mattinen, J. and Pihlaja, K.**, ^{17}O NMR spectra of methyl substitued 2-oxo-1,3,2-dioxathianes, *Magn. Reson. Chem.*, 25, 569, 1987.

32. **Pihlaja, K., Rossi, K. and Nikander, H.,** Conformational analysis 25. ^{13}C NMR chemical shift-sensitive detectors in structure determination 3. The proposal for non-chair conformations in methyl substituted 2-oxo-1,3,2-dioxathianes, *J. Org. Chem.,* 50, 644, 1985.

33. **Hellier, D. G. and Liddy, H. G.,** Chemistry of the S=O bond. 11*-Carbon-13 and oxygen-17 nuclear magnetic resonance studies of stereoisomerization in 1,3,2-dioxathianes, *Magn. Reson. Chem.,* 27, 431, 1989.

34. **Bellamy, L. J.,** *Organic Sulfur Compounds,* Vol. 1, Kharash, N., Ed., Pergamon Press, Oxford, 1961, 47.

35. (a) **Craig, D. P., Maccol, A., Nyholm, R. S., Orgel, L. E., and Sutton, L. E.,** Chemical bonds involving d-orbitals Part I, *J. Chem. Soc.,* p. 332, 1954; (b) **Craig, D. P. and Magnusson, E. A.,** d-Orbital contraction in chemical bonding, *J. Chem. Soc.,* p. 4895, 1956; (c) **Janssen, M. J.,** *Organic Sulphur Chemistry,* Stirling, C. J. M., Ed., Butterworths, London, 1975, 19.

36. (a) **Delseth, C. and Kintzinger, J.-P.,** ^{17}O NMR aliphatic aldehydes and ketones, additivity of substituent effects and correlation with ^{13}C NMR, *Helv. Chim. Acta,* 59, 466, 1976; (b) **Harris, R. K. and Kimberd, B. J.,** d-Orbital contraction in chemical bonding, *Org. Magn. Reson.,* 59, 460, 1975.

37. **Fehnel, E. A. and Carmack, M.,** The ultraviolet absorption spectra of organic sulfur compounds. II. Compounds containing the sulfone function, *J. Am. Chem. Soc.,* 71, 231, 1949.

38. **Kelly, J. W. and Evans, S. A., Jr.,** Oxygen-17 NMR spectral studies of selected aromatic sulfones, *Magn. Reson. Chem.,* 25, 305, 1987.

39. **Koch, H. P. and Moffitt, W. E.,** Conjugation in sulphones, *Trans. Faraday Soc.,* 47, 7, 1951.

40. **Barbarella, G., Chatgilialoglu, C., Rossini, S., and Tugnoli, V.,** [33]S and ^{17}O NMR of compounds containing the SO_2 moiety. The chlorine effect, *J. Magn. Reson.,* 70, 204, 1986.

41. **Huheey, J. E.,** The electronegativity of groups, *J. Phys. Chem.,* 69, 3284, 1965.

42. **Ternay, A. L., Jr., Ens, L., Herrmann, J., and Evans, S.,** The stereochemistry of thioxanthene sulfoxide and its C-9 methylated analogs. The control of conformational distribution on folded molecules by peri substitution, *J. Org. Chem.,* 34, 940, 1969.

43. **Hakkinen, A.-M. and Ruostesuo, P.,** Carbon-13, nitrogen-15, oxygen-17, and sulphur-33 NMR chemical shifts of some sulphonamides and related compounds, *Magn. Reson. Chem.,* 23, 424, 1985.

44. **Eliel, E. L. and Evans, S. A.,** An unusually strong intramolecular interaction between the sulfone or sulfoxide and the alkoxide function, *J. Am. Chem. Soc.,* 94, 8587, 1972.

45. **Juaristi, E., Martinez, R., Mendez, R., Toscano, R. A., Soriano-Garcia, M., Eliel, E. L., Petsom, A., and Glass, R. S.,** Conformational analysis of 1,3-dioxanes with sulfide, sulfoxide, and sulfone substitution at C(5). Finding an eclipsed conformation in *cis*-2-*tert*-butyl-s-(*tert*-butylsulfonyl)1,3-dioxane, *J. Org. Chem.,* 52, 3806, 1987.

46. **Cinquini, G. B. M., Colonna, S., and Montanari, F.,** The reaction of sulphides with iodobenzene dichloride in aqueous pyridine. Synthesis of sulphoxides free from sulphone and ^{18}O-labeled sulphoxides, *J. Chem. Soc. (C),* p. 659, 1968.

47. **Block E., Bassi, A. A., Lambert, J. B., Wharry, S. M., Andersen, K. K., Dittmer, D. L., Patwarden, B. H., and Smith, D. J. H.,** Carbon-13 and oxygen-17 nuclear magnetic resonance studies of organosulfur compounds: the four-membered-ring sulphone effect, *J. Org. Chem.,* 45, 4807, 1980.

48. (a) **Sammakia, T. H., Harris, D. L., and Evans, S. A., Jr.,** Oxygen-17 NMR spectral investigation of 3-alkoxy-*trans*-3,4 disubstituted thiolene 1,1-dioxides and related compounds, *Org. Magn. Reson.,* 22, 747, 1984; (b) For additional insights, See **Bakke, J. M., Ronneberg, H., and Chadwick, D. J.,** NMR and theoretical (MNDO and *ab initio*) studies of lone pair electron-π-electron interactions: the conformation of 9-methoxyfluorene, *Magn. Reson. Chem.,* 25, 251, 1987.

49. **Iwamura, H., Sugawara, T., Kawada, Y., Tori, K., Muneyuki, R., and Noyori, R.,** ^{17}O NMR chemical shifts versus structure relationships in oxiranes, *Tetrahedron Lett.,* p. 3449, 1979.

50. **Inagaki, F. and Miyazawa, T.,** NMR analysis of molecular conformations and conformational equilibrium with the lanthanide probe method, *Prog. Nucl. Magn. Reson. Spectrosc.,* 14, 67, 1981.

51. **Lowe, G. and Salamone, S. J.,** Application of a lanthanide shift reagent in [17]O N.M.R. spectroscopy to determine the stereochemical course of oxidation of cyclic sulphite diesters to cyclic sulphate diesters with ruthenium, *J. Chem. Soc. Chem. Commun.,* p. 1392, 1983.

52. **Virtanen, T., Nikander, H., Pihlaja, K., and Rahkamaa, E.,** Conformational analysis XXI ^{1}H NMR conformational study of alkyl substituted 2-OXO-1,3,2-dioxathianes, *Tetrahedron,* 38, 2821, 1982.

53. **Hellier, D. G. and Liddy, H. G.,** Chemistry of the S=O bond. 10 - Conformational analysis of cyclic sulphates via ^{13}C and [17]O N.M.R. spectroscopy, *Magn. Reson. Chem.,* 26, 671, 1988.

54. (a) **Tsai, M.-D., Huang, S. L., Kozlowski, J. F., and Chang, C. C.,** Applicability of the phosphorus-31 (oxygen-17) nuclear magnetic resonance method in the study of enzyme mechanism involving phosphorus, *Biochemistry,* 19, 3531, 1980; (b) **Gerlt, J. A., Reynolds, M. A., Demou, P. C., and Kenyon, G. L.,** ^{17}O NMR spectral properties of pyrophosphate, simple phosphonates, and thiophosphate and phosphanate analogues of ATP, *J. Am. Chem. Soc.,* 105, 6469, 1983.

55. **Tsai, M.-D. and Bruzik, K.**, NMR methods involving oxygen isotopes in biophosphates, in *Biological Magnetic Resonance*, Vol. 5, Berliner, L. J. and Reuben, J., Eds., Plenum Press, New York, 1983, 129.

56. **Gray, G. A. and Albright, T. A.**, ^{15}N and ^{17}O nuclear magnetic resonance of organophosphorus compounds. Experimental and theoretical determinations of ^{15}N-^{31}P and ^{17}O-^{31}P nuclear spin coupling constants, *J. Am. Chem. Soc.*, 99, 3243, 1977.

57. **Cheng, C. P. and Brown, T. L.**, Oxygen-17 nuclear quadrupole double resonance. 3. Results for N-O, P-O, and S-O bonds, *J. Am. Chem. Soc.*, 102, 6418, 1980.

58. **Sammons, D., Frey, P. A., Bruzik, K., and Tsai, M.-D.**, Effects of ^{17}O and ^{18}O on ^{31}P NMR: further investigation and applications, *J. Am. Chem. Soc.*, 105, 5455, 1983.

59. **Gerlt, J. A., Demou, P. C., and Mehdi, S.**, ^{17}O NMR spectral properties of simple phosphate esters and adenine nucleotides, *J. Am. Chem. Soc.*, 104, 423, 1982.

60. **Gerlt, J. A., Demou, P. C., and Mehdi, S.**, High field ^{17}O NMR studies of adenine nucleotides, *Nucleic Acids Res. Symp.*, Series 9, 1981, 11.

61. **Quin, L. D., Szewczyk, J., Lineham, K., and Harris, D. L.**, Structural effects on ^{17}O NMR shifts of cyclic phosphine oxides, *Magn. Reson. Chem.*, 25, 271, 1987.

62. **Duddeck, H. and Hanna, A. G.**, Structural effects on ^{17}O NMR shifts of cyclic phosphine oxides, *Magn. Reson. Chem.*, 23, 533, 1985.

63. **Grossman, G., Gruner, M., and Seifert, G.**, ^{17}O NMR spectroscopic investigation of phosphoryl compounds; $OPCl_3$, $OPCl_2X$, $OPClX_2$, OPX_3[X: F, Br, OCH_3, NMe_2], *Z. Chem.*, 16, 362, 1976.

64. **Sugawara, T., Kawada, Y., Katsh, M., and Iwamura, H.**, Oxygen-17 nuclear magnetic resonance. III. Oxygen atoms with a coordination number of two, *Bull. Chem. Soc. Jpn.*, 52, 3391, 1979.

65. **Hibbert, R. C. and Logan, N.**, Multinuclear magnetic resonance study of the interaction of some phosphorous (V) compounds with inorganic acids. The protonating abilities of NO_3, $MeSO_3H$, and HPO_2F_2 towards the phosphoryl group, *J. Chem. Soc. Dalton Trans.*, p. 865, 1985.

66. **Coderre, J. A., Mehdi, S., Demou, P. C., Weber, R., Traficante, D. D., and Gerlt, J. A.**, Oxygen chiral phosphodiesters. 3. Use of ^{17}O NMR spectroscopy to demonstrate differences of cyclic 2'-deoxyadenosine 3',5' - [^{17}O, ^{18}O] monophosphate, *J. Am. Chem. Soc.*, 103, 1870, 1981.

67. **Bock, P. L., Mosbo, J. A., and Redmon, J. L.**, ^{17}O NMR spectra of equatorial and axial hydrocarbons and 5-hydroxy-1,3-dioxanes and their methyl ethers. Cyclic 3', 5' cyclic-(R_p)-phosphoranilidate for conformational assessment, *Org. Magn. Reson.*, 21, 491, 1983.

68. **Lesnikowski, Z. J., Stec, W. J., Zielinski, W. S., Adamiak, D., and Saenger, W.**, Crystallographic assessment of absolute configuration in 2'-deoxyadenosine, *J. Am. Chem. Soc.*, 103, 2862, 1981.

69. **Eliel, E. L., Chandrasekaran, S., Carpenter, L. E., and Verkade, J. G.**, ^{17}O NMR spectra of cyclic phosphites, phosphates, and thiophosphates, *J. Am. Chem. Soc.*, 108, 6651, 1986.

70. **Nifant'ev, E. E., Borisenko, A. A., Nasonovoskii, I. S., and Matrosov, E. I.**, Stereochemistry of 1,3-butylene phosphites, *Dokl. Akad. Nauk SSSR*, 196, 28, 1971.

71. **Sopchik, A. E., Cairns, S. M., and Bentrude, W. G.**, ^{17}O NMR of diastereomeric 3',5'-cyclic thymidine methyl phosphates, methylphosphates, and *N,N*-dimethyl phosphoramides. Phosphorus configuration of P[^{17}O,^{18}O]-nucleoside phosphate diesters, *Tetrahedron Lett.*, 30, 1221, 1989.

72. **Quin, L. D., Yao, E.-Y., and Szewczyk, J.**, Special properties imparted to the 9-phosphabicyclo[6.1.0]nonatriere system by a P-(2,4,6-tri-*t*-butylphenyl) substituent; ^{17}O NMR spectrum of a bicyclic phosphirane oxide, *Tetrahedron Lett.*, 28, 1077, 1987.

73. **Fritzgerald, A., Campbell, J. A., Smith, G. D., Caughlan, C. N., and Cremer, S. E.**, Solid state studies on crowded molecules. Crystal and molecular structures of 2,2,3-trimethyl-1-phenylphosphatane 1-oxide and 2,2,3,3,4-pentamethyl-1-phenylphosphatane 1-oxide, *J. Org. Chem.*, 43, 3513, 1978.

INDEX

Printed and bound by CPI Group (UK) Ltd, Croydon, CR0 4YY

23/10/2024

01778245-0010